Almost Periodic Solutions of Differential Equations in Banach Spaces

Stability and Control: Theory, Methods and Applications
A series of books and monographs on the theory of stability and control
Edited by A.A. Martynyuk, Institute of Mechanics, Kiev, Ukraine and V. Lakshmikantham, Florida Institute of Technology, USA

Volume 1
Theory of Integro-Differential Equations
V. Lakshmikantham and M. Rama Mohana Rao

Volume 2
Stability Analysis: Nonlinear Mechanics Equations
A.A. Martynyuk

Volume 3
Stability of Motion of Nonautonomous Systems (Method of Limiting Equations)
J. Kato, A.A. Martynyuk and A.A. Shestakov

Volume 4
Control Theory and its Applications
E.O. Roxin

Volume 5
Advances in Nonlinear Dynamics
edited by S. Sivasundaram and A.A. Martynyuk

Volume 6
Solving Differential Problems by Multistep Initial and Boundary Value Methods
L. Brugnano and D. Trigiante

Volume 7
Dynamics of Machines with Variable Mass
L. Cveticanin

Volume 8
Optimization of Linear Control Systems: Analytical Methods and Computational Algorithms
F.A. Aliev and V.B. Larin

Volume 9
Dynamics and Control
edited by G. Leitmann, F.E. Udwadia and A.V. Kryazhimskii

Volume 10
Volterra Equations and Applications
edited by C. Corduneanu and I.W. Sandberg

Volume 11
Nonlinear Problems in Aviation and Aerospace
edited by S. Sivasundaram

Volume 12
Stabilization of Programmed Motion
E.Ya. Smirnov

Please see the back of this book for other titles in the Stability and Control: Theory, Methods and Applications series.

Almost Periodic Solutions of Differential Equations in Banach Spaces

Yoshiyuki Hino
Department of Mathematics and Informatics
Chiba University, Chiba, Japan

Toshiki Naito
Department of Mathematics
University of Electro-Communications, Tokyo, Japan

Nguyen Van Minh
Department of Mathematics
Hanoi University of Science, Hanoi, Vietnam

Jong Son Shin
Department of Mathematics
Korea University, Tokyo, Japan

London and New York

First published 2002 by Taylor & Francis
11 New Fetter Lane, London EC4P 4EE

Simultaneously published in the USA and Canada
by Taylor & Francis Inc,
29 West 35th Street, New York, NY 10001

Taylor & Francis is an imprint of the Taylor & Francis Group

© 2002 Taylor & Francis

Publisher's Note
This book has been prepared from camera-ready copy provided by the authors.

Printed and bound in Great Britain by TJ International Ltd, Padstow, Cornwall

All rights reserved. No part of this book may be reprinted or reproduced or utilised in any form or by any electronic, mechanical, or other means, now known or hereafter invented, including photocopying and recording, or in any information storage or retrieval system, without permission in writing from the publishers.

Every effort has been made to ensure that the advice and information in this book is true and accurate at the time of going to press. However, neither the publisher nor the authors can accept any legal responsibility or liability for any errors or omissions that may be made. In the case of drug administration, any medical procedure or the use of technical equipment mentioned within this book, you are strongly advised to consult the manufacturer's guidelines.

British Library Cataloguing in Publication Data
A catalogue record for this book is available from the British Library

Library of Congress Cataloging in Publication Data
A catalog record has been requested.

ISBN 0-415-27266-1

Contents

Introduction to the Series	1
Preface	3
Acknowledgements	5

1 C_0-SEMIGROUPS, WELL POSED EVOLUTION EQUATIONS, SPECTRAL THEORY AND ALMOST PERIODICITY OF FUNCTIONS 7
  1.1. STRONGLY CONTINUOUS SEMIGROUPS OF LINEAR OPERATORS 7
    1.1.1. Definition and Basic Properties 7
    1.1.2. Compact Semigroups and Analytic Strongly Continuous Semigroups 10
    1.1.3. Spectral Mapping Theorems 11
  1.2. EVOLUTION EQUATIONS 15
    1.2.1. Well-Posed Evolution Equations 15
    1.2.2. Functional Differential Equations with Finite Delay 18
    1.2.3. Equations with Infinite Delay 20
  1.3. SPECTRAL THEORY AND ALMOST PERIODICITY OF BOUNDED UNIFORMLY CONTINUOUS FUNCTIONS 24
    1.3.1. Spectrum of a Bounded Function 24
    1.3.2. Almost Periodic Functions 26
    1.3.3. Spectrum of an Almost Periodic Function 27
    1.3.4. A Spectral Criterion for Almost Periodicity of a Function 28

2 SPECTRAL CRITERIA FOR PERIODIC AND ALMOST PERIODIC SOLUTIONS 31
  2.1. EVOLUTION SEMIGROUPS AND ALMOST PERIODIC SOLUTIONS OF PERIODIC EQUATIONS 31
    2.1.1. Evolution Semigroups 31

	2.1.2. Almost Periodic Solutions and Applications	35
2.2.	EVOLUTION SEMIGROUPS, SUMS OF COMMUTING OPERATORS AND SPECTRAL CRITERIA FOR ALMOST PERIODICITY	45
	2.2.1. Differential Operator $d/dt - \mathcal{A}$ and Notions of Admissibility	48
	2.2.2. Admissibility for Abstract Ordinary Differential Equations	53
	2.2.3. Higher Order Differential Equations	55
	2.2.4. Abstract Functional Differential Equations	62
	2.2.5. Examples and Applications	66
2.3.	DECOMPOSITION THEOREM AND PERIODIC, ALMOST PERIODIC SOLUTIONS	77
	2.3.1. Spectral Decomposition	79
	2.3.2. Spectral Oriteria For Almost Periodic Solutions	85
	2.3.3. When Does Boundedness Yield Uniform Continuity?	89
	2.3.4. Periodic Solutions of Partial Functional Differential Equations	91
	2.3.5. Almost Periodic Solutions of Partial Functional Differential Equations	95
2.4.	FIXED POINT THEOREMS AND FREDHOLM OPERATORS	109
	2.4.1. Fixed Point Theorems	109
	2.4.2. Decomposition of Solution Operators	110
	2.4.3. Periodic Solutions and Fixed Point Theorems	113
	2.4.4. Existence of Periodic Solutions: Bounded Perturbations	116
	2.4.5. Existence of Periodic Solutions: Compact Perturbations	120
	2.4.6. Uniqueness of Periodic Solutions I	125
	2.4.7. Uniqueness of Periodic Solutions II	127
	2.4.8. An Example	129
	2.4.9. Periodic Solutions in Equations with Infinite Delay	130
2.5.	BOUNDEDNESS AND ALMOST PERIODICITY IN DISCRETE SYSTEMS	132
	2.5.1. Spectrum of Bounded Sequences and Decomposition	133
	2.5.2. Almost Periodic Solutions of Discrete Systems	137
	2.5.3. Applications to Evolution Equations	139
2.6.	BOUNDEDNESS AND ALMOST PERIODIC SOLUTIONS OF SEMILINEAR EQUATIONS	143
	2.6.1. Evolution Semigroups and Semilinear Evolution Equations	143
	2.6.2. Bounded and Periodic Solutions to Abstract Functional Differential Equations with Finite Delay	151
2.7.	ALMOST PERIODIC SOLUTIONS OF NONLINEAR EVOLUTION EQUATIONS	153
	2.7.1. Nonlinear Evolution Semigroups in $AP(\Delta)$	153
	2.7.2. Almost Periodic Solutions of Dissipative Equations	157
	2.7.3. An Example	160
2.8.	NOTES	161

3 STABILITY METHODS FOR SEMILINEAR EVOLUTION EQUATIONS AND NONLINEAR EVOLUTION EQUATIONS 163
   3.1. SKEW PRODUCT FLOWS OF PROCESSES AND QUASI-PROCESSES AND STABILITY OF INTEGRALS 163

CONTENTS

- 3.2. EXISTENCE THEOREMS OF ALMOST PERIODIC INTEGRALS — 168
  - 3.2.1. Asymptotic Almost Periodicity and Almost Periodicity — 168
  - 3.2.2. Uniform Asymptotic Stability and Existence of Almost Periodic Integrals — 171
  - 3.2.3. Separation Condition and Existence of Almost Periodic Integrals — 172
  - 3.2.4. Relationship between the Uniform Asymptotic Stability and the Separation Condition — 175
  - 3.2.5. Existence of an Almost Periodic Integral of Almost Quasi-Processes — 176
- 3.3. PROCESSES AND QUASI-PROCESSES GENERATED BY ABSTRACT FUNCTIONAL DIFFERENTIAL EQUATIONS — 176
  - 3.3.1. Abstract Functional Differential Equations with Infinite Delay — 176
  - 3.3.2. Processes and Quasi-Processes Generated by Abstract Functional Differential Equations with Infinite Delay — 180
  - 3.3.3. Stability Properties for Abstract Functional Differential Equations with Infinite Delay — 185
- 3.4. EQUIVALENT RELATIONSHIPS BETWEEN BC-STABILITIES AND ρ-STABILITIES — 190
  - 3.4.1. BC-Stabilities in Abstract Functional Differential Equations with Infinite Delay — 190
  - 3.4.2. Equivalent Relationship between BC-Uniform Asymptotic Stability and ρ-Uniform Asymptotic Stability — 192
  - 3.4.3. Equivalent Relationship Between BC-Total Stability and ρ-Total Stability — 195
  - 3.4.4. Equivalent Relationships of Stabilities for Linear Abstract Functional Differential Equations with Infinite Delay — 198
- 3.5. EXISTENCE OF ALMOST PERIODIC SOLUTIONS — 202
  - 3.5.1. Almost Periodic Abstract Functional Differential Equations with Infinite Delay — 202
  - 3.5.2. Existence Theorems of Almost Periodic Solutions for Nonlinear Systems — 203
  - 3.5.3. Existence Theorems of Almost Periodic Solutions for Linear Systems — 204
- 3.6. APPLICATIONS — 207
  - 3.6.1. Damped Wave Equation — 207
  - 3.6.2. Integrodifferential Equation with Diffusion — 210
  - 3.6.3. Partial Functional Differential Equation — 214
- 3.7. NOTES — 217

4 APPENDICES — 221
- 4.1. FREDHOLM OPERATORS AND CLOSED RANGE THEOREMS — 221
- 4.2. ESSENTIAL SPECTRUM AND MEASURES OF NONCOMPACTNESS — 224
- 4.3. SUMS OF COMMUTING OPERATORS — 231
- 4.4. LIPSCHITZ OPERATORS — 232

REFERENCES — 235

INDEX — 249

Introduction to the Series

The problems of modern society are both complex and interdisciplinary. Despite the apparent diversity of problems, tools developed in one context are often adaptable to an entirely different situation. For example, consider the Lyapunov's well known second method. This interesting and fruitful technique has gained increasing significance and has given a decisive impetus for modern development of the stability theory of differential equations. A manifest advantage of this method is that it does not demand the knowledge of solutions and therefore has great power in application. It is now well recognized that the concept of Lyapunov-like functions and the theory of differential and integral inequalities can be utilized to investigate qualitative and quantitative properties of nonlinear dynamic systems. Lyapunov-like functions serve as vehicles to transform the given complicated dynamic systems into a relatively simpler system and therefore it is sufficient to study the properties of this simpler dynamic system. It is also being realized that the same versatile tools can be adapted to discuss entirely different nonlinear systems, and that other tools, such as the variation of parameters and the method of upper and lower solutions provide equally effective methods to deal with problems of a similar nature. Moreover, interesting new ideas have been introduced which would seem to hold great potential.

Control theory, on the other hand, is that branch of application-oriented mathematics that deals with the basic principles underlying the analysis and design of control systems. To control an object implies the influence of its behavior so as to accomplish a desired goal. In order to implement this influence, practitioners build devices that incorporate various mathematical techniques. The study of these devices and their interaction with the object being controlled is the subject of control theory. There have been, roughly speaking, two main lines of work in control theory which are complementary. One is based on the idea that a good model of the object to be controlled is available and that we wish to optimize its behavior, and the other is based on the constraints imposed by uncertainty about the model in which the object operates. The control tool in the latter is the use of feedback in order to correct for deviations from the desired behavior. Mathematically, stability theory, dynamic systems and functional analysis have had a strong influence on this approach.

Volume 1, *Theory of Integro-Differential Equations*, is a joint contribution by V. Lakshmikantham (USA) and M. Rama Mohana Rao (India).

Volume 2, *Stability Analysis: Nonlinear Mechanics Equations*, is by A.A. Martynyuk (Ukraine).

Volume 3, *Stability of Motion of Nonautonomous Systems: The Method of Limiting Equations*, is a collaborative work by J. Kato (Japan), A.A. Martynyuk (Ukraine) and A.A. Shestakov (Russia).

Volume 4, *Control Theory and its Applications*, is by E.O. Roxin (USA).

Volume 5, *Advances in Nonlinear Dynamics*, is edited by S. Sivasundaram (USA) and A.A. Martynyuk (Ukraine) and is a multiauthor volume dedicated to Professor S. Leela.

Volume 6, *Solving Differential Problems by Multistep Initial and Boundary Value Methods*, is a joint contribution by L. Brugnano (Italy) and D. Trigiante (Italy).

Volume 7, *Dynamics of Machines with Variable Mass*, is by L. Cveticanin (Yugoslavia).

Volume 8, *Optimization of Linear Control Systems: Analytical Methods and Computational Algorithms*, is a joint work by F.A. Aliev (Azerbaijan) and V.B. Larin (Ukraine).

Volume 9, *Dynamics and Control*, is edited by G. Leitmann (USA), F.E. Udwadia (USA) and A.V. Kryazhimskii (Russian) and is a multiauthor volume.

Volume 10, *Volterra Equations and Applications*, is edited by C. Corduneanu (USA) and I.W. Sandberg (USA) and is a multiauthor volume.

Volume 11, *Nonlinear Problems in Aviation and Aerospace*, is edited by S. Sivasundaram (USA) and is a multiauthor volume.

Volume 12, *Stabilization of Programmed Motion*, is by E.Ya. Smirnov (Russia).

Volume 13, *Advances in Stability Theory at the end of the 20th Century*, is edited by A.A. Martynyuk.

Volume 14, *Dichotomies and Stability in Nonautonomous Linear Systems*, is by Yu.A. Mitropolskii, A.M. Samoilenko and V.L. Kulik.

Volume 15, *Almost Periodic Solutions of Differential Equations in Banach Spaces*, is by Yoshiyuki Hino, Toshiki Naito, Nguyen Van Minh and Jong Son Shin

Due to the increased interdependency and cooperation among the mathematical sciences across the traditional boundaries, and the accomplishments thus far achieved in the areas of stability and control, there is every reason to believe that many breakthroughs await us, offering existing prospects for these versatile techniques to advance further. It is in this spirit that we see the importance of the 'Stability and Control' series, and we are immensely thankful to Gordon and Breach Science Publishers for their interest and cooperation in publishing this series.

Preface

Almost periodic solutions of differential equations have been studied since the very beginning of this century. The theory of almost periodic solutions has been developed in connection with problems of differential equations, dynamical systems, stability theory and its applications to control theory and other areas of mathematics. The classical books by C. Corduneanu [50], A.M. Fink [67], T. Yoshizawa [231], L. Amerio and G. Prouse [7], B.M. Levitan and V.V. Zhikov [137] gave a very nice presentation of methods as well as results in the area. In recent years, there has been an increasing interest in extending certain classical results to differential equations in Banach spaces. In this book we will make an attempt to gather systematically certain recent results in this direction.

We outline briefly the contents of our book. The main results presented here are concerned with conditions for the existence of periodic and almost periodic solutions and its connection with stability theory. In the qualitative theory of differential equations there are two classical results which serve as models for many works in the area. Namely,

> **Theorem A** *A periodic inhomogeneous linear equation has a unique periodic solution (vith the same period) if 1 is not an eigenvalue of its monodromy operator.*

> **Theorem B** *A periodic inhomogeneous linear equation has a periodic solution (uith the same period) if and only if it has a bounded solution.*

In our book, a main part will be devoted to discuss the question as how to extend these results to the case of almost periodic solutions of (linear and nonlinear) equations in Banach spaces. To this end, in the first chapter we present introductions to the theory of semigroups of linear operators (Section 1), its applications to evolution equations (Section 2) and the harmonic analysis of bounded functions on the real line (Section 3). In Chapter 2 we present the results concerned with autonomous as well as periodic evolution equations, extending Theorems A and B to the infinite dimensional case. In contrast to the finite dimensional case, in general one cannot treat periodic evolution equations as autonomous ones. This is due to the fact that in the infinite dimensional case there is no Floquet representation, though one can prove many similar assertions to the autonomous case (see e.g. [78], [90], [131]). Sections 1, 2 of this chapter are devoted to the investigation by means of evolution semigroups in translation invariant subspaces of $BUC(\mathbf{R}, \mathbf{X})$ (of bounded uniformly continuous $\mathbf{X}$-valued functions on the real line). A new technique of spectral decomposition is presented in Section 3. Section 4 presents various results extending Theorem B to periodic solutions of abstract functional differential equations. In Section 5 we prove analogues of results in Sections 1, 2. 3 for discrete systems and discuss an alternative method to extend Theorems A and B to periodic and almost periodic solutions of differential equations. In Sections 6 and 7 we extend the method used in the previous ones to semilinear and fully nonlinear equations. The conditions are given in terms of the dissipativeness of the equations under consideration.

In Chapter 3 we present the existence of almost periodic solutions of almost periodic evolution equations by using stability properties of nonautonomous dynamical systems. Sections 1 and 2 of this chapter extend the concept of skew product flow of processes to a more general concept which is called skew product flow of quasi-processes and investigate the existence of almost periodic integrals for almost periodic quasi-processes. For abstract functional differential equations uith infinite delay, there are three kinds of definitions of stabilities. In Sections 3 and 4. we prove some equivalence of these definitions of stabilities and show that these stabilities fit in with quasiprocesses. By using results in Section 2, we discuss the existence of almost periodic solutions for abstract almost periodic evolution equations in Section 5. Concrete applications for functional partial differential equations are given in Section 6.

Acknowledgements

We wish to thank Professors T.A. Burton and J. Kato for their kind interest, encouragement, and especially for reading the manuscript and making valuable comments on the contents as well as on the presentation of this book. It is also our pleasure to acknowledge our indebtedness to Professor S. Murakami for his interest, encouragement and remarks to improve several results as well as their presentation. The main part of the book was written during the visit of the third author, N.V. Minh, to the University of Electro-Communications (Tokyo) supported by a fellow-ship of the Japan Society for the Promotion of Science. He wishes to thank the University for its warm hospitality and the Society for the generous support.

Yoshiyuki Hino
Toshiki Naito
Nguyen Van Minh
Jong Son Shin

CHAPTER 1

C_0-SEMIGROUPS, WELL POSED EVOLUTION EQUATIONS, SPECTRAL THEORY AND ALMOST PERIODICITY OF FUNCTIONS

1.1. STRONGLY CONTINUOUS SEMIGROUPS OF LINEAR OPERATORS

In this section we collect some well-known facts from the theory of strongly continuous semigroups of operators on a Banach space for the reader's convenience. We will focus the reader's attention on several important classes of semigroups such as analytic and compact semigroups which will be discussed later in the next chapters. Among the basic properties of strongly continuous semigroups we will put emphasis on the spectral mapping theorem. Since the materials of this section as well as of the chapter in the whole can be found in any standard book covering the area, here we aim at freshening up the reader's memory rather than giving a logically self contained account of the theory.

Throughout the book we will denote by **X** a complex Banach space. The set of all real numbers and the set of nonnegative real numbers will be denoted by **R** and **R**$^+$, respectively. $BC(\mathbf{R}, \mathbf{X})$, $BUC(\mathbf{R}, \mathbf{X})$ stand for the spaces of bounded, continuous functions and bounded, uniformly continuous functions, respectively.

1.1.1. Definition and Basic Properties

Definition 1.1 A family $(T(t))_{t\geq 0}$ of bounded linear operators acting on a Banach space **X** is a *strongly continuous semigroup of bounded linear operators*, or briefly, a C_0-*semigroup* if the following three properties are satisfied:

i) $T(0) = I$, the identity operator on **X**;

ii) $T(t)T(s) = T(t+s)$ for all $t, s \geq 0$;

iii) $\lim_{t\downarrow 0} \|T(t)x - x\| = 0$ for all $x \in \mathbf{X}$.

The *infinitesimal generator* of $(T(t))_{t\geq 0}$, or briefly, the *generator*, is the linear operator A with domain $D(A)$ defined by

$$D(A) = \{x \in \mathbf{X} : \lim_{t\downarrow 0} \frac{1}{t}(T(t)x - x) \text{ exists}\},$$

$$Ax = \lim_{t\downarrow 0} \frac{1}{t}(T(t)x - x), \quad x \in D(A).$$

The generator is always a closed, densely defined operator.

Theorem 1.1 *Let $(T(t))_{t\geq 0}$ be a C_0-semigroup. Then there exist constants $\omega \geq 0$ and $M \geq 1$ such that*
$$\|T(t)\| \leq Me^{\omega t}, \ \forall t \geq 0.$$

Proof. For the proof see e.g. [179, p. 4].

Corollary 1.1 *If $(T(t))_{t\geq 0}$ is a C_0-semigroup, then the mapping $(x,t) \mapsto T(t)x$ is a continuous function from $\mathbf{X} \times \mathbf{R}^+ \to \mathbf{X}$.*

Proof. For any $x, y \in \mathbf{X}$ and $t \leq s \in \mathbf{R}^+ := [0, \infty)$,

$$\begin{aligned}
\|T(t)x - T(s)y\| &\leq \|T(t)x - T(s)x\| + \|T(s)x - T(s)y\| \\
&\leq Me^{\omega s}\|x-y\| + \|T(t)\|\|T(s-t)x - x\| \\
&\leq Me^{\omega s}\|x-y\| + Me^{\omega t}\|T(s-t)x - x\|. \quad (1.1)
\end{aligned}$$

Hence, for fixed x, t ($t \leq s$) if $(y, s) \to (x, t)$, then $\|T(t)x - T(s)y\| \to 0$. Similarly, for $s \leq t$

$$\begin{aligned}
\|T(t)x - T(s)y\| &\leq \|T(t)x - T(s)x\| + \|T(s)x - T(s)y\| \\
&\leq Me^{\omega s}\|x-y\| + \|T(s)\|\|T(t-s)x - x\| \\
&\leq Me^{\omega s}\|x-y\| + Me^{\omega s}\|T(t-s)x - x\|. \quad (1.2)
\end{aligned}$$

Hence, if $(y, s) \to (x, t)$, then $\|T(t)x - T(s)y\| \to 0$.

Other basic properties of a C_0-semigroup and its generator are listed in the following:

Theorem 1.2 *Let A be the generator of a C_0-semigroup $(T(t))_{t\geq 0}$ on $\mathbf{X}$. Then*

i) For $x \in \mathbf{X}$,
$$\lim_{h\to 0} \frac{1}{h} \int_t^{t+h} T(s)x\,ds = T(t)x.$$

ii) For $x \in \mathbf{X}$, $\int_0^t T(s)x\,ds \in D(A)$ and
$$A\left(\int_0^t T(s)x\,ds\right) = T(t)x - x.$$

iii) For $x \in D(A), T(t)x \in D(A)$ and
$$\frac{d}{dt}T(t)x = AT(t)x = T(t)Ax.$$

iv) For $x \in D(A)$,
$$T(t)x - T(s)x = \int_s^t T(\tau)Ax\,d\tau = \int_s^t AT(\tau)x\,d\tau.$$

Proof. For the proof see e.g. [179, p. 5]. ∎

We continue with some useful fact about semigroups that will be used throughout this book. The first of these is the *Hille-Yosida theorem*, which characterizes the generators of C_0-semigroups among the class of all linear operators.

Theorem 1.3 *Let A be a linear operator on a Banach space $\mathbf{X}$, and let $\omega \in \mathbf{R}$ and $M \geq 1$ be constants. Then the following assertions are equivalent:*

i) *A is the generator of a C_0-semigroup $(T(t))_{t \geq 0}$ satisfying $\|T(t)\| \leq M e^{\omega t}$ for all $t \geq 0$;*

ii) *A is closed, densely defined, the half-line (ω, ∞) is contained in the resolvent set $\rho(A)$ of A, and we have the estimates*

$$\|R(\lambda, A)^n\| \leq \frac{M}{(\lambda - \omega)^n}, \quad \forall \lambda > \omega, \quad n = 1, 2, \ldots \tag{1.3}$$

Here, $R(\lambda, A) := (\lambda - A)^{-1}$ denotes the resolvent of A at λ. If one of the equivalent assertions of the theorem holds, then actually $\{Re\lambda > \omega\} \subset \rho(A)$ and

$$\|R(\lambda, A)^n\| \leq \frac{M}{(Re\lambda - \omega)^n}, \quad \forall Re\lambda > \omega, \quad n = 1, 2, \ldots \tag{1.4}$$

Moreover, for $Re\lambda > \omega$ the resolvent is given explicitly by

$$R(\lambda, A)x = \int_0^\infty e^{-\lambda t} T(t)x\, dt, \quad \forall x \in \mathbf{X}. \tag{1.5}$$

We shall mostly need the implication (i)$\Rightarrow$(ii), which is the easy part of the theorem. In fact, one checks directly from the definitions that

$$R_\lambda x := \int_0^\infty e^{-\lambda t} T(t)x\, dt$$

defines a two-sided inverse for $\lambda - A$. The estimate (1.4) and the identity (1.5) follow trivially from this.

A useful consequence of (1.3) is that

$$\lim_{\lambda \to \infty} \|\lambda R(\lambda, A)x - x\| = 0, \quad \forall x \in X. \tag{1.6}$$

This is proved as follows. Fix $x \in D(A)$ and $\mu \in \rho(A)$, and let $y \in X$ be such that $x = R(\mu, A)y$. By (1.3) we have $\|R(\lambda, A)\| = O(\lambda^{-1})$ as $\lambda \to \infty$. Therefore, the *resolvent identity*

$$R(\lambda, A) - R(\mu, A) = (\mu - \lambda) R(\lambda, A) R(\mu, A) \tag{1.7}$$

implies that

$$\lim_{\lambda \to \infty} \|\lambda R(\lambda, A)x - x\| = \lim_{\lambda \to \infty} \|R(\lambda, A)(\mu R(\mu, A)y - y)\| = 0.$$

This proves (1.6) for elements $x \in D(A)$. Since $D(A)$ is dense in X and the operators $\lambda R(\lambda, A)$ are uniformly bounded as $\lambda \to \infty$ by (1.3), (1.6) holds for all $x \in \mathbf{X}$.

1.1.2. Compact Semigroups and Analytic Strongly Continuous Semigroups

Definition 1.2 A C_0-semigroup $(T(t))_{t\geq 0}$ is called *compact* for $t > t_0$ if for every $t > t_0$, $T(t)$ is a compact operator. $(T(t))_{t\geq 0}$ is called *compact* if it is compact for $t > 0$.

If a C_0-semigroup $(T(t))_{t\geq 0}$ is compact for $t > t_0$, then it is continuous in the uniform operator topology for $t > t_0$.

Theorem 1.4 *Let A be the generator of a C_0-semigroup $(T(t))_{t\geq 0}$. Then $(T(t))_{t\geq 0}$ is a compact semigroup if and only if $T(t)$ is continuous in the uniform operator topology for $t > 0$ and $R(\lambda; A)$ is compact for $\lambda \in \rho(A)$.*

Proof. For the proof see e.g. [179, p. 49].

In this book we distinguish the notion of analytic C_0-semigroups from that of analytic semigroups in general. To this end we recall several notions. Let A be a linear operator $D(A) \subset \mathbf{X} \to \mathbf{X}$ with *not necessarily dense domain*.

Definition 1.3 A is said to be *sectorial* if there are constants $\omega \in \mathbf{R}, \theta \in (\pi/2, \pi), M > 0$ such that the following conditions are satisfied:

$$\begin{cases} i) & \rho(A) \supset S_{\theta,\omega} = \{\lambda \in \mathbf{C} : \lambda \neq \omega, |arg(\lambda - \omega)| < \theta\}, \\ ii) & \|R(\lambda, A)\| \leq M/|\lambda - \omega| \ \forall \lambda \in S_{\theta,\omega}. \end{cases}$$

If we assume in addtion that $\rho(A) \neq \emptyset$, then A is closed. Thus, $D(A)$, endowed with the graph norm

$$\|x\|_{D(A)} := \|x\| + \|Ax\|,$$

is a Banach space. For a sectorial operator A, from the definition, we can define a linear bounded operator e^{tA} by means of the Dunford integral

$$e^{tA} := \frac{1}{2\pi i} \int_{\omega + \gamma_{r,\eta}} e^{t\lambda} R(\lambda, A) d\lambda, t > 0, \tag{1.8}$$

where $r > 0, \eta \in (\pi/2, \theta)$ and $\gamma_{r,\eta}$ is the curve

$$\{\lambda \in \mathbf{C} : |arg\lambda| = \eta, |\lambda| \geq r\|\} \cup \{\lambda \in \mathbf{C} : |arg\lambda| \leq \eta, |\lambda| = r\},$$

oriented counterclockwise. In addition, set $e^{0A}x = x, \forall x \in \mathbf{X}$.

Theorem 1.5 *Under the above notation, for a sectorial operator A the following assertions hold true:*

i) $e^{tA}x \in D(A^k)$ for every $t > 0, x \in \mathbf{X}, k \in \mathbf{N}$. If $x \in D(A^k)$, then

$$A^k e^{tA} x = e^{tA} A^k x, \ \forall t \geq 0;$$

ii) $e^{tA}e^{sA} = e^{(t+s)A}$, $\forall t, s \geq 0$;

iii) *There are positive constants* $M_0, M_1, M_2, ...$, *such that*

$$\begin{cases} (a) & \|e^{tA}\| \leq M_0 e^{\omega t}, \ t \geq 0, \\ (b) & \|t^k(A-\omega I)^k e^{tA}\| \leq M_k e^{\omega t}, \ t \geq 0, \end{cases}$$

where ω *is determined from Definition 1.3. In particular, for every* $\varepsilon > 0$ *and* $k \in \mathbf{N}$ *there is* $C_{k,\varepsilon}$ *such that*

$$\|t^k A^k e^{tA}\| \leq C_{k,\varepsilon} e^{(\omega+\varepsilon)t}, \ t > 0;$$

iv) *The function* $t \mapsto e^{tA}$ *belongs to* $C^\infty((0,+\infty), L(\mathbf{X}))$, *and*

$$\frac{d^k}{dt^k} e^{tA} = A^k e^{tA}, \ t > 0,$$

moreover it has an analytic extension in the sector

$$S = \{\lambda \in \mathbf{C} : |arg\lambda| < \theta - \pi/2\}.$$

Proof. For the proof see [140, pp. 35-37].

Definition 1.4 For every sectorial operator A the semigroup $(e^{tA})_{t \geq 0}$ defined in Theorem 1.5 is called *the analytic semigroup* generated by A in $\mathbf{X}$. An analytic semigroup is said to be an *analytic strongly continuous semigroup* if in addition, it is strongly continuous.

There are analytic semigroups which are not strongly continuous, for instance, the analytic semigroups generated by nondensely defined sectorial operators. From the definition of sectorial operators it is obvious that for a sectorial operator A the intersection of the spectrum $\sigma(A)$ with the imaginary axis is bounded.

1.1.3. Spectral Mapping Theorems

If A is a bounded linear operator on a Banach space $\mathbf{X}$, then by the Dunford Theorem [63] $\sigma(exp(tA)) = exp(t\sigma(A))$, $\forall t \geq 0$. It is natural to expect this relation holds for any C_0-semigroups on a Banach space. However, this is not true in general as shown by the following counterexample

Example 1.1

For $n = 1, 2, 3, ...$, let A_n be the $n \times n$ matrix acting on $\mathbf{C}^n$ defined by

$$A_n := \begin{pmatrix} 0 & 1 & 0 & 0 & \cdots \\ 0 & 0 & 1 & 0 & \cdots \\ \vdots & \vdots & \ddots & \ddots & \ddots \end{pmatrix}$$

Each matrix A_n is nilpotent and therefore $\sigma(A_n) = \{0\}$. Let X be the Hilbert space consisting of all sequences $x = (x_n)_{n \in \mathbf{N}}$ with $x_n \in \mathbf{C}^n$ such that

$$\|x\| := \left(\sum_{n=1}^{\infty} \|x_n\|_{\mathbf{C}^n}^2 \right)^{\frac{1}{2}} < \infty.$$

Let $(T(t))_{t \geq 0}$ be the semigroup on X defined coordinatewise by

$$(T(t)) = (e^{int} e^{tA_n})_{n \in \mathbf{N}}.$$

It is easily checked that $(T(t))_{t \geq 0}$ is a C_0-semigroup on X and that $(T(t))_{t \geq 0}$ extends to a C_0-group. Since $\|A_n\| = 1$ for $n \geq 2$, we have $\|e^{tA_n}\| \leq e^t$ and hence $\|T(t)\| \leq e^t$, so $\omega_0((T(t))_{t \geq 0}) \leq 1$, where

$$\omega_0((T(t))_{t \geq 0}) := \inf\{\alpha : \exists N \geq 1 \text{ such that } \|T(t)\| \leq Ne^{\alpha t}, \, \forall t \geq 0\}.$$

First, we show that $s(A) = 0$, where A is the generator of $(T(t))_{t \geq 0}$ and $s(A) := \{\sup \mathrm{Re}\lambda, \lambda \in \sigma(A)\}$. To see this, we note that A is defined coordinatewise by

$$A = (in + A_n)_{n \geq 1}.$$

An easy calculation shows that for all $\mathrm{Re}\lambda > 0$,

$$\lim_{n \to \infty} \|R(\lambda, A_n + in)\|_{\mathbf{C}^n} = 0.$$

It follows that the operator $(R(\lambda, A_n + in))_{n \geq 1}$ defines a bounded operator on X, and clearly this operator is a two-sided inverse of $\lambda - A$. Therefore $\{\mathrm{Re}\lambda > 0\} \subset rho(A)$ and $s(A) \leq 0$. On the other hand, $in \in \sigma(in + A_n) \subset \sigma(A)$ for all $n \geq 1$, so $s(A) = 0$.

Next, we show that $\omega_0((T(t))_{t \geq 0}) = 1$. In view of $\omega_0((T(t))_{t \geq 0}) \leq 1$ it suffices to show that $\omega_0((T(t))_{t \geq 0}) \geq 1$. For each n we put

$$x_n := n^{-\frac{1}{2}}(1, 1, ..., 1) \in \mathbf{C}^n.$$

Then, $\|x_n\|_{\mathbf{C}^n} = 1$ and

$$\|e^{tA_n} x_n\|_{\mathbf{C}^n}^2 = \frac{1}{n} \sum_{m=0}^{n-1} \left(\sum_{j=0}^{m} \frac{t^j}{j!} \right)^2$$

$$= \frac{1}{n} \sum_{m=0}^{n-1} \left(\sum_{j,k=0}^{m} \frac{t^{j+k}}{j!k!} \right)$$

$$= \frac{1}{n} \sum_{m=0}^{n-1} \sum_{i=0}^{2m} t^i \sum_{j+k=i} \frac{1}{j!k!}$$

$$= \frac{1}{n} \sum_{m=0}^{n-1} \sum_{i=0}^{2m} \frac{t^i}{i!} \sum_{j=0}^{i} \frac{i!}{j!(i-j)!}$$

$$= \frac{1}{n}\sum_{m=0}^{n-1}\sum_{i=0}^{2m}\frac{2^i t^i}{i!}$$

$$\geq \frac{1}{n}\sum_{i=0}^{2n-2}\frac{2^i t^i}{i!}.$$

For $0 < q < 1$, we define $x_q \in X$ by $x_q := (n^{\frac{1}{2}}q^n x_n)_{n\geq 1}$. It is easy to check that $x_q \in D(A)$ and

$$\begin{aligned}\|T(t)x_q\|^2 &= \sum_{n=1}^{\infty} nq^{2n}\|e^{tA_n}x_n\|^2 \\ &\geq \sum_{n=1}^{\infty} nq^{2n}\left(\frac{1}{n}\sum_{i=0}^{2n-2}\frac{2^i t^i}{i!}\right) \\ &= \sum_{i=0}^{\infty}\frac{2^i t^i}{i!}\sum_{n=\{i/2\}+1}^{\infty}q^{2n} \\ &= \sum_{i=0}^{\infty}\frac{q^{2\{i/2\}+2}}{1-q^2}\frac{2^i t^i}{i!} \\ &\geq \frac{q^3}{1-q^2}e^{2tq}.\end{aligned}$$

Here $\{a\}$ denotes the least integer greater than or equal to a; we used that $2\{k/2\}+2 \leq k+3$ for all $k = 0, 1, \ldots$ Thus, $\omega_0((T(t))_{t\geq 0}) \geq q$ for all $0 < q < 1$, so $\omega_0((T(t))_{t\geq 0}) \geq 1$. Hence, the relation $\sigma(T(t)) = e^{t\sigma(A)}$ does not holds for the semigroup $(T(t))_{t\geq 0}$.

In this section we prove the spectral inclusion theorem:

Theorem 1.6 *Let $(T(t))_{t\geq 0}$ be a C_0-semigroup on a Banach space X, with generator A. Then we have the spectral inclusion relation*

$$\sigma(T(t)) \supset e^{t\sigma(A)}, \ \forall t \geq 0.$$

Proof. By Theorem 1.2 for the semigroup $(T_\lambda(t))_{t\geq 0} := \{e^{-\lambda t}T(t)\}_{t\geq 0}$ generated by $A - \lambda$, for all $\lambda \in \mathbf{C}$ and $t \geq 0$

$$(\lambda - A)\int_0^t e^{\lambda(t-s)}T(s)x\,ds = (e^{\lambda t} - T(t))x, \quad \forall x \in X,$$

and

$$\int_0^t e^{\lambda(t-s)}T(s)(\lambda - A)x\,ds = (e^{\lambda t} - T(t))x, \quad \forall x \in D(A). \tag{2.1.1}$$

Suppose $e^{\lambda t} \in \rho(T(t))$ for some $\lambda \in \mathbf{C}$ and $t \geq 0$, and denote the inverse of $e^{\lambda t} - T(t)$ by $Q_{\lambda,t}$. Since $Q_{\lambda,t}$ commutes with $T(t)$ and hence also with A, we have

$$(\lambda - A) \int_0^t e^{\lambda(t-s)} T(s) Q_{\lambda,t} x \, ds = x, \quad \forall x \in X,$$

and

$$\int_0^t e^{\lambda(t-s)} T(s) Q_{\lambda,t} (\lambda - A) x \, ds = x, \quad \forall x \in D(A).$$

This shows the boundedness of the operator B_λ defined by

$$B_\lambda x := \int_0^t e^{\lambda(t-s)} T(s) Q_{\lambda,t} x \, ds$$

is a two-sided inverse of $\lambda - A$. It follows that $\lambda \in \varrho(A)$.

As shown by Example 1.1 the converse inclusion

$$\exp(t\sigma(A)) \supset \sigma(T(t)) \backslash \{0\}$$

in general fails. For certain parts of the spectrum, however, the spectral mapping theorem holds true. To make it more clear we recall that for a given closed operator A on a Banach space $\mathbf{X}$ the *point spectrum* $\sigma_p(A)$ is the set of all $\lambda \in \sigma(A)$ for which there exists a non-zero vector $x \in D(A)$ such that $Ax = \lambda x$, or equivalently, for which the operator $\lambda - A$ is not injective; the *residual spectrum* $\sigma_r(A)$ is the set of all $\lambda \in \sigma(A)$ for which $\lambda - A$ does not have dense range; the *approximate point spectrum* $\sigma_a(A)$ is the set of all $\lambda \in \sigma(A)$ for which there exists a sequence (x_n) of norm one vectors in X, $x_n \in D(A)$ for all n, such that

$$\lim_{n \to \infty} \|Ax_n - \lambda x_n\| = 0.$$

Obviously, $\sigma_p(A) \subset \sigma_a(A)$.

Theorem 1.7 *Let $(T(t))_{t \geq 0}$ be a C_0-semigroup on a Banach space $\mathbf{X}$, with generator A. Then*

$$\sigma_p(T(t)) \backslash \{0\} = e^{t\sigma_p(A)}, \quad \forall t \geq 0.$$

Proof. For the proof see e.g. [179, p. 46]. ∎

Recall that a family of bounded linear operators $(T(t))_{t \in \mathbf{R}}$ is said to be a *strongly continuous group* if it satisfies

i) $T(0) = I$,

ii) $T(t+s) = T(t)T(s)$, $\forall t, s \in \mathbf{R}$,

iii) $\lim_{t \to 0} T(t)x = x$, $\forall x \in \mathbf{X}$.

Similarly to C_0-semigroups, the generator of a strongly continuous group $(T(t))_{t\in\mathbf{R}}$ is defined to be the operator

$$Ax := \lim_{t\to 0} \frac{T(t)x - x}{t},$$

with the domain $D(A)$ consisting of all elements $x \in \mathbf{X}$ such that the above limit exists. For bounded strongly continuous groups of linear operators the following weak spectral mapping theorem holds:

Theorem 1.8 *Let $(T(t))_{t\in\mathbf{R}}$ be a bounded strongly continuous group, i.e., there exists a positive M such that $\|T(t)\| \leq M$, $\forall t \in \mathbf{R}$ with generator A. Then*

$$\sigma(T(t)) = \overline{e^{t\sigma(A)}}, \ \forall t \in \mathbf{R}. \qquad (1.9)$$

Proof. For the proof see e.g. [163] or [173, Chapter 2].

Example 1.2 *Let $\mathcal{M}$ be a closed translation invariant subspace of the space of $\mathbf{X}$-valued bounded uniformly continuous functions on the real line $BUC(\mathbf{R}, \mathbf{X})$, i.e., $\mathcal{M}$ is closed and $S(t)\mathcal{M} \subset \mathcal{M}$, $\forall t$, where $(S(t))_{t\in\mathbf{R}}$ is the translation group on $BUC(\mathbf{R}, \mathbf{X})$. Then*

$$\sigma(S(t)|_{\mathcal{M}}) = \overline{e^{t\sigma(\mathcal{D}_{\mathcal{M}})}}, \ \forall t \in \mathbf{R},$$

where $\mathcal{D}_{\mathcal{M}}$ is the generator of $(S(t)|_{\mathcal{M}})_{t\in\mathbf{R}}$ (the restriction of the group $(S(t))_{t\in\mathbf{R}}$ to $\mathcal{M}$).

In the next chapter we will again consider situations similar to this example which arise in connection with invariant subspaces of so-called evolution semigroups.

1.2. EVOLUTION EQUATIONS

1.2.1. Well-Posed Evolution Equations

Homogeneous and inhomogeneous equations

For a densely defined linear operator A let us consider the *abstract Cauchy problem*

$$\begin{cases} \frac{du(t)}{dt} = Au(t), \ \forall t > 0, \\ u(0) = x \in D(A). \end{cases} \qquad (1.10)$$

The problem (1.10) is called *well posed* if $\rho(A) \neq \emptyset$ and for every $x \in D(A)$ there is a unique (classical) solution $u : [0, \infty) \to D(A)$ of (1.10) in $C^1([0, \infty), \mathbf{X})$. The well posedness of (1.10) involves the existence, uniqueness and continuous dependence on the initial data. The following result is fundamental.

Theorem 1.9 *The problem (1.10) is well posed if and only if A generates a C_0-semigroup on $\mathbf{X}$. In this case the solution of (1.10) is given by $u(t) = T(t)x, t > 0$.*

Proof. The detailed proof of this theorem can be found in [71, p. 83].

In connection with the well posed problem (1.10) we consider the following Cauchy problem

$$\begin{cases} \frac{du(t)}{dt} = Au(t) + f(t), \ \forall t > 0, \\ u(0) = u_0. \end{cases} \quad (1.11)$$

Theorem 1.10 *Let the problem (1.10) be well posed and $u_0 \in D(A)$. Assume either*

i) $f \in C([0,\infty), \mathbf{X})$ *takes values in $D(A)$ and $Af(\cdot) \in C([0,\infty), \mathbf{X})$, or*

ii) $f \in C^1([0,\infty), \mathbf{X})$.

Then the problem (1.11) has a unique solution $u \in C^1([0,\infty), \mathbf{X})$ with values in $D(A)$.

Proof. The detailed proof of this theorem can be found in [71, pp. 84-85].

Even when the conditions of Theorem 1.10 are not satisfied we can speak of *mild solutions* by which we mean continuous solutions of the equation

$$\begin{cases} u(t) = T(t-s)u(s) + \int_s^t T(t-\xi)f(\xi)d\xi, \ \forall t \geq s \geq 0 \\ u(0) = u_0, \ u_0 \in \mathbf{X}, \end{cases} \quad (1.12)$$

where $(T(t))_{t \geq 0}$ is the semigroup generated by A and f is assumed to be continuous. It is easy to see that there exists a unique mild solution of Eq.(1.12) for every $x \in \mathbf{X}$.

Nonautonomous equations

To a time-dependent equation

$$\begin{cases} \frac{du(t)}{dt} = A(t)u(t), \ \forall t \geq s \geq 0, \\ u(s) = x, \end{cases} \quad (1.13)$$

where $A(t)$ is in general unbounded linear operator, the notion of well posedness can be extended, roughly speaking, as follows: if the initial data x is in a dense set of the phase space $\mathbf{X}$, then there exists a unique (classical) solution of (1.13) which depends continuously on the initial data. Let us denote by $U(t,s)x$ the solution of (1.13). By the uniqueness we see that $(U(t,s))_{t \geq s \geq 0}$ is a family of bounded linear operators on $\mathbf{X}$ with the properties

i) $U(t,s)U(s,r) = U(t,r), \ \forall t \geq s \geq r \geq 0$;

ii) $U(t,t) = I, \ \forall t \geq 0$;

iii) $U(\cdot, \cdot)x$ is continuous for every fixed $x \in \mathbf{X}$.

In the next chapter we will deal with families $(U(t,s))_{t \geq s \geq 0}$ rather than with the equations of the form (1.13) which generate such families. This general setting enables us to avoid stating complicated sets of conditions imposed on the coefficient-operators $A(t)$. We refer the reader to [71, pp. 140-147] and [179, Chapter 5] for more information on this subject.

Semilinear evolution equations

The notion of well posedness discussed above can be extended to semilinear equations of the form
$$\frac{dx}{dt} = Ax + Bx, \quad x \in \mathbf{X} \qquad (1.14)$$
where $\mathbf{X}$ is a Banach space, A is the infinitesimal generator of a C_0-semigroup $S(t)$, $t \geq 0$ of linear operators of type ω, i.e.
$$\|S(t)x - S(t)y\| \leq e^{\omega t}\|x - y\|, \ \forall \, t \geq 0, \ x, y \in \mathbf{X},$$
and B is an everywhere defined continuous operator from $\mathbf{X}$ to $\mathbf{X}$. Hereafter, by a mild solution $x(t), t \in [s, \tau]$ of equation (1.14) we mean a continuous solution of the integral equation
$$x(t) = S(t-s)x + \int_s^t S(t-\xi)Bx(\xi)d\xi, \ \forall s \leq t \leq \tau. \qquad (1.15)$$

Before proceeding we recall some notions and results which will be frequently used later on. We define the bracket $[\cdot, \cdot]$ in a Banach space $\mathbf{Y}$ as follows (see e.g. [142] for more information)
$$[x, y] = \lim_{h \to +0} \frac{\|x + hy\| - \|y\|}{h} = \inf_{h > 0} \frac{\|x + hy\| - \|y\|}{h}$$

Definition 1.5 Suppose that F is a given operator on a Banach space $\mathbf{Y}$. Then $(F + \gamma I)$ is said to be *accretive* if and only if for every $\lambda > 0$ one of the following equivalent conditions is satisfied

i) $(1 - \lambda\gamma)\|x - y\| \leq \|x - y + \lambda(Fx - Fy)\|, \ \forall x, y \in D(F)$,

ii) $[x - y, Fx - Fy] \geq -\gamma\|x - y\|, \ \forall x, y \in D(F)$.

In particular, if $\gamma = 0$, then F is said to be accretive.

Remark 1.1 From this definition we may conclude that $(F + \gamma I)$ is accretive if and only if
$$\|x - y\| \leq \|x - y + \lambda(Fx - Fy)\| + \lambda\gamma\|x - y\| \qquad (1.16)$$
for all $x, y \in D(F), \lambda > 0, 1 \geq \lambda\gamma$.

Theorem 1.11 *Let the above conditions hold true. Then for every fixed $s \in \mathbf{R}$ and $x \in \mathbf{X}$ there exists a unique mild solution $x(\cdot)$ of Eq.(1.14) defined on $[s, +\infty)$. Moreover, the mild solutions of Eq.(1.14) give rise to a semigroup of nonlinear operators $T(t), t \geq 0$ having the following properties:*

i) $\quad T(t)x = S(t)x + \int_0^t S(t-\xi)BT(\xi)xd\xi, \ \forall t \geq 0, x \in \mathbf{X}, \qquad (1.17)$

ii) $\quad \|T(t)x - T(t)y\| \leq e^{(\omega+\gamma)t}\|x - y\|, \ \forall t \geq 0, x, y \in \mathbf{X}. \qquad (1.18)$

More detailed information on this subject can be found in [142].

1.2.2. Functional Differential Equations with Finite Delay

Let E be a Banach space with norm $|\cdot|$. Denote by $\mathcal{C} := C([-r, 0], E)$ the Banach space of continuous functions on $[-r, 0]$ taking values in the Banach space E with the maximum norm. Let A be the generator of a C_0-semigroup $(T(t))_{t\geq 0}$ on E. If $(T(t))_{t\geq 0}$ is a C_0-semigroup, then there exist constants $M_w \geq 1, w$ such that

$$\|T(t)\| \leq M_w e^{wt} \quad \text{for } t \geq 0. \tag{1.19}$$

Suppose that $F(t, \phi)$ is an E-valued continuous function defined for $t \geq \sigma, \phi \in \mathcal{C}$, and that there exists a locally integrable function $N(t)$ such that

$$|F(t, \phi) - F(t, \psi)| \leq N(t)|\phi - \psi|, \quad t \geq \sigma, \phi, \psi \in \mathcal{C}.$$

If a continuous function $u : [\sigma - r, \sigma + a) \to E$ satisfies the following equation

$$u(t) = T(t - \sigma)u(\sigma) + \int_\sigma^t T(t-s)F(s, u_s)ds \quad \sigma \leq t < \sigma + a,$$

it is called a mild solution of the functional differential equation

$$u'(t) = Au(t) + F(t, u_t) \tag{1.20}$$

on the interval $[\sigma, \sigma + a)$.

We will need the following lemma to prove the existence and uniqueness of mild solutions.

Lemma 1.1 *Suppose that $a(t)$ and $f_n(t), n \geq 0$, are nonnegative continuous functions for $t \geq \sigma$ such that, for $n \geq 1$*

$$f_n(t) \leq \int_\sigma^t a(s)f_{n-1}(s)ds \quad \sigma \leq t.$$

Then $\sum_{n\geq 1} f_n(t)$ converges uniformly on $[\sigma, \tau]$ for any $\tau \geq \sigma$ and

$$\sum_{n\geq 1} f_n(t) \leq \int_\sigma^t f_0(s)a(s) \exp\left(\int_s^t a(r)dr\right) ds$$

Proof. By induction we have the following inequality for $n \geq 1$:

$$f_n(t) \leq \int_\sigma^t \frac{a(s)}{(n-1)!}\left(\int_\sigma^t a(r)dr\right)^{n-1} f_0(s)ds \quad \sigma \leq t.$$

The lemma follows immediately from this result.

Theorem 1.12 *For every $\phi \in \mathcal{C}$, Eq.(1.20) has a unique mild solution $u(t) = u(t, \sigma, \phi)$ on the interval $[\sigma, \infty)$ such that $u_\sigma = \phi$. Moreover, it satisfies the following inequality:*

$$|u_t| \leq |\phi| M_w e^{\max\{0,w\}(t-\sigma)} \exp\left(\int_\sigma^t M_w N(r) dr\right)$$
$$+ \int_\sigma^t |F(s,0)| M_w e^{\max\{0,w\}(t-s)} \exp\left(\int_s^t M_w N(r) dr\right) ds.$$

Proof. Set $z = \max\{0, w\}$. The successive approximations $\{u^n(t)\}, n \geq 0$, are defined as follows: for $\sigma - r \leq t \leq \sigma$, $u^n(t) = \phi(t - \sigma), n \geq 0$; and for $\sigma < t$, $u^0(t) = T(t - \sigma)\phi(0)$ and

$$u^n(t) = T(t-\sigma)\phi(0) + \int_\sigma^t T(t-s) F(s, u_s^{n-1}) ds$$

for $n \geq 1$, successively. For $n \geq 2$, we have that, for $t \geq \sigma$,

$$|u^n(t) - u^{n-1}(t)| \leq \int_\sigma^t \|T(t-s)\| N(s) |u_s^{n-1} - u_s^{n-2}| ds$$
$$\leq \int_\sigma^t M_w e^{z(t-s)} N(s) |u_s^{n-1} - u_s^{n-2}| ds.$$

The last term is nondecreasing for $t \geq \sigma$. Hence, the continuous functions $f_n(t) := e^{-zt}|u_t^{n+1} - u_t^n|, n \geq 0$, satisfy the inequality of the above lemma with $a(s) = M_w N(s)$. It follows that

$$\sum_{n \geq 1} |u_t^{n+1} - u_t^n| \leq \int_\sigma^t e^{z(t-s)} |u_s^1 - u_s^0| M_w N(s) \exp\left(\int_s^t M_w N(r) dr\right) ds.$$

Thus, for every $a > 0$ the sequence $\{u_t^n\}$ converges uniformly with respect to $t \in [\sigma, \sigma + a]$, and $u(t) = \lim_{n \to \infty} u^n(t)$ is a mild solution on $t \in [\sigma, \infty)$ with $u_\sigma = \phi$. Furthermore,

$$|u_t - u_t^1| \leq \int_\sigma^t e^{z(t-s)} |u_s^1 - u_s^0| M_w N(s) \exp\left(\int_s^t M_w N(r) dr\right) ds.$$

Notice that $u^1(t) - u^0(t) = 0$ for $t \in [\sigma - r, \sigma]$ and

$$u^1(t) - u^0(t) = \int_\sigma^t T(t-s) F(s, u_s^0) ds$$

for $t \geq \sigma$. Set $g(t) = F(t, 0)$ for $t \geq \sigma$. Then

$$|F(t, \phi)| \leq N(t)|\phi| + |g(t)| \quad t \geq \sigma, \phi \in \mathcal{C}.$$

This implies that

$$|u_t^1 - u_t^0| \leq \int_\sigma^t M_w e^{z(t-s)} [N(s)|u_s^0| + |g(s)|] ds.$$

Hence

$$
\begin{aligned}
|u_t - u_t^1| &\leq \int_\sigma^t e^{z(t-s)} \left(\exp\left(\int_s^t M_w N(r) dr\right) - 1 \right) M_w[N(s)|u_s^0| + |g(s)|] ds \\
&= \int_\sigma^t e^{z(t-s)} \exp\left(\int_s^t M_w N(r) dr\right) M_w[N(s)|u_s^0| + |g(s)|] ds \\
&\quad - |u_t^1 - u_t^0|.
\end{aligned}
$$

Notice that $|u_t| \leq |u_t - u_t^1| + |u_t^1 - u_t^0| + |u_t^0|$ and that

$$|u_t^0| \leq M_w e^{z(t-\sigma)} |\phi|.$$

Finally, these prove the estimate of the solution in the theorem.

The following result will be used later whose proof can be found in [216].

Theorem 1.13 *Let $T(t)$ be compact for $t > 0$ and $F(t, \cdot)$ be Lipshitz continuous uniformly in t. Then for every $s > r$ the solution operator $\mathcal{C} \ni \phi \mapsto u_s \in \mathcal{C}$, $(u(t) := u(t, 0, \phi))$, is a compact operator.*

Suppose that $L : R \times \mathcal{C} \to E$ is a continuous function such that, for each $t \in R$, $L(t, \cdot)$ is a continuous linear operator from $\mathcal{C}$ to E. Notice that $\|L(t)\| := \|L(t, \cdot)\|$ is a locally bounded, lower semicontinuous function. For every $\phi \in \mathcal{C}$ and $\sigma \in R$, and for every continuous function $f : R \to E$, the inhomogeneous linear equation

$$u'(t) = Au(t) + L(t, u_t) + f(t) \tag{1.21}$$

with the initial condition $u_\sigma = \phi$ has a unique mild solution $u(t, \sigma, \phi, f)$.

Corollary 1.2 *If $u(t) = u(t, \sigma, \phi, f)$ is a mild solution of Eq.(1.21), then, for $t \geq \sigma$,*

$$
\begin{aligned}
|u_t| &\leq |\phi| M_w e^{\max\{0,w\}(t-\sigma)} \exp\left(\int_\sigma^t M_w \|L(r)\| dr\right) \\
&\quad + \int_\sigma^t |f(s)| M_w e^{\max\{0,w\}(t-s)} \exp\left(\int_s^t M_w \|L(r)\| dr\right) ds.
\end{aligned}
$$

1.2.3. Equations with Infinite Delay

As the phase space for equations with finite delay one usually takes the space of continuous functions. This is justifiable because the section x_t of the solution becomes a continuous function for $t \geq \sigma + r$, where σ is the initial time and r is the delay time of the equation, even if the initial function x_σ is not continuous. However, the situation is different for equations with infinite delay. The section x_t contains the initial function x_σ as its part for every $t \geq \sigma$. There are many candidates for the phase space of equations with infinite delay. However, we can discuss many problems independently of the choice of the phase space. This can be done by extracting the common properties of phase spaces as the axioms of an abstract phase space $\mathcal{B}$. We will use the following fundamental axioms, due to Hale and Kato [79].

The space $\mathcal{B}$ is a Banach space consisting of E-valued functions $\phi, \psi, \cdots$, on $(-\infty, 0]$ satisfying the following axioms.

(B) If a function $x : (-\infty, \sigma + a) \to E$ is continuous on $[\sigma, \sigma + a)$ and $x_\sigma \in \mathcal{B}$, then, for $t \in [\sigma, \sigma + a)$,

i) $x_t \in \mathcal{B}$ and x_t is continuous in $\mathcal{B}$,

ii) $H^{-1}|x(t)| \leq |x_t| \leq K(t-\sigma)\sup\{|x(s)| : \sigma \leq s \leq t\} + M(t-\sigma)|x_\sigma|$, where $H > 0$ is constant, $K, M : [0, \infty) \to [0, \infty)$ are independent of x, K is continuous, M is measurable, and locally bounded.

Now we consider several examples for the space $\mathcal{B}$ of functions $\phi : (-\infty, 0] \to E$. Let $g(\theta), \theta \leq 0$, be a positive, continuous function such that $g(\theta) \to \infty$ as $\theta \to -\infty$. The space UC_g is a set of continuous functions ϕ such that $\phi(\theta)/g(\theta)$ is bounded and uniformly continuous for $\theta \leq 0$. Set

$$|\phi| = \sup\{|\phi(\theta)|/g(\theta) : \theta \in (-\infty, 0]\}.$$

Then this space satsifies the above axioms. The space C_g is the set of continuous functions ϕ such that $\phi(\theta)/g(\theta)$ has a limit in E as $\theta \to -\infty$. Thus, C_g is a closed subspace of UC_g and satisfies the above axioms with respect to the same norm. The space $\mathcal{L}_g$ is the set of strongly measurable functions ϕ such that $|\phi(\theta)|/g(\theta)$ is integrable over $(-\infty, 0]$. Set

$$|\phi| = |\phi(0)| + \int_{-\infty}^{0} |\phi(\theta)|/g(\theta) d\theta.$$

Then this space satisfies the above axioms.

Next, we present several fundamental properties of $\mathcal{B}$. Let BC be the set of bounded, continuous functions on $(-\infty, 0]$ to E, and C_{00} be its subset consisting of functions with compact support. For $\phi \in BC$ put

$$|\phi|_\infty = \sup\{|\phi(\theta)| : \theta \in (-\infty, 0]\}.$$

Every function $\phi \in C_{00}$ is obtained as $x_r = \phi$ for some $r \geq 0$ and for some continuous function $x : R \to E$ such that $x(\theta) = 0$ for $\theta \leq 0$. Since $x_0 = 0 \in \mathcal{B}$, Axiom (B) i) implies that $x_t \in \mathcal{B}$ for $t \geq 0$. As a result C_{00} is a subspace of $\mathcal{B}$, and

$$\|\phi\|_\mathcal{B} \leq K(r)\|\phi\|_\infty, \quad \phi \in C_{00},$$

provided $\mathrm{supp}\phi \subset [-r, 0]$. For every $\phi \in BC$, there is a sequence $\{\phi^n\}$ in C_{00} such that $\phi^n(\theta) \to \phi(\theta)$ uniformly for θ on every compact interval, and that $\|\phi^n\|_\infty \leq \|\phi\|_\infty$. From this observation, the space BC is contained in $\mathcal{B}$ under the additional axiom (C).

(C) If a uniformly bounded sequence $\{\phi^n(\theta)\}$ in C_{00} converges to a function $\phi(\theta)$ uniformly on every compact set of $(-\infty, 0]$, then $\phi \in \mathcal{B}$ and $\lim_{n\to\infty} |\phi^n - \phi| = 0$.

In fact, BC is continuously imbedded into $\mathcal{B}$. The following result is found in [107].

Lemma 1.2 *If the phase space $\mathcal{B}$ satisfies the axiom (C), then $BC \subset \mathcal{B}$ and there is a constant $J > 0$ such that $|\phi|_\mathcal{B} \leq J\|\phi\|_\infty$ for all $\phi \in BC$.*

For each $b \in E$, define a constant function $\bar{b}$ by $\bar{b}(\theta) = b$ for $\theta \in (-\infty, 0]$; then $|\bar{b}|_\mathcal{B} \leq J|b|$ from Lemma 1.2. Define operators $S(t) : \mathcal{B} \to \mathcal{B}, t \geq 0$, as follows:

$$[S(t)\phi](\theta) = \begin{cases} \phi(0) & -t \leq \theta \leq 0, \\ \phi(t+\theta) & \theta \leq -t. \end{cases}$$

Let $S_0(t)$ be the restriction of $S(t)$ to $\mathcal{B}_0 := \{\phi \in \mathcal{B} : \phi(0) = 0\}$. If $x : R \to E$ is continuous on $[\sigma, \infty)$ and $x_\sigma \in \mathcal{B}$, we take a function $y : R \to E$ defined by $y(t) = x(t), t \geq \sigma; y(t) = x(\sigma), t \leq \sigma$. From Lemma 1.2 $y_t \in \mathcal{B}$ for $t \geq \sigma$, and x_t is decomposed as

$$x_t = y_t + S_0(t - \sigma)[x_\sigma - \overline{x(\sigma)}] \quad \text{for } t \in [\sigma, \infty).$$

Using Lemma 1.2 and this equation, we have an inequality

$$|x_t| \leq J \sup\{|x(s)| : \sigma \leq s \leq t\} + |S_0(t-\sigma)[x_\sigma - \overline{x(\sigma)}]|.$$

The phase space $\mathcal{B}$ is called a fading memory space [107] if the axiom (C) holds and $S_0(t)\phi \to 0$ as $t \to \infty$ for each $\phi \in \mathcal{B}_0$. If $\mathcal{B}$ is such a space, then $\|S_0(t)\|$ is bounded for $t \geq 0$ by the Banach Steinhaus theorem, and

$$|x_t| \leq J \sup\{|x(s)| : \sigma \leq s \leq t\} + M|x_\sigma|,$$

where $M = (1 + HJ)\sup_{t\geq 0} \|S_0(t)\|$. As a result, we have the following property.

Proposition 1.1 *Assume that $\mathcal{B}$ is a fading memory space. If $x : R \to E$ is bounded, and continuous on $[\sigma, \infty)$ and $x_\sigma \in \mathcal{B}$, then x_t is bounded in $\mathcal{B}$ for $t \geq \sigma$.*

In addition, if $\|S_0(t)\| \to 0$ as $t \to \infty$, then $\mathcal{B}$ is called a uniform fading memory space. It is shown in [107, p.190], that the phase space $\mathcal{B}$ is a uniform fading memory space if and only if the axiom (C) holds and $K(t)$ is bounded and $\lim_{t\to\infty} M(t) = 0$ in the axiom (B).

For the space UC_g, we have that

$$\|S_0(t)\| = \sup\{g(s)/g(s-t) : s \leq 0\},$$

and this space is a uniform fading memory space if and only if it is a fading memory space, cf. [107, p.191].

Let A be the infinitesimal generator of a C_0-semigroup on E such that $\|T(t)\| \leq Me^{wt}, t \geq 0$. Suppose that $F(t,\phi)$ is an E-valued continuous function defined for $t \geq \sigma, \phi \in \mathcal{B}$, and that there exists a locally integrable function $N(t)$ such that

$$|F(t,\phi) - F(t,\psi)| \leq N(t)|\phi - \psi|, \quad t \geq \sigma, \phi, \psi \in \mathcal{B}.$$

Every continuous solution $u : [\sigma - r, \sigma + a) \to E$ of the equation

$$u(t) = T(t-\sigma)u(\sigma) + \int_\sigma^t T(t-s)F(s, u_s)ds \quad \sigma \leq t < \sigma + a, \qquad (1.22)$$

will be called a mild solution of the functional differential equation

$$u'(t) = Au(t) + F(t, u_t) \qquad (1.23)$$

on the interval $[\sigma, \sigma + a)$.

As in the equations with finite delay, the mild solution exists uniquely for $\phi \in \mathcal{B}$, and the norm $|u_t(\phi)|$ is estimated in the similar manner in terms of the functions $K(r), M(r)$ appearing in the axiom (B). We refer the reader to [206], [108] for more details on the results of this section. The compact property of the orbit in $\mathcal{B}$ of a bounded solution follows from the following lemmas (for the proofs see [108]).

Lemma 1.3 *Let S be a compact subset of a fading memory space $\mathcal{B}$. Let $W(S)$ be a set of functions $x : R \to E$ having the following properties :*

i) $x_0 \in S$.

ii) The family of the restrictions of x to $[0, \infty)$ is equicontinuous.

iii) The set $\{x(t) : t \leq 0, x \in W(S)\}$ is relatively compact in E.

Then the set $V(S) := \{x_t : t \geq 0, x \in W(S)\}$ is relatively compact in $\mathcal{B}$.

Lemma 1.4 *In Eq.(1.22) let $\mathcal{B}$ be a fading memory space, $(T(t))_{t \geq 0}$ be a compact C_0-semigroup, and $F(t,\phi)$ be such that for every $B > 0$*

$$\sup_{t \geq 0, |\phi| \leq B}\{|F(t,\phi)|\} < +\infty.$$

Then, for every solution $u(t)$ of Eq.(1.22) bounded on $[0,+\infty)$, the orbit $\{u_t : t \geq 0\}$ is relatively compact in $\mathcal{B}$.

1.3. SPECTRAL THEORY AND ALMOST PERIODICITY OF BOUNDED UNIFORMLY CONTINUOUS FUNCTIONS

1.3.1. Spectrum of a Bounded Function

We denote by $\mathcal{F}$ the Fourier transform, i.e.

$$(\mathcal{F}f)(s) := \int_{-\infty}^{+\infty} e^{-ist} f(t) dt \qquad (1.24)$$

($s \in \mathbf{R}, f \in L^1(\mathbf{R})$). Then the *Beurling spectrum* of $u \in BUC(\mathbf{R}, \mathbf{X})$ is defined to be the following set

$$sp(u) := \{\xi \in \mathbf{R} : \forall \varepsilon > 0 \ \exists f \in L^1(\mathbf{R}), supp\mathcal{F}f \subset (\xi - \varepsilon, \xi + \varepsilon), f * u \neq 0\} \qquad (1.25)$$

where

$$f * u(s) := \int_{-\infty}^{+\infty} f(s-t) u(t) dt.$$

Example 1.3 *If $f(t)$ is a 1-periodic function with Fourier series*

$$\sum_{k \in \mathbf{Z}} f_k e^{2i\pi kt},$$

then

$$sp(f) = \{2\pi k : f_k \neq 0\}.$$

Proof. For every $\lambda \neq 2k_0\pi$, $k_0 \in \mathbf{Z}$ or $\lambda = 2k_0\pi$ at which $f_{k_0} = 0$, where f_n is the Fourier coefficients of f, and for every positive ε, let $\phi \in L^1(\mathbf{R})$ be a complex valued continuous function such that the support of its Fourier transform $supp\mathcal{F}\phi \subset [\lambda - \varepsilon, \lambda + \varepsilon]$. Put

$$u(t) = f * \phi(t) = \int_{-\infty}^{\infty} f(t-s) \phi(s) ds.$$

Since f is periodic, there is a sequence of trigonometric polynomials

$$P_n(t) = \sum_{k=1}^{N(n)} a_{n,k} e^{2ik\pi t}$$

convergent uniformly to f with respect to $t \in \mathbf{R}$ such that $\lim_{n \to \infty} a_{n,k} = f_n$. We have

$$\begin{aligned} u(t) &= f * \phi(t) = \lim_{n \to \infty} P_n * \phi(t) \\ &= \lim_{n \to \infty} \sum_{k=1}^{N(n)} a_{n,k} e^{2ik\pi \cdot} * \phi(t) \end{aligned}$$

$$\begin{aligned}
&= \lim_{n\to\infty} \sum_{k=1}^{N(n)} a_{n,k} e^{2ik\pi t} \int_{-\infty}^{\infty} e^{-2ik\pi s}\phi(s)ds \\
&= \lim_{n\to\infty} \sum_{k=1}^{N(n)} a_{n,k} e^{2ik\pi t} \mathcal{F}\phi(2k\pi) \\
&= 0.
\end{aligned}$$

This, by definition, shows that $sp(f) \subset \{m \in 2\pi\mathbf{Z} : f_m \neq 0\}$. Conversely, for $\lambda \in \{m \in 2\pi\mathbf{Z} : f_m \neq 0\}$ and for every sufficiently small positive ε we can choose a complex function $\varphi \in L^1(\mathbf{R})$ such that $\mathcal{F}\varphi(\xi) = 1$, $\forall \xi \in [\lambda - \varepsilon, \lambda + \varepsilon]$ and $\mathcal{F}\varphi(\xi) = 0$, $\forall \xi \notin [\lambda - \varepsilon, \lambda + \varepsilon]$. Repeating the above argument, we have

$$\begin{aligned}
w(t) &= f * \varphi(t) = \lim_{n\to\infty} P_n(t) * \varphi(t) \\
&= \lim_{n\to\infty} \sum_{k=1}^{N(n)} a_{n,k} e^{2ik\pi t} \mathcal{F}\varphi(2k\pi) \\
&= \lim_{n\to\infty} a_{n,k_0} e^{2ik_0\pi t}.
\end{aligned} \qquad (1.26)$$

Since $\lim_{n\to\infty} a_{n,k_0} = f_{k_0}$ this shows that $w \neq 0$. Thus, $\lambda \in sp(f)$.

Theorem 1.14 *Under the notation as above, $sp(u)$ coincides with the set consisting of $\xi \in \mathbf{R}$ such that the Fourier-Carleman transform of u*

$$\hat{u}(\lambda) = \begin{cases} \int_0^\infty e^{-\lambda t} u(\lambda) dt, & (Re\lambda > 0); \\ -\int_0^\infty e^{\lambda t} u(-t) dt, & (Re\lambda < 0) \end{cases} \qquad (1.27)$$

has no holomorphic extension to any neighborhood of $i\xi$.

Proof. For the proof we refer the reader to [185, Proposition 0.5, p.22].

We collect some main properties of the spectrum of a function, which we will need in the sequel.

Theorem 1.15 *Let $f, g_n \in BUC(\mathbf{R}, \mathbf{X}), n \in \mathbf{N}$ such that $g_n \to f$ as $n \to \infty$. Then*

i) *$sp(f)$ is closed,*

ii) *$sp(f(\cdot + h)) = sp(f)$,*

iii) *If $\alpha \in \mathbf{C}\backslash\{0\}$ $sp(\alpha f) = sp(f)$,*

iv) *If $sp(g_n) \subset \Lambda$ for all $n \in \mathbf{N}$ then $sp(f) \subset \overline{\Lambda}$,*

v) *If A is a closed operator, $f(t) \in D(A) \forall t \in \mathbf{R}$ and $Af(\cdot) \in BUC(\mathbf{R}, \mathbf{X})$, then, $sp(Af) \subset sp(f)$,*

vi) *$sp(\psi * f) \subset sp(f) \cap supp\mathcal{F}\psi, \forall \psi \in L^1(\mathbf{R})$.*

Proof. For the proof we refer the reader to [221, Proposition 0.4, p. 20, Theorem 0.8 , p. 21] and [185, p. 20-21].

As an immediate consequence of the above theorem we have the following.

Corollary 1.3 *Let Λ be a closed subset of $\mathbf{R}$. Then the set*

$$\Lambda(\mathbf{X}) := \{g \in BUC(\mathbf{R}, \mathbf{X}) : sp(g) \subset \Lambda\}$$

is a closed subspace of $BUC(\mathbf{R}, \mathbf{X})$ which is invariant under translations.

We consider the translation group $(S(t))_{t \in \mathbf{R}}$ on $BUC(\mathbf{R}, \mathbf{X})$. One of the frequently used properties of the spectrum of a function is the following:

Theorem 1.16 *Under the notation as above,*

$$i\, sp(u) = \sigma(\mathcal{D}_u), \tag{1.28}$$

where $\mathcal{D}_u$ is the generator of the restriction of the group $S(t)$ to $\mathcal{M}_u$.

Proof. For the proof see [61, Theorem 8.19, p. 213].

We will need also the following result (see e.g. [8], [173, Lemma 5.1.7]) in the next chapter.

Lemma 1.5 *Let A be the generator of a C_0-group $\mathcal{U} = (U(t))_{t \in \mathbf{R}}$ of isometries on a Banach space $\mathbf{Y}$. Let $z \in \mathbf{Y}$ and $\xi \in \mathbf{R}$ and suppose that there exist a neighborhood V of $i\xi$ in $\mathbf{C}$ and a holomorphic function $h : V \to \mathbf{Y}$ such that $h(\lambda) = R(\lambda, A)z$ whenever $\lambda \in V$ and $\Re\lambda > 0$. Then $i\xi \in \rho(A_z)$, where A_z is the generator of the restriction of $\mathcal{U}$ to the closed linear span of $\{U(t)z, t \in \mathbf{R}\}$ in $\mathbf{Y}$.*

1.3.2. Almost Periodic Functions

Definition and basic properties

A subset $E \subset \mathbf{R}$ is said to be *relatively dense* if there exists a number $l > 0$ (*inclusion length*) such that every interval $[a, a+l]$ contains at least one point of E. Let f be a continuous function on $\mathbf{R}$ taking values in a complex Banach space $\mathbf{X}$. f is said to be *almost periodic* if to every $\varepsilon > 0$ there corresponds a relatively dense set $T(\varepsilon, f)$ (*of ε-translations, or ε-periods*) such that

$$\sup_{t \in \mathbf{R}} \|f(t+\tau) - f(t)\| \leq \varepsilon, \ \forall \tau \in T(\varepsilon, f).$$

Example 1.4 *All trigonometric polynomials*

$$P(t) = \sum_{k=1}^{n} a_k e^{i\lambda_k t}, \ (a_k \in \mathbf{X}, \lambda_k \in \mathbf{R})$$

are almost periodic.

We collect some basic properties of almost periodic functions in the following:

Theorem 1.17 *Let f and $f_n, n \in \mathbf{R}$ be almost periodic functions with values in $\mathbf{X}$. Then the following assertions hold true:*

i) The range of f is precompact, i.e., the set $\overline{\{f(t), t \in \mathbf{R}\}}$ is a compact subset of $\mathbf{X}$, so f is bounded;

ii) f is uniformly continuous on $\mathbf{R}$;

iii) If $f_n \to g$ as $n \to \infty$ uniformly, then g is almost periodic;

iv) If f' is uniformly continuous, then f' is almost periodic.

Proof. For the proof see e.g. [7, pp. 5-6].

As a consequence of Theorem 1.17 the space of all almost periodic functions taking values in $\mathbf{X}$ with sup-norm is a Banach space which will be denoted by $AP(\mathbf{X})$. For almost periodic functions the following criterion holds (*Bochner's criterion*):

Theorem 1.18 *Let f be a continuous function taking values in $\mathbf{X}$. Then f is almost periodic if and only if given a sequence $\{c_n\}_{n \in \mathbf{N}}$ there exists a subsequence $\{c_{n_k}\}_{k \in \mathbf{N}}$ such that the sequence $\{f(\cdot + c_{n_k})\}_{k \in \mathbf{N}}$ converges uniformly in $BC(\mathbf{R}, \mathbf{X})$.*

Proof. For the proof see e.g. [7, p. 9].

1.3.3. Spectrum of an Almost Periodic Function

There is a natural extension of the notion of Fourier exponents of periodic functions to almost periodic functions. In fact, if f is almost periodic function taking values in $\mathbf{X}$, then for every $\lambda \in \mathbf{R}$ the average

$$a(f, \lambda) := \lim_{T \to \infty} \frac{1}{2T} \int_{-T}^{T} e^{-i\lambda t} f(t) dt$$

exists and is different from 0 at most at countably many points λ. The set $\{\lambda \in \mathbf{R} : a(f, \lambda) \neq 0\}$ is called *Bohr spectrum* of f which will be denoted by $\sigma_b(f)$. The following Approximation Theorem of almost periodic functions holds

Theorem 1.19 *(Approximation Theorem) Let f be an almost periodic function. Then for every $\varepsilon > 0$ there exists a trigonometric polynomial*

$$P_\varepsilon(t) = \sum_{j=1}^{N} a_j e^{i\lambda_j t}, \ a_j \in \mathbf{X}, \lambda_j \in \sigma_b(f)$$

such that

$$\sup_{t \in \mathbf{R}} \|f(t) - P_\varepsilon(t)\| < \varepsilon.$$

Proof. For the proof see e.g. [137, pp. 17-24].

Remark 1.2 The trigonometric polynomials $P_\varepsilon(t)$ in Theorem 1.19 can be chosen as an element of the space

$$\mathcal{M}_f := \overline{span\{S(\tau)f, \tau \in \mathbf{R}\}}$$

(see [137, p. 29]. Moreover, without loss of generality by assuming that $\sigma_b(f) = \{\lambda_1, \lambda_2, \cdots\}$ one can choose a sequence of trigonometric polynomials, called *trigonometric polynomials of Bochner-Fejer*, approximating f such that

$$P_m(t) = \sum_{j=1}^{N(m)} \gamma_{m,j} a(\lambda_j, f) e^{i\lambda_j t}, m \in \mathbf{N},$$

where $\lim_{m\to\infty} \gamma_{m,j} = 1$. As a consequence we have:

Corollary 1.4 *let f be almost periodic. Then*

$$\mathcal{M}_f = \overline{span\{a(\lambda, f)e^{i\lambda \cdot}, \lambda \in \sigma_b(f)\}}.$$

Proof. By Theorem 1.19,

$$\mathcal{M}_f \subset \overline{span\{a(f, \lambda)e^{i\lambda \cdot}, \lambda \in \sigma_b(f)\}}.$$

On the other hand, it is easy to prove by induction that if P is any trigonometric polynomial with different exponetnts $\{\lambda_1, \cdots, \lambda_k\}$, such that

$$P(t) = \sum_{j=1}^{k} x_j e^{i\lambda_k t},$$

then $x_j e^{i\lambda_j} \in \mathcal{M}_P$, $\forall j = 1, \cdots, k$. Hence by Remark 1.2, obviously, $a(\lambda_j, f)e^{i\lambda_j} \in \mathcal{M}_f$, $\forall j \in \mathbf{N}$.

The relation between the spectrum of an almost periodic function f and its Bohr spectrum is stated in the following:

Proposition 1.2 *If f is an almost periodic function, then $sp(f) = \overline{\sigma_b(f)}$.*

Proof. Let $\lambda \in \sigma_b(f)$. Then there is a $x \in \mathbf{X}$ such that $xe^{i\lambda \cdot} \in \mathcal{M}_f$. Obviously, $\lambda \in \sigma(\mathcal{D}|_{\mathcal{M}_f})$. By Theorem 1.16 $\lambda \in sp(f)$. Conversely, by Theorem 1.19 f can be approximated by a sequnece of trigonometric polynomials with exponents contained in $\sigma_b(f)$. In view of Theorem 1.15 $sp(f) \subset \overline{\sigma_b(f)}$.

1.3.4. A Spectral Criterion for Almost Periodicity of a Function

Suppose that we know beforehand that $f \in BUC(\mathbf{R}, \mathbf{X})$. It is often possible to establish the almost periodicity of this function starting from certain *a priori* information about its spectrum.

Theorem 1.20 *Let $\mathcal{E}$ and $\mathcal{G}$ be closed, translation invariant subspaces of the space $BUC(\mathbf{R}, \mathbf{X})$ and suppose that*

i) $\mathcal{G} \subset \mathcal{E}$;

ii) $\mathcal{G}$ *contains all constant functions which belong to $\mathcal{E}$;*

iii) $\mathcal{E}$ *and $\mathcal{G}$ are invariant under multiplications by $e^{i\xi \cdot}$ for all $\xi \in \mathbf{R}$;*

iv) whenever $f \in \mathcal{G}$ and $F \in \mathcal{E}$, where $F(t) = \int_0^t f(s)ds$, then $F \in \mathcal{G}$.

Let $u \in \mathcal{E}$ have countable reduced spectrum

$$sp_\mathcal{G}(u) := \{\xi \in \mathbf{R} : \forall \varepsilon > 0 \ \exists f \in L^1(\mathbf{R}) \text{ such that}$$
$$\text{supp}\mathcal{F}f \subset (\xi - \varepsilon, \xi + \varepsilon) \text{ and } f * u \notin \mathcal{G}\}.$$

Then $u \in \mathcal{G}$.

Proof. For the proof see [8, p. 371].

Remark 1.3 In the case where $\mathcal{G} = AP(\mathbf{X})$ the condition iv) in Theorem 1.20 can be replaced by the condition that $\mathbf{X}$ does not contain c_0 (see [8, Proposition 3.1, p. 369]). Another alternative of the condition iv) is the total ergodicity of u which is defined as follows: $u \in BUC(\mathbf{R}, \mathbf{X})$ is called *totally ergodic* if

$$M_\eta u := \lim_{\tau \to \infty} \frac{1}{2\tau} \int_{-\tau}^{\tau} e^{i\eta s} S(s) ds$$

exists in $BUC(\mathbf{R}, \mathbf{X})$ for all $\eta \in \mathbf{R}$ (see [26, Theorem 2.6], [27, Theorem 4.2.6], [192]). Another case in which the condition iv) can be dropped is that $sp(u)$ is discrete (see [12]). From this remark the following example is obvious:

Example 1.5 *A function $f \in BUC(\mathbf{R}, \mathbf{X})$ is 2π-periodic if and only if $sp(f) \subset \mathbf{Z}$.*

CHAPTER 2

SPECTRAL CRITERIA FOR PERIODIC AND ALMOST PERIODIC SOLUTIONS

The problem of our primary concern in this chapter is to find spectral conditions for the existence of almost periodic solutions of periodic equations. Although the theory for periodic equations can be carried out parallelly to that for autonomous equations, there is always a difference between them. This is because that in general there is no Floquet representation for the monodromy operators in the infinite dimensional case. Section 1 will deal with evolution semigroups acting on invariant function spaces of $AP(\mathbf{X})$. Since, originally, this technique is intended for nonautonomous equations we will treat equations with as much nonautonomousness as possible, namely, periodic equations. The spectral conditions are found in terms of spectral properties of the monodromy operators. Meanwhile, for the case of autonomous equations these conditions will be stated in terms of spectral properties of the operator coefficients. This can be done in the framework of evolution semigroups and sums of commuting operators in Section 2. Section 3 will be devoted to the critical case in which a fundamental technique of decomposition is presented. In Section 4 we will present another, but traditional, approach to periodic solutions of abstract functional differential equations. The remainder of the chapter will be devoted to several extensions of these methods to discrtete systems and nonlinear equations. As will be shown in Section 5, many problems of evolution equations can be studied through discrete systems with less sophisticated notions.

2.1. EVOLUTION SEMIGROUPS AND ALMOST PERIODIC SOLUTIONS OF PERIODIC EQUATIONS

2.1.1. Evolution Semigroups

Let us consider the following linear evolution equations

$$\frac{dx}{dt} = A(t)x, \qquad (2.1)$$

and

$$\frac{dx}{dt} = A(t)x + f(t), \qquad (2.2)$$

where $x \in \mathbf{X}$, $\mathbf{X}$ is a complex Banach space, $A(t)$ is a (unbounded) linear operator acting on $\mathbf{X}$ for every fixed $t \in \mathbf{R}$ such that $A(t) = A(t+1)$ for all $t \in \mathbf{R}$, $f : \mathbf{R} \to \mathbf{X}$ is an almost periodic function. Under suitable conditions Eq.(2.1) is well-posed (see e.g. [179]), i.e., one can associate with this equation an evolutionary process

$(U(t,s))_{t\geq s}$ which satisfies, among other things, the conditions in the following definition.

Definition 2.1 A family of bounded linear operators $(U(t,s))_{t\geq s}, (t,s \in \mathbf{R})$ from a Banach space $\mathbf{X}$ to itself is called *1-periodic strongly continuous evolutionary process* if the following conditions are satisfied:

i) $U(t,t) = I$ for all $t \in \mathbf{R}$,

ii) $U(t,s)U(s,r) = U(t,r)$ for all $t \geq s \geq r$,

iii) The map $(t,s) \mapsto U(t,s)x$ is continuous for every fixed $x \in \mathbf{X}$,

iv) $U(t+1, s+1) = U(t,s)$ for all $t \geq s$,

v) $\|U(t,s)\| < Ne^{\omega(t-s)}$ for some positive N, ω independent of $t \geq s$.

If it does not cause any danger of confusion, for the sake of simplicity, we shall often call 1-periodic strongly continuous evolutionary process *(evolutionary) process*. We emphasize that in this chapter, for the sake of simplicity of the notations we assume the 1-periodicity of the processes under consideration, and this does not mean any restriction on the period of the processes.

Once the well-posedness of the equations in question is assumed in stead of the equations with operator-coefficient $A(t)$ we are in fact concerned with the evolutionary processes generated by these equations. In light of this, throughout the book we will deal with the asymptotic behavior of evolutionary processes as defined in Definition 2.1. Our main tool to study the asymptotic behavior of evolutionary processes is to use the notion of *evolution semigroups* associated with given evolutionary processes, which is defined in the following:

Definition 2.2 For a given 1-periodic strongly continuous evolutionary process $(U(t,s))_{t\geq s}$, the following formal semigroup associated with it

$$(T^h u)(t) := U(t, t-h)u(t-h), \forall t \in \mathbf{R}, \qquad (2.3)$$

where u is an element of some function space, is called *evolutionary semigroup* associated with the process $(U(t,s))_{t\geq s}$.

Below, for a given strongly continuous 1-periodic evolutionary process $(U(t,s))_{t\geq s}$ we will be concerned with the following inhomogeneous equation

$$x(t) = U(t,s)x(s) + \int_s^t U(t,\xi)f(\xi)d\xi, \forall t \geq s \qquad (2.4)$$

associated with it. A continuous solution $u(t)$ of Eq.(2.4) will be called *mild solution* to Eq.(2.2). The following lemma will be the key tool to study spectral criteria for almost periodicity in this section which relates the evolution semigroup (2.3) with the operator defined by Eq.(2.4).

Lemma 2.1 *Let $(U(t,s))_{t\geq s}$ be a 1-periodic strongly continuous evolutionary process. Then its associated evolutionary semigroup $(T^h)_{h\geq 0}$ is strongly continuous in $AP(\mathbf{X})$. Moreover, the infinitesimal generator of $(T^h)_{h\geq 0}$ is the operator L defined as follows: $u \in D(L)$ and $Lu = -f$ if and only if $u, f \in AP(\mathbf{X})$ and u is the solution to Eq.(2.4).*

Proof. Let $v \in AP(\mathbf{X})$. First we can see that T^h acts on $AP(\mathbf{X})$. To this end, we will prove the following assertion: Let $Q(t) \in L(\mathbf{X})$ be a family of bounded linear operators which is periodic in t and strongly continuous, i.e., $Q(t)x$ is continuous in t for every given $x \in \mathbf{X}$. Then if $f(\cdot) \in AP(\mathbf{X})$, $Q(\cdot)f(\cdot) \in AP(\mathbf{X})$. The fact that $\sup_t \|Q(t)\| < \infty$ follows from the uniform boundedness principle. By the Approximation Theorem of almost periodic functions we can choose sequences of trigonometric polynomials $f_n(t)$ which converges uniformly to $f(t)$ on the real line. For every $n \in \mathbf{N}$ it is obvious that $Q(\cdot)f_n(\cdot) \in AP(\mathbf{X})$. Hence

$$\sup_t \|Q(t)f_n(t) - Q(t)f(t)\| \leq \sup_t \|Q(t)\| \sup_t \|f_n(t) - f(t)\|$$

implies the assertion. We continue our proof of Lemma 2.1. By definition we have to prove that

$$\limsup_{h\downarrow 0}{}_t \|U(t, t-h)v(t-h) - v(t)\| = 0. \tag{2.5}$$

Since $v \in AP(\mathbf{X})$ the range of $v(\cdot)$ which we denote by K (consisting of $x \in \mathbf{X}$ such that $x = v(t)$ for some real t) is a relatively compact subset of $\mathbf{X}$. Hence the map $(t, s, x) \mapsto U(t, s)x$ is uniformly continuous in the set $\{1 \geq t \geq s \geq -1, x \in K\}$. Now let ε be any positive real. In view of the uniform continuity of the map $(t, s, x) \mapsto U(t, s)x$ in the above-mentioned set there is a positive real $\delta = \delta(\varepsilon)$ such that

$$\|U(t - [t], t - [t] - h)x - x\| < \varepsilon \tag{2.6}$$

for all $0 < h < \delta < 1$ and $x \in K$, where $[t]$ denotes the integer n such that $n \leq t < n+1$. Since $(U(t,s))_{t\geq s}$ is 1-periodic from (2.6) this yields

$$\limsup_{h\downarrow 0}{}_t \|U(t, t-h)v(t-h) - v(t-h)\| = 0. \tag{2.7}$$

Now we have

$$\limsup_{h\downarrow 0}{}_t \|U(t, t-h)v(t-h) - v(t)\| \leq$$
$$\leq \lim_{h\to 0^+} \sup_t \|U(t, t-h)v(t-h) - v(t-h)\| +$$
$$+ \lim_{h\to 0^+} \sup_t \|v(t-h) - v(t)\|. \tag{2.8}$$

Since v is uniformly continuous this estimate and (2.8) imply (2.5), i.e., the evolutionary semigroup $(T^h)_{h\geq 0}$ is strongly continuous in $AP(\mathbf{X})$.

Now let $u \in AP(\mathbf{X})$ be a solution of Eq.(2.4). Then we will show that $u \in D(L)$ and $Lu = -f$. In fact, in view of the strong continuity of $(T^h))_{h\geq 0}$ in $AP(\mathbf{X})$ the integral $\int_0^h T^\xi f d\xi$ exists as an element of $AP(\mathbf{X})$. Hence, by definition,

$$u(t) = U(t, t-h)u(t-h) + \int_{t-h}^{t} U(t,\eta)f(\eta)d\eta,$$
$$= [T^h u](t) + [\int_0^h T^\xi f d\xi](t), \quad \forall h \geq 0, t \in \mathbf{R}.$$

Thus, $u = T^h u + \int_0^h T^\xi f d\xi$, $\forall h \geq 0$. This yields that

$$\lim_{h \downarrow 0} \frac{T^h u - u}{h} = -\lim_{h \downarrow 0} \frac{1}{h} \int_0^h T^\xi f d\xi = -f,$$

i.e., $u \in D(L)$ and $Lu = -f$. Conversely, let $u \in D(L)$ and $Lu = -f$. Then we will show that $u(\cdot)$ is a solution of Eq.(2.4). In fact, this can be done by reversing the above argument, so the details are omitted.

Remark 2.1 It may be noted that in the proof of Lemma 2.1 the precompactness of the ranges of u and f are essiential. Hence, in the same way, we can show the strong continuity of the evolution semigroup $(T^h)_{h\geq 0}$ in $C_0(\mathbf{R}, \mathbf{X})$. Finally, combining this remark and Lemma 2.1 we get immediately the following corollary.

Corollary 2.1 *Let $(U(t,s))_{t\geq s}$ be a 1-periodic strongly continuous process. Then its associated evolutionary semigroup $(T^h)_{h\geq 0}$ is a C_0-semigroup in*

$$AAP(\mathbf{X}) := AP(\mathbf{X}) \oplus C_0(\mathbf{R}, \mathbf{X}).$$

One of the interesting applications of Corollary 2.1 is the following.

Corollary 2.2 *Let $(U(t,s))_{t\geq s}$ be a 1-periodic strongly continuous evolutionary process. Moreover, let $u, f \in AAP(\mathbf{X})$ such that u is a solution of Eq.(2.4). Then the almost periodic component u_{ap} of u satisfies Eq.(2.4) with $f := f_{ap}$, where f_{ap} is the corresponding almost periodic component of f.*

Proof. The evolution semigroup $(T^h)_{h\geq 0}$ leaves $AP(\mathbf{X})$ and $C_0(\mathbf{R}, \mathbf{X})$ invariant. Let us denote by P_{ap}, P_0 the projections on these function spaces, respectively. Then since u is a solution to Eq.(2.4), by Lemma 2.1,

$$\lim_{h \downarrow 0} \frac{T^h u - u}{h} = -f.$$

Hence,

$$P_{ap} \lim_{h \downarrow 0} \frac{T^h u - u}{h} = \lim_{h \downarrow 0} \frac{T^h P_{ap} u - P_{ap} u}{h} = -P_{ap} f.$$

This, by Lemma 2.1, shows that $P_{ap} u := u_{ap}$ is a solution of Eq.(2.4) with $f := P_{ap} f := f_{ap}$.

2.1.2. Almost Periodic Solutions and Applications

Monodromy operators

First we collect some results which we shall need in the book. Recall that for a given 1-periodic evolutionary process $(U(t,s))_{t\geq s}$ the following operator

$$P(t) := U(t, t-1), t \in \mathbf{R} \qquad (2.9)$$

is called *monodromy operator* (or sometime, *period map*, *Poincaré map*). Thus we have a family of monodromy operators. Throughout the book we will denote $P := P(0)$. The nonzero eigenvalues of $P(t)$ are called *characteristic multipliers*. An important property of monodromy operators is stated in the following lemma.

Lemma 2.2 *Under the notation as above the following assertions hold:*

i) $P(t+1) = P(t)$ *for all t; characteristic multipliers are independent of time, i.e. the nonzero eigenvalues of $P(t)$ coincide with those of P,*

ii) $\sigma(P(t))\setminus\{0\} = \sigma(P)\setminus\{0\}$, *i.e., it is independent of t ,*

iii) *If $\lambda \in \rho(P)$, then the resolvent $R(\lambda, P(t))$ is strongly continuous.*

Proof. The periodicity of $P(t)$ is obvious. In view of this property we will consider only the case $0 \leq t \leq 1$. Suppose that $\mu \neq 0, Px = \mu x \neq 0$, and let $y = U(t,0)x$, so $U(1,t)y = \mu y \neq 0$, $y \neq 0$ and $P(t)y = \mu y$. By the periodicity this shows the first assertion.

Let $\lambda \neq 0$ belong to $\rho(P)$. We consider the equation

$$\lambda x - P(t)x = y, \qquad (2.10)$$

where $y \in \mathbf{X}$ is given. If x is a solution to Eq.(2.10), then $\lambda x = y + w$, where $w = U(t,0)(\lambda - P)^{-1}U(1,t)y$. Conversely, defining x by this equation, it follows that $(\lambda - P(t))x = y$ so $\rho(P(t)) \supset \rho(P)\setminus\{0\}$. The second assertion follows by the periodicity. Finally, the above formula involving x proves the third assertion.

Remark 2.2 In view of the above lemma, below, in connection with spectral properties of the monodromy operators, the terminology "monodromy operators" may be referred to as the operator P if this does not cause any danger of confusion.

Invariant functions spaces of evolution semigroups

Below we shall consider the evolutionary semigroup $(T^h)_{h\geq 0}$ in some special invariant subspaces **M** of $AP(\mathbf{X})$.

Definition 2.3 The subspace **M** of $AP(\mathbf{X})$ is said to satisfy *condition H* if the following conditions are satisfied:

i) **M** is a closed subspace of $AP(\mathbf{X})$,

ii) There exists $\lambda \in \mathbf{R}$ such that $\mathbf{M}$, contains all functions of the form $e^{i\lambda \cdot}x$, $x \in \mathbf{X}$,

iii) If $C(t)$ is a strongly continuous 1-periodic operator valued function and $f \in \mathbf{M}$, then $C(\cdot)f(\cdot) \in \mathbf{M}$,

iv) $\mathbf{M}$ is invariant under the group of translations.

In the sequel we will be mainly concerned with the following concrete examples of subspaces of $AP(\mathbf{X})$ which satisfy condition H:

Example 2.1 *Let us denote by $\mathcal{P}(1)$ the subspace of $AP(\mathbf{X})$ consisting of all 1-periodic functions. It is clear that $\mathcal{P}(1)$ satisfies condition H.*

Example 2.2 *Let $(U(t,s))_{t \geq s}$ be a strongly continuous 1-periodic evolutionary process. Hereafter, for every given $f \in AP(\mathbf{X})$, we shall denote by $\mathcal{M}(f)$ the subspace of $AP(\mathbf{X})$ consisting of all almost periodic functions u such that $sp(u) \subset \overline{\{\lambda + 2\pi n, n \in \mathbf{Z}, \lambda \in sp(f)\}}$. Then $\mathcal{M}(f)$ satisfies condition H.*

In fact, obviously, it is a closed subspace of $AP(\mathbf{X})$, and moreover it satisfies conditions ii), iv) of the definition. We now check that condition iii) is also satisfied by proving the following lemma:

Lemma 2.3 *Let $Q(t)$ be a 1-periodic operator valued function such that the map $(t,x) \mapsto Q(t)x$ is continuous. Then for every $u(\cdot) \in AP(\mathbf{X})$, the following spectral estimate holds true:*

$$sp(Q(\cdot)u(\cdot)) \subset \Lambda, \qquad (2.11)$$

where $\Lambda := \overline{\{\lambda + 2k\pi, \lambda \in sp(u), k \in \mathbf{Z}\}}$.

Proof. Using the Approximation Theorem of almost periodic functions we can choose a sequence of trigonometric polynomials

$$u^{(m)}(t) = \sum_{k=1}^{N(m)} e^{i\lambda_{k,m}t} a_{k,m}, a_{k,m} \in \mathbf{X}$$

such that $\lambda_{k,m} \in \sigma_b(u)$ (:= Bohr spectrum of u), $\lim_{m \to \infty} u^{(m)}(t) = u(t)$ uniformly in $t \in \mathbf{R}$. The lemma is proved if we have shown that

$$sp(Q(\cdot)u^{(m)}(\cdot)) \subset \Lambda. \qquad (2.12)$$

In turn, to this end, it suffices to show that

$$sp(Q(\cdot)e^{i\lambda_{k,m}\cdot}a_{k,m}) \subset \Lambda. \qquad (2.13)$$

In fact, since $Q(\cdot)a_{k,m}$ is 1-periodic in t, there is a sequence of trigonometric polynomials

$$P_n(t) = \sum_{k=-N(n)}^{N(n)} e^{i2\pi kt} p_{k,n}, p_{k,n} \in \mathbf{X}$$

converging to $Q(\cdot)a_{k,m}$ uniformly as n tends to ∞. Obviously,

$$sp(e^{i\lambda_{k,m}\cdot}P_n(\cdot)) \subset \Lambda. \tag{2.14}$$

Hence,

$$sp(e^{i\lambda_{k,m}\cdot}Q(\cdot)a_{k,m}) \subset \Lambda.$$

The following corollary will be the key tool to study the unique solvability of the inhomogeneous equation (2.4) in various subspaces $\mathbf{M}$ of $AP(\mathbf{X})$ satisfying condition H.

Corollary 2.3 *Let $\mathbf{M}$ satisfy condition H. Then, if $1 \in \rho(T^1|_{\mathbf{M}})$, the inhomogeneous equation (2.4) has a unique solution in $\mathbf{M}$ for every $f \in \mathbf{M}$.*

Proof. Under the assumption, the evolutionary semigroup $(T^h)_{h\geq 0}$ leaves $\mathbf{M}$ invariant. The generator A of $(T^h|_{\mathbf{M}})_{h\geq 0}$ can be defined as the part of L in $\mathbf{M}$. Thus, the corollary is an immediate consequence of Lemma 2.1 and the spectral inclusion $e^{\sigma(A)} \subset \sigma(T^1|_{\mathbf{M}})$.

Let $\mathbf{M}$ be a subspace of $AP(\mathbf{X})$ invariant under the evolution semigroup $(T^h)_{h\geq 0}$ associated with the given 1-periodic evolutionary process $(U(t,s))_{t\geq s}$ in $AP(\mathbf{X})$. Below we will use the following notation

$$\hat{P}_{\mathbf{M}}v(t) := P(t)v(t), \forall t \in \mathbf{R}, v \in \mathbf{M}.$$

If $\mathbf{M} = AP(\mathbf{X})$ we will denote $\hat{P}_{\mathbf{M}} = \hat{P}$.

In the sequel we need the following lemma:

Lemma 2.4 *Let $(U(t,s))_{t\geq s}$ be a 1-periodic strongly continuous evolutionary process and $\mathbf{M}$ be an invariant subspace of the evolution semigroup $(T^h)_{h\geq 0}$ associated with it in $AP(\mathbf{X})$. Then for all invariant subspaces $\mathbf{M}$ satisfying condition H,*

$$\sigma(\hat{P}_{\mathbf{M}})\backslash\{0\} = \sigma(P)\backslash\{0\}.$$

Proof. For $u, v \in \mathbf{M}$, consider the equation $(\lambda - \hat{P}_{\mathbf{M}})u = v$. It is equivalent to the equation $(\lambda - P(t))u(t) = v(t), t \in \mathbf{R}$. If $\lambda \in \rho(\hat{P}_{\mathbf{M}})\backslash\{0\}$, for every v the first equation has a unique solution u, and $\|u\| \leq \|R(\lambda, \hat{P}_{\mathbf{M}})\|\|v\|$. Take a function $v \in \mathbf{M}$ of the form $v(t) = ye^{i\mu t}$, for some $\mu \in \mathbf{R}$; the existence of such a μ is guaranteed by the axioms of condition H. Then the solution u satisfies $\|u\| \leq \|R(\lambda, \hat{P}_{\mathbf{M}})\|\|y\|$. Hence, for every $y \in \mathbf{X}$ the solution of the equation $(\lambda - P(0))u(0) = y$ has a unique solution $u(0)$ such that

$$\|u(0)\| \leq \sup_t \|u(t)\| \leq \|R(\lambda, \hat{P}_{\mathbf{M}})\| \sup_t \|v(t)\| \leq \|R(\lambda, \hat{P}_{\mathbf{M}})\|\|y\|.$$

This implies that $\lambda \in \rho(P)\backslash\{0\}$ and $\|R(\lambda, P(t))\| \leq \|R(\lambda, \hat{P}_{\mathbf{M}})\|$.

Conversely, suppose that $\lambda \in \rho(P)\backslash\{0\}$. By Lemma 2.2 for every v the second equation has a unique solution $u(t) = R(\lambda, P(t))v(t)$ and the map taking t into $R(\lambda, P(t))$ is strongly continuous. By definition of condition H, the function taking t into $(\lambda - P(t))^{-1}v(t)$ belongs to $\mathbf{M}$. Since $R(\lambda, P(t))$ is a strongly continuous, 1-periodic function, by the uniform boundedness principle it holds that $r := \sup\{\|R(\lambda, P(t))\| : t \in R\} < \infty$. This means that $\|u(t)\| \leq r\|v(t)\| \leq r \sup_t \|v(t)\|$, or $\|u\| \leq r\|v\|$. Hence $\lambda \in \rho(\hat{P}_{\mathbf{M}})$, and $\|R(\lambda, \hat{P}_{\mathbf{M}})\| \leq r$.

Unique solvability of the inhomogeneous equations in $\mathcal{P}(1)$

We now illustrate Corollary 2.2 in some concrete situations. First we will consider the unique solvability of Eq.(2.4) in $\mathcal{P}(1)$.

Proposition 2.1 *Let $(U(t,s))_{t \geq s}$ be 1-periodic strongly continuous. Then the following assertions are equivalent:*

i) $1 \in \rho(P)$,

ii) Eq.(2.4) is uniquely solvable in $\mathcal{P}(1)$ for a given $f \in \mathcal{P}(1)$.

Proof. Suppose that i) holds true. Then we show that ii) holds by applying Corollary 2.2. To this end, we show that $\sigma(T^1|_{\mathcal{P}(1)})\backslash\{0\} \subset \sigma(P)\backslash\{0\}$. To see this, we note that
$$T^1|_{\mathcal{P}(1)} = \hat{P}_{\mathcal{P}(1)}.$$
In view of Lemma 2.4 $1 \in \rho(T^1|_{\mathcal{P}(1)})$. By Example 2.1 and Corollary 2.2 ii) holds also true.

Conversely, we suppose that Eq.(2.4) is uniquely solvable in $\mathcal{P}(1)$. We now show that $1 \in \rho(P)$. For every $x \in \mathbf{X}$ put $f(t) = U(t,0)g(t)x$ for $t \in [0,1]$, where $g(t)$ is any continuous function of t such that $g(0) = g(1) = 0$, and
$$\int_0^1 g(t)dt = 1.$$
Thus $f(t)$ can be continued to a 1-periodic function on the real line which we denote also by $f(t)$ for short. Put $Sx = [L^{-1}(-f)](0)$. Obviously, S is a bounded operator. We have
$$[L^{-1}(-f)](1) = U(1,0)[L^{-1}(-f)](0) + \int_0^1 U(1,\xi)U(\xi,0)g(\xi)xd\xi$$
$$Sx = PSx + Px.$$
Thus
$$(I - P)(Sx + x) = Px + x - Px = x.$$
So, $I - P$ is surjective. From the uniqueness of solvability of (2.4) we get easily the injectiveness of $I - P$. In other words, $1 \in \rho(P)$.

Unique solvability in $AP(\mathbf{X})$ and exponential dichotomy

This subsection will be devoted to the unique solvability of Eq.(2.4) in $AP(\mathbf{X})$ and its applications to the study of exponential dichotomy. Let us begin with the following lemma which is a consequence of Proposition 2.1.

Lemma 2.5 *Let $(U(t,s))_{t\geq s}$ be 1-periodic strongly continuous. Then the following assertions are equivalent:*

i) $S^1 \cap \sigma(P) = \emptyset$,

ii) For every given $\mu \in \mathbf{R}, f \in \mathcal{P}(1)$ the following equation has a unique solution in $AP(\mathbf{X})$

$$x(t) = U(t,s)x(s) + \int_s^t U(t,\xi)e^{i\mu\xi}f(\xi)d\xi, \forall t \geq s. \qquad (2.15)$$

Proof. Suppose that i) holds, i.e $S^1 \cap \sigma(P) = \emptyset$. Then, since

$$T^1 = S(-1) \cdot \hat{P} = \hat{P} \cdot S(-1)$$

in view of the commutativeness of two operators $\hat{P}$ and $S(-1)$ (see e.g. [193], Theorem 11.23, p.193)

$$\sigma(T^1) \subset \sigma(S(-1)).\sigma(\hat{P}).$$

It may be noted that $\sigma(S(-1)) = S^1$. Thus

$$\sigma(T^1) \subset \{e^{i\mu}\lambda, \mu \in \mathbf{R}, \lambda \in \sigma(\hat{P})\}.$$

Hence, in view of Lemma 2.4

$$\sigma(T^1) \cap S^1 = \emptyset.$$

Let us consider the process $(V(t,s))_{t\geq s}$ defined by

$$V(t,s)x := e^{-i\mu(t-s)}U(t,s)x$$

for all $t \geq s, x \in \mathbf{X}$. Let $Q(t)$ denote its monodromy operator, i.e. $Q(t) = e^{-i\mu}V(t, t-1)$ and $(T^h_\mu)_{h\geq 0}$ denote the evolution semigroup associated with the evolutionary process $(V(t,s))_{t\geq s}$. Then by the same argument as above we can show that since $\sigma(T^h_\mu) = e^{-i\mu}\sigma(T^h)$,

$$\sigma(T^h_\mu) \cap S^1 = \emptyset.$$

By Lemma 2.1 and Corollary 2.2, the following equation

$$y(t) = V(t,s)y(s) + \int_s^t V(t,\xi)f(\xi)d\xi, \ \forall t \geq s$$

has a unique almost periodic solution $y(\cdot)$. Let $x(t) := e^{i\mu t}y(t)$. Then

$$x(t) = e^{i\mu t}y(t) = U(t,s)e^{i\mu s}y(s) + \int_s^t U(t,\xi)e^{i\mu\xi}f(\xi)d\xi$$
$$= U(t,s)x(s) + \int_s^t U(t,\xi)e^{i\mu\xi}f(\xi)d\xi \; \forall t \geq s.$$

Thus $x(\cdot)$ is an almost periodic solution of Eq.(2.15). The uniqueness of $x(\cdot)$ follows from that of the solution $y(\cdot)$.

We now prove the converse. Let $y(t)$ be the unique almost periodic solution to the equation

$$y(t) = U(t,s)y(s) + \int_s^t U(t,\xi)e^{i\mu\xi}f(\xi)d\xi, \; \forall t \geq s. \tag{2.16}$$

Then $x(t) := e^{-i\mu t}y(t)$ must be the unique solution to the following equation

$$x(t) = e^{-i\mu(t-s)}U(t,s)x(s) + \int_s^t e^{-i\mu(t-\xi)}U(t,\xi)f(\xi)d\xi), \; \forall t \geq s. \tag{2.17}$$

And vice versa. We show that $x(t)$ should be periodic. In fact, it is easily seen that $x(1+\cdot)$ is also an almost periodic solution to Eq.(2.16). From the uniqueness of $y(\cdot)$ (and then that of $x(\cdot)$) we have $x(t+1) = x(t)$, $\forall t$. By Proposition 2.1 this yields that $1 \in \rho(Q(0))$, or in other words, $e^{i\mu} \in \rho(P)$. From the arbitrary nature of μ, $S^1 \cap \sigma(P) = \emptyset$.

Remark 2.3 From Lemma 2.5 it follows in particular that *the inhomogeneous equation (2.4) is uniquely solvable in the function space* $AP(\mathbf{X})$ *if and only if* $S^1 \cap \sigma(P) = \emptyset$. This remark will be useful to consider the asymptotic behavior of the solutions to the homogeneous equation (2.1).

Before applying the above results to study the exponential dichotomy of 1-periodic strongly continuous processes we recall that a given 1-periodic strongly continuous evolutionary process $(U(t,s))_{t \geq s}$ is said to have an *exponential dichotomy* if there exist a family of projections $Q(t), t \in \mathbf{R}$ and positive constants M, α such that the following conditions are satisfied:

i) For every fixed $x \in \mathbf{X}$ the map $t \mapsto Q(t)x$ is continuous,

ii) $Q(t)U(t,s) = U(t,s)Q(s), \; \forall t \geq s,$

iii) $\|U(t,s)x\| \leq Me^{-\alpha(t-s)}\|x\|, \; \forall t \geq s, x \in ImQ(s),$

iv) $\|U(t,s)y\| \geq M^{-1}e^{\alpha(t-s)}\|y\|, \; \forall t \geq s, y \in KerQ(s),$

v) $U(t,s)|_{KerQ(s)}$ is an isomorphism from $KerQ(s)$ onto $KerQ(t), \; \forall t \geq s.$

Theorem 2.1 *Let* $(U(t,s))_{t \geq s}$ *be given 1-periodic strongly continuous evolutionary process. Then the following assertions are equivalent:*

i) The process $(U(t,s))_{t \geq s}$ *has an exponential dichotomy;*

ii) For every given bounded and continuous f the inhomogeneous equation (2.4) has a unique bounded solution;

iii) The spectrum of the monodromy operator P does not intersect the unit circle;

iv) For every given $f \in AP(\mathbf{X})$ the inhomogeneous equation (2.4) is uniquely solvable in the function space $AP(\mathbf{X})$.

Proof. The equivalence of i) and ii) has been established by Zikov (for more general conditions , see e.g. [137, Chap. 10, Theorem 1]. Now we show the equivalence between i), ii) and iii). Let the process have an exponential dichotomy. We now show that the spectrum of the monodromy operator P does not intersect the unit circle. In fact, from ii) it follows that for every 1-periodic function f on the real line there is a unique bounded solution $x(\cdot)$ to Eq.(2.4). This solution should be 1-periodic by the periodicity of the process $(U(t,s))_{t \geq s}$. According to Lemma 2.5, $1 \in \rho(P)$. By the same argument as in the proof of Lemma 2.5 we can show that $e^{i\mu} \in \rho(P), \forall \mu \in \mathbf{R}$. Conversely, suppose that the spectrum of the monodromy operator P does not intersect the unit circle. The assertion follows readily from [90, Theorem 7.2.3 , p. 198]. The equivalence of iv) and iii) is clear from Lemma 2.5.

Unique solvability of the inhomogeneous equations in $\mathcal{M}(f)$

Now let us return to the more general case where the spectrum of the monodromy operator may intersect the unit circle. In the sequel we shall need the following basic property of the translation group on $\Lambda(\mathbf{X})$ which proof can be done in a standard manner.

Lemma 2.6 *Let Λ be a closed subset of the real line. Then*

i) $\sigma(\mathcal{D}_{\Lambda(\mathbf{X})}) = i\Lambda$,

ii) $\sigma(\mathcal{D}_{\Lambda(\mathbf{X}) \cap AP(\mathbf{X})}) = i\Lambda$.

Proof. (i) First, we note that for every $\lambda \in \Lambda, i\lambda \in \sigma(\mathcal{D}_{\Lambda(\mathbf{X})})$. In fact, $\mathcal{D}e^{i\lambda \cdot}x = i\lambda e^{i\lambda \cdot}x$. Now suppose that $\lambda_0 \notin \Lambda$. Then we shall show that $i\lambda_0 \in \rho(\mathcal{D}_{\Lambda(\mathbf{X})})$. To this end, we consider the following equation

$$\frac{du}{dt} = i\lambda_0 u + g(t), g \in \Lambda(\mathbf{X}). \qquad (2.18)$$

By Theorem 1.16 we have that $isp(g) = \sigma(\mathcal{D}_{\mathcal{M}_g})$, where $\mathcal{M}_g$ is the closed subspace of $BUC(\mathbf{R}, \mathbf{X})$, spanned by all translations of g. Thus, $i\lambda_0 \notin \sigma(\mathcal{D}_{\mathcal{M}_g})$; and hence the above equation has a unique solution $h \in \mathcal{M}_g \subset \Lambda(\mathbf{X})$. If k is another solution to Eq.(2.18) in $\Lambda(\mathbf{X})$, then $h - k$ is a solution in $\Lambda(\mathbf{X})$ to the homogeneous equation associated with Eq.(2.18). Thus, a computation via Carleman transform shows that $sp(h-k) \subset \{\lambda_0\}$. On the one hand, we get $\lambda_0 \notin sp(h-k)$ because of $sp(h-k) \subset \Lambda$. Hence, $sp(h-k) = \emptyset$, and then $h - k = 0$. In other words, Eq.(2.18) has a unique solution in $\Lambda(\mathbf{X})$. This shows that the above equation has a unique solution in $\Lambda(\mathbf{X})$, i.e. $i\lambda_0 \in \rho(\mathcal{D}_{\Lambda(\mathbf{X})})$.
(ii) The second assertion can be proved in the same way.

Theorem 2.2 *Let $(U(t,s))_{t\geq s}$ be a 1-periodic strongly continuous evolutionary process. Moreover, let $f \in AP(\mathbf{X})$ such that $\sigma(P) \cap \overline{\{e^{i\lambda}, \lambda \in sp(f)\}} = \emptyset$. Then the inhomogeneous equation (2.4) has an almost periodic solution which is unique in $\mathcal{M}(f)$.*

Proof. From Example 2.2 it follows that the function space $\mathcal{M}(f)$ satisfies condition H. Since $(S(t))_{t\in \mathbf{R}}$ is an isometric C_0-group, by the weak spectral mapping theorem for isometric groups (Theorem 1.8) we have

$$\sigma(S(1)|_{\mathcal{M}(f)}) = \overline{e^{\sigma(\mathcal{D}|_{\mathcal{M}(f)})}},$$

where $\mathcal{D}|_{\mathcal{M}(f)}$ is the generator of $(S(t)|_{\mathcal{M}(f)})_{t\geq 0}$. By Lemma 2.6 we have

$$\sigma(\mathcal{D}|_{\mathcal{M}(f)}) = i\Lambda,$$

where $\Lambda = \overline{\{\lambda + 2\pi k, \lambda \in sp(f), k \in \mathbf{Z}\}}$. Hence, since

$$e^{\sigma(\mathcal{D}|_{\mathcal{M}(f)})} = e^{i\Lambda} \subset \overline{e^{isp(f)}} \subset \overline{e^{i\Lambda}},$$

we have

$$\sigma(S(1)|_{\mathcal{M}(f)}) = \overline{e^{\sigma(\mathcal{D}|_{\mathcal{M}(f)})}} = \overline{e^{isp(f)}}.$$

Thus, the condition

$$\sigma(P) \cap \overline{e^{isp(f)}} = \emptyset$$

is equivalent to the following

$$1 \notin \sigma(P).\sigma(S(-1)|_{\mathcal{M}(f)}).$$

In view of the inclusion

$$\begin{aligned}\sigma(T^1|_{\mathcal{M}(f)})\backslash\{0\} &\subset \sigma(\hat{P}_{\mathcal{M}(f)}).\sigma(S(-1)|_{\mathcal{M}(f)})\backslash\{0\} \\ &\subset \sigma(P).\sigma(S(-1)|_{\mathcal{M}(f)})\backslash\{0\}\end{aligned}$$

which follows from the commutativeness of the operator $\hat{P}_{\mathcal{M}(f)}$ with the operator $S(-1)|_{\mathcal{M}(f)}$, the above inclusion implies that

$$1 \notin \sigma(T^1|_{\mathcal{M}(f)}).$$

Now the assertion of the theorem follows from Corollary 2.2.

Unique solvability of nonlinearly perturbed equations

Let us consider the semilinear equation

$$x(t) = U(t,s)x(s) + \int_s^t U(t,\xi)g(\xi, x(\xi))d\xi. \qquad (2.19)$$

We shall be interested in the unique solvability of (2.19) for a larger class of the forcing term g. We shall show that the generator of evolutionary semigroup is still useful in studying the perturbation theory in the critical case in which the spectrum of the monodromy operator P may intersect the unit circle. We suppose that $g(t,x)$ is Lipschitz continuous with coefficient k and the Nemystky operator F defined by $(Fv)(t) = g(t,v(t)), \forall t \in \mathbf{R}$ acts in **M**. Below we can assume that **M** *is any closed subspace of* the space of all bounded continuous functions $BC(\mathbf{R},\mathbf{X})$. We consider the operator L in $BC(\mathbf{R},\mathbf{X})$. If $(U(t,s))_{t \geq s}$ is strongly continuous, then L is a single-valued operator from $D(L) \subset BC(\mathbf{R},\mathbf{X})$ to $BC(\mathbf{R},\mathbf{X})$.

Lemma 2.7 *Let **M** be any closed subspace of $BC(\mathbf{R},\mathbf{X})$, $(U(t,s))_{t \geq s}$ be strongly continuous and Eq.(2.4) be uniquely solvable in **M**. Then for sufficiently small k, Eq.(2.19) is also uniquely solvable in this space.*

Proof. First, we observe that under the assumptions of the lemma we can define a single-valued operator L acting in **M** as follows: $u \in D(L)$ if and only if there is a function $f \in \mathbf{M}$ such that Eq.(2.4) holds. From the strong continuity of the evolutionary process $(U(t,s))_{t \geq s}$ one can easily see that there is at most one function f such that Eq.(2.4) holds (the proof of this can be carried out in the same way as in that of Lemma 2.9 in the next section). This means L is single-valued. Moreover, one can see that L is closed. Now we consider the Banach space $[D(L)]$ with graph norm, i.e. $|v| = \|v\| + \|Lv\|$. By assumption it is seen that L is an isomorphism from $[D(L)]$ onto **M**. In view of the Lipschitz Inverse Mapping Theorem for Lischitz mappings (see e.g. Theorem 4.11) for sufficiently small k the operator $L - F$ is invertible. Hence there is a unique $u \in \mathbf{M}$ such that $Lu - Fu = 0$. From the definition of operator L we see that u is a unique solution to Eq.(2.19).

Corollary 2.4 *Let **M** be any closed subspace of $AP(\mathbf{X})$, $(U(t,s))_{t \geq s}$ be 1-periodic strongly continuous evolutionary process and for every $f \in \mathbf{M}$ the inhomogeneous equation (2.4) be uniquely solvable in **M**. Moreover let the Nemytsky operator F induced by the nonlinear function g in Eq.(2.19) act on **M**. Then for sufficiently small k, the semilinear equation (2.19) is uniquely solvable in **M**.*

Proof. The corollary is an immediate consequence of Lemma 2.7.

Example 1

In this example we shall consider the abstract form of parabolic partial differential equations (see e.g. [90] for more details) and apply the results obtained above to study the existence of almost periodic solutions to these equations. It may be noted that a necessary condition for the existence of Floquet representation is that the process under consideration is invertible. It is known for the bounded case (see e.g. [55, Chap. V, Theorem 1.2]) that if the spectrum of the monodromy operator does not circle the origin (of course, it should not contain the origin), then the evolution operators admit Floquet representation. In the example below, in general, Floquet representation does not exist. For instance, if the sectorial operator A has compact resolvent, then monodromy operator is compact (see [90] for more details). Thus, if $dim\mathbf{X} = \infty$, then monodromy operators cannot be invertible. However, the above results can apply.

Let A be sectorial operator in a Banach space $\mathbf{X}$, and the mapping taking t into $B(t) \in L(\mathbf{X}^\alpha, \mathbf{X})$ be Hölder continuous and 1-periodic. Then there is a 1-periodic evolutionary process $(U(t,s))_{t \geq s}$ associated with the equation

$$\frac{du}{dt} = (-A + B(t))u. \qquad (2.20)$$

We have the following:

Claim 1 *For any $x_0 \in \mathbf{X}$ and τ there exists a unique (strong) solution $x(t) := x(t; \tau, x_0)$ of Eq.(2.20) on $[\tau, +\infty)$ such that $x(\tau) = x_0$. Moreover, if we write $x(t; \tau, x_0) := T(t, \tau)x_0, \forall t \geq \tau$, then $(T(t,\tau))_{t \geq \tau}$ is a strongly continuous 1-periodic evolutionary process. In addition, if A has compact resolvent, then the monodromy operator $P(t)$ is compact.*

Proof. This claim is an immediate consequence of [90, Theorem 7.1.3, p. 190-191]. In fact, it is clear that $(T(t,\tau))_{t \geq \tau}$ is strongly continuous and 1-periodic. The last assertion is contained in [90, Lemma 7.2.2, p. 197]).

Thus, in view of the above claim if $dim\mathbf{X} = \infty$, then Floquet representation does not exist for the process. This means that the problem cannot reduced to the autonomous and bounded case. To apply our results, let the function f taking t into $f(t) \in \mathbf{X}$ be almost periodic and the spectrum of the monodromy operator of the process $(U(t,s))_{t \geq s}$ be separated from the set $\overline{e^{isp(f)}}$. Then the following inhomogeneous equation

$$\frac{du}{dt} = (-A + B(t))u + f(t)$$

has a unique almost periodic solution u such that

$$sp(u) \subset \overline{\{\lambda + 2\pi k, k \in \mathbf{Z}, \lambda \in sp(f)\}}.$$

We now show

Claim 2 *Let the conditions of Claim 1 be satisfied except for the compactness of the resolvent of A . Then*

$$\frac{dx}{dt} = (-A + B(t))x \qquad (2.21)$$

has an exponential dichotomy if and only if the spectrum of the monodromy operator does not intersect the unit circle. Moreover, if A has compact resolvent, it has an exponential dichotomy if and only if all multipliers have modulus different from one. In particular, it is asymptotically stable if and only if all characteristic multipliers have modulus less than one.

Proof. The operator $T(t,s), t > s$ is compact if A has compact resolvent (see e.g. [90, p. 196]). The claim is an immediate consequence of Theorem 2.1.

Example 2

We examine in this example how the condition of Theorem 2.2 cannot be droped. In fact we consider the simplest case with $A = 0$

$$\frac{dx}{dt} = f(f), x \in \mathbf{R}, \qquad (2.22)$$

where f is continuous and 1-periodic. Obviously,

$$\sigma(e^A) = \{1\} = \overline{e^{i\ sp(f)}}.$$

We assume further that the integral $\int_0^t f(\xi)d\xi$ is bounded. Then every solution to Eq.(2.22) can be extended to a periodic solution defined on the whole line of the form

$$x(t) = c + \int_0^t f(\xi)d\xi, t \in \mathbf{R}.$$

Thus the uniqueness of a periodic solution to Eq.(2.22) does not hold.

Now let us consider the same Eq.(2.22) but with 1-anti-periodic f, i.e., $f(t+1) = f(t), \forall t \in \mathbf{R}$. Clearly,

$$\overline{e^{i\ sp(f)}} = \{-1\} \cap \sigma(e^A) = \varnothing.$$

Hence the conditions of Theorem 2.2 are satisfied. Recall that in this theorem we claim that the uniqueness of the almost periodic solutions is among the class of almost periodic functions g with $\overline{e^{i\ sp(g)}} \subset \overline{e^{i\ sp(f)}}$. Now let us have a look at our example. Every solution to Eq.(2.22) is a sum of the unique 1-anti-periodic solution, which existence is guaranteed by Theorem 2.2, and a solution to the corresponding homogeneous equation, i.e., in this case a constant function. Hence, Eq.(2.22) has infinitely many almost periodic solutions.

2.2. EVOLUTION SEMIGROUPS, SUMS OF COMMUTING OPERATORS AND SPECTRAL CRITERIA FOR ALMOST PERIODICITY

Let $\mathbf{X}$ be a given complex Banach space and $\mathcal{M}$ be a translation invariant subspace of the space of $\mathbf{X}$-valued bounded uniformly continuous functions on the real line $BUC(\mathbf{R}, \mathbf{X})$. The problem of our primary concern in this section is to find conditions for $\mathcal{M}$ to be admissible with respect to differential equations of the form

$$\frac{du}{dt} = Au + f(t), \qquad (2.23)$$

where A is a (unbounded) linear operator with nonempty resolvent set on the Banach space $\mathbf{X}$. By a tradition, by admissibility here we mean that for every $f \in \mathcal{M}$ Eq.(2.23) has a unique solution (in a suitable sense) which belongs to $\mathcal{M}$ as well.

The admissibility theory of function spaces is a classical and well-studied subject of the qualitative theory of differential equations which goes back to a fundamental study of O. Perron on characterization of exponential dichotomy of linear ordinary differential equations.

The next problem we will deal with in the section is concerned with the situation in which one fails to solve Eq.(2.23) uniquely in $\mathcal{M}$. Even in this case one can still find conditions on A so that a given solution $u_f \in BUC(\mathbf{R}, \mathbf{X})$ belongs to $\mathcal{M}$. In this direction, one interesting criterion is the countability of the imaginary spectrum of the operator A which is based on the spectral inclusion

$$isp_{AP}(u_f) \subset i\mathbf{R} \cap \sigma(A),$$

where $sp_{AP}(u_f)$ is called (in terminology of [137]) *the set of points of non-almost periodicity*). In the next section we will discuss another direction dealing with such conditions that if Eq.(2.23) has a solution $u_f \in BUC(\mathbf{R}, \mathbf{X})$, then it has a solution in $\mathcal{M}$ (which may be different from u_f).

In this section we will propose a new approach to the admissibility theory of function spaces of Eq.(2.23) by considering the sum of two commuting operators $-d/dt := -\mathcal{D}_\mathcal{M}$ and the operator of multiplication by A on $\mathcal{M}$. As a result we will give simple proofs of recent results on the subject. Moreover, by this approach the results can be naturally extended to general classes of differential equations, including higher order and abstract functional differential equations. Various spectral criteria of the type $\sigma(\mathcal{D}_\mathcal{M}) \cap \sigma(A) = \oslash$ for the admissibility of the function space $\mathcal{M}$ and applications will be discussed.

We now describe more detailedly our approach which is based on the notion of evolution semigroups and the method of sums of commuting operators. Let A be the infinitesimal generator of a C_0-semigroup $(e^{tA})_{t \geq 0}$. Then the evolution semigroup $(T^h)_{h \geq 0}$ on the function space $\mathcal{M}$ associated with Eq.(2.23) is defined by $T^h g(t) := e^{h\overline{A}} g(t-h), \forall t \in \mathbf{R}, h \geq 0, g \in \mathcal{M}$. Under certain conditions on $\mathcal{M}$, the evolution semigroup $(T^h)_{h \geq 0}$ is strongly continuous. On the one hand, its infinitesimal generator G is the closure of the operator $-d/dt + A$ on $\mathcal{M}$. On the other hand, the generator G relates a mild solution u of Eq.(2.23) to the forcing term f by the rule $Gu = -f$. The conditions for which the closure of $-d/dt + A$, as a sum of two commuting operators, is invertible, are well studied in the theory of sums of commuting operators. We refer the reader to a summary of basic notions and results in the appendices of this book, and the references for more information on the theory and applications of sums of commuting operators method to differential equations. So far this method is mainly applied to study the existence and regularity of solutions to the Cauchy problem corresponding to Eq.(2.23) on a *finite interval*. In this context, it is natural to extend this method to the admissibility theory of function spaces.

Recall that by $(S(t))_{t \geq 0}$ we will denote the translation group on the function space $BUC(\mathbf{R}, \mathbf{X})$, i.e., $S(t)v(s) := v(t+s), \forall t, s \in \mathbf{R}, v \in BUC(\mathbf{R}, \mathbf{X})$ with infinitesimal generator $\mathcal{D} := d/dt$ defined on $D(\mathcal{D}) := BUC^1(\mathbf{R}, \mathbf{X})$. Let $\mathcal{M}$ be a subspace of $BUC(\mathbf{R},\mathbf{X})$, A be a linear operator on $\mathbf{X}$. We shall denote by $\mathcal{A}_\mathcal{M}$ the operator $f \in \mathcal{M} \mapsto Af(\cdot)$ with $D(\mathcal{A}_\mathcal{M}) = \{f \in \mathcal{M} | \forall t \in \mathbf{R}, f(t) \in D(A), Af(\cdot) \in \mathcal{M}\}$. When $\mathcal{M} = BUC(\mathbf{R},\mathbf{X})$ we shall use the notation $\mathcal{A} := \mathcal{A}_\mathcal{M}$. Throughout the paragraph we always assume that A is a given operator on $\mathbf{X}$ with $\rho(A) \neq \emptyset$, (so it is closed).

In this paragraph we will use the notion of translation-invariance of a function space, which we recall in the following definition, and additional conditions on it.

Definition 2.4 A closed and translation invariant subspace $\mathcal{M}$ of the function space $BUC(\mathbf{R},\mathbf{X})$, i.e., $S(\tau)\mathcal{M} \subset \mathcal{M}$ for all $\tau \in \mathbf{R}$, is said to satisfy

i) *condition H1* if the following condition is fulfilled:
$$\forall C \in L(\mathbf{X}), \forall f \in \mathcal{M} \Rightarrow Cf \in \mathcal{M},$$

ii) *condition H2* if the following condition is fulfilled:
For every closed linear operator A, if $f \in \mathcal{M}$ such that $f(t) \in D(A), \forall t$, $Af \in BUC(\mathbf{R},\mathbf{X})$, then $Af \in \mathcal{M}$,

iii) *condition H3* if the following condition is fulfilled: For every bounded linear operator $B \in L(BUC(\mathbf{R},\mathbf{X}))$ which commutes with the translation group $(S(t))_{t \in \mathbf{R}}$ one has $B\mathcal{M} \subset \mathcal{M}$.

Remark 2.4 As remarked in [222, p.401], condition H3 is equivalent to the assertion that
$$\forall B \in L(\mathcal{M}, \mathbf{X}) \ \ \forall f \in \mathcal{M} \Rightarrow BS(\cdot)f \in \mathcal{M}.$$

Obviously, conditions H2, H3 are stronger than condition H1. In the sequel, we will define the autonomousness of a functional operator via condition H3.

In connection with the translation-invariant subspaces we need the following simple spectral properties.

Lemma 2.8 *i) Let $\mathcal{M}$ satisfy condition H1. Then*
$$\sigma(\mathcal{A}_\mathcal{M}) \subset \sigma(\mathcal{A}) = \sigma(A)$$

and
$$\|R(\lambda, \mathcal{A}_\mathcal{M})\| \leq \|R(\lambda, \mathcal{A})\| = \|R(\lambda, A)\|, \forall \lambda \in \rho(A);$$

ii) Let $\mathcal{M}$ satisfy condition H3 and $\mathcal{B}$ be a bounded linear operator on $BUC(\mathbf{R},\mathbf{X})$ which commutes with the translation group. Then $\sigma(\mathcal{B}_\mathcal{M}) \subset \sigma(\mathcal{B})$ and
$$\|R(\lambda, \mathcal{B}_\mathcal{M})\| \leq \|R(\lambda, \mathcal{B})\|, \forall \lambda \in \rho(\mathcal{B}).$$

Proof. i) Let $\lambda \in \rho(A)$. We show that $\lambda \in \rho(\mathcal{A}_\mathcal{M})$. In fact, as $\mathcal{M}$ satisfies condition H1, $\forall f \in \mathcal{M}, R(\lambda, A)f(\cdot) := (\lambda - A)^{-1}f(\cdot) \in \mathcal{M}$. Thus the function $R(\lambda, A)f(\cdot)$ is a solution to the equation $(\lambda - \mathcal{A}_\mathcal{M})u = f$. Moreover, since $\lambda \in \rho(A)$ it is seen that the above equation has at most one solution. Hence $\lambda \in \rho(\mathcal{A}_\mathcal{M})$. Moreover, it is seen that $\|R(\lambda, \mathcal{A}_\mathcal{M})\| \leq \|R(\lambda, A)\|$. Similarly, we can show that if $\lambda \in \rho(\mathcal{A})$, then $\lambda \in \rho(A)$ and $\|R(\lambda, A)\| \leq \|R(\lambda, \mathcal{A})\|$.
ii) The proof of the second assertion can be done in the same way.

In the section, as a model of the translation - invariant subspaces, which satisfy all conditions H1, H2, H3 we can take the spectral spaces

$$\Lambda(\mathbf{X}) := \{u \in BUC(\mathbf{R}, \mathbf{X}) : \text{sp}(u) \subset \Lambda\},$$

where Λ is a given closed subset of the real line.

2.2.1. Differential Operator $d/dt - \mathcal{A}$ and Notions of Admissibility

We start the main subsection of this ection by discussing various notions of admissibility and their inter-relations via the differential operator $d/dt - \mathcal{A}$, or more precisely its closed extensions, for the following equation

$$\frac{dx}{dt} = Ax + f(t), x \in \mathbf{X}, t \in \mathbf{R}, \qquad (2.24)$$

where A is a linear operator acting on $\mathbf{X}$.

Using the notation $BUC^1(\mathbf{R}, \mathbf{X}) := \{g \in BUC(\mathbf{R}, \mathbf{X}) : \exists g', g' \in BUC(\mathbf{R}, \mathbf{X})\}$ we recall that

Definition 2.5 i) An $\mathbf{X}$-valued function u on $\mathbf{R}$ is said to be a *solution on* $\mathbf{R}$ to Eq.(2.24) for given linear operator A and $f \in BUC(\mathbf{R}, \mathbf{X})$ (or sometime, *classical solution*) if $u \in BUC^1(\mathbf{R}, \mathbf{X}), u(t) \in D(A), \forall t$ and u satisfies Eq.(2.24) for all $t \in \mathbf{R}$.

ii) Let A be the generator of a C_0 semigroup of linear operators. An $\mathbf{X}$-valued continuous function u on $\mathbf{R}$ is said to be a *mild solution on* $\mathbf{R}$ to Eq.(2.24) for a given $f \in BUC(\mathbf{R}, \mathbf{X})$ if u satisfies

$$u(t) = e^{(t-s)A}u(s) + \int_s^t e^{(t-r)A}f(r)dr, \forall t \geq s.$$

Definition 2.6 i) (cf. [186]) A closed translation invariant subspace $\mathcal{M} \subset BUC(\mathbf{R}, \mathbf{X})$ is said to be *admissible for Eq.(2.24)* if for each $f \in \mathcal{M}_0 := \mathcal{M} \cap BUC^1(\mathbf{R}, \mathbf{X})$ there is a unique solution $u \in \mathcal{M}_0$ of Eq.(2.24) and if $f_n \in \mathcal{M}_0, n \in \mathbf{N}, f_n \to 0$ as $n \to \infty$ in $\mathcal{M}_0$ imply $u_n \to 0$ as $n \to \infty$.

ii) Let $\mathcal{M}$ satisfy condition H1. $\mathcal{M}$ is said to be *weakly admissible for Eq.(2.24)* if $\mathcal{D}_\mathcal{M} - \mathcal{A}_\mathcal{M}$ is $\mathcal{T}_A$-closable and $0 \in \rho(\overline{\mathcal{D}_\mathcal{M} - \mathcal{A}_\mathcal{M}}^A)$.

iii) Let A be the generator of a C_0-semigroup. A translation - invariant closed subspace $\mathcal{M}$ of $BUC(\mathbf{R}, \mathbf{X})$ is said to be *mildly admissible for Eq.(2.24)* if for every $f \in \mathcal{M}$ there exists a unique mild solution $x_f \in \mathcal{M}$ to Eq.(2.24).

Remark 2.5 By definition it is obvious that admissibility and the closability of $\mathcal{D}_\mathcal{M} - \mathcal{A}_\mathcal{M}$ imply that $0 \in \rho(\overline{\mathcal{D}_\mathcal{M} - \mathcal{A}_\mathcal{M}})$.

We now discuss the relationship between the notions of admissibility, weak admissibility and mild admissibility if A is the generator of a C_0-semigroup. To this end, we introduce the following operator $L_\mathcal{M}$ which will be the key tool in our construction.

Definition 2.7 Let $\mathcal{M}$ be a translation invariant closed subspace of $BUC(\mathbf{R}, \mathbf{X})$. We define the operator $L_\mathcal{M}$ on $\mathcal{M}$ as follows: $u \in D(L_\mathcal{M})$ if and only if $u \in \mathcal{M}$ and there is $f \in \mathcal{M}$ such that

$$u(t) = e^{(t-s)A}u(s) + \int_s^t e^{(t-r)A}f(r)dr, \forall t \geq s \qquad (2.25)$$

and in this case $L_\mathcal{M} u := f$.

Let A be a given operator and $\mathcal{M}$ be a translation invariant closed subspace of $BUC(\mathbf{R}, \mathbf{X})$. We recall that in $\mathcal{M}$ the topology $\mathcal{T}_A$ is defined by the norm $\|f\|_A := \|R(\lambda, \mathcal{A}_\mathcal{M})f\|$ for $\lambda \in \rho(A) \subset \rho(\mathcal{A}_\mathcal{M})$. The following lemma will play an important role below.

Lemma 2.9 *Let A be the generator of a C_0-semigroup and $\mathcal{M}$ be a translation invariant closed subspace of $BUC(\mathbf{R}, \mathbf{X})$ satisfying condition H1. Then the operator $L_\mathcal{M}$ is well-defined single valued, $\mathcal{T}_A$-closed, and thus it is a closed and $\mathcal{T}_A$-closed extension of the sum of two commuting operators $\mathcal{D}_\mathcal{M} - \mathcal{A}_\mathcal{M}$.*

Proof. First we show that $L_\mathcal{M}$ is a well defined singled valued operator on $\mathcal{M}$. To this purpose, we suppose that there are $u, f_1, f_2 \in \mathcal{M}$ such that

$$L_\mathcal{M} u = f_1, L_\mathcal{M} u = f_2.$$

By definition this means that Eq.(2.25) holds for $f = f_i, i = 1, 2$. We now show that $f_1 = f_2$. In fact we have

$$\begin{aligned} u(t) &= e^{(t-s)A}u(s) + \int_s^t e^{(t-r)A}f_1(r)dr, \forall t \geq s, \\ &= e^{(t-s)A}u(s) + \int_s^t e^{(t-r)A}f_2(r)dr, \forall t \geq s. \end{aligned}$$

This yields that

$$\int_s^t e^{(t-r)A}(f_1(r) - f_2(r))dr = 0, \forall t \geq s. \qquad (2.26)$$

Since $(e^{tA})_{t\geq 0}$ is strongly continuous, the integrand in the left hand side of (2.26) is continuous with respect to $r, (r \leq t)$. Thus,

$$\begin{aligned} 0 &= \lim_{s\uparrow\uparrow t} \frac{1}{t-s} \int_s^t e^{(t-r)A}(f_1(r) - f_2(r))dr = e^{(t-t)A}f_1(t) - f_2(t) = \\ &= f_1(t) - f_2(t), \forall t. \end{aligned}$$

Now we show the $\mathcal{T}_\mathcal{A}$-closedness of the operator $L_\mathcal{M}$. Let $u_n, f_n \in \mathcal{M} \subset BUC(\mathbf{R}, \mathbf{X})$, $n \in \mathbf{N}$ such that

$$\begin{aligned} \lim_{n\to\infty} R(\lambda, \mathcal{A}_\mathcal{M})u_n &= R(\lambda, \mathcal{A}_\mathcal{M})u, u \in \mathcal{M}, \\ \lim_{n\to\infty} R(\lambda, \mathcal{A}_\mathcal{M})f_n &= R(\lambda, \mathcal{A}_\mathcal{M})f, f \in \mathcal{M}, \\ L_\mathcal{M} u_n &= f_n. \end{aligned}$$

By assumption we have

$$u_n(t) = e^{(t-s)A}u_n(s) + \int_s^t e^{(t-\xi)A}f_n(\xi)d\xi, \forall t \geq s.$$

Thus

$$\begin{aligned} u(t) + (u_n(t) - u(t)) &= e^{(t-s)A}u(s) + (e^{(t-s)A}u_n(s) - e^{(t-s)A}u(s)) + \\ &+ \int_s^t e^{(t-\xi)A}f(\xi)d\xi + (\int_s^t e^{(t-\xi)A}(f_n(\xi) - f(\xi))d\xi, \\ &\forall t \geq s. \end{aligned}$$

Since $R(\lambda, A)e^{tA} = e^{tA}R(\lambda, A)$ we have

$$\begin{aligned} \|R(\lambda, A)e^{(t-s)A}(u_n(s) - u(s))\| &= \|e^{(t-s)A}R(\lambda, A)(u_n(s) - u(s))\| \\ &\leq Ne^{(t-s)\omega}\|R(\lambda, A)(u_n(s) - u(s))\| \\ &\leq Ne^{(t-s)\omega}\|R(\lambda, \mathcal{A}_\mathcal{M})(u_n - u)\|_{\mathcal{T}_\mathcal{A}}. \end{aligned}$$

Thus,

$$\lim_{n\to\infty} \|R(\lambda, A)e^{(t-s)A}(u_n(s) - u(s))\| = 0. \tag{2.27}$$

On the other hand,

$$\begin{aligned} \|R(\lambda, A)\int_s^t e^{(t-\xi)A}(f_n(\xi) - f(\xi))d\xi\| &= \\ &= \|\int_s^t e^{(t-\xi)A}R(\lambda, A)(f_n(\xi) - f(\xi))d\xi\| \\ &\leq \int_s^t Ne^{(t-s)\omega}\|R(\lambda, \mathcal{A}_\mathcal{M})(f_n - f)\|d\xi. \end{aligned}$$

This yields that

$$\lim_{n\to\infty} R(\lambda, A) \int_s^t e^{(t-\xi)A}(f_n(\xi) - f(\xi))d\xi = 0 \forall t \geq s. \qquad (2.28)$$

Combining (2.27) and (2.28) we see that

$$R(\lambda, A)u(t) = R(\lambda, A)(e^{(t-s)A}u(s) + \int_s^t e^{(t-\xi)A}f(\xi)d\xi), \forall t \geq s.$$

From the injectiveness of $R(\lambda, A)$ we get

$$u(t) = e^{(t-s)A}u(s) + \int_s^t e^{(t-\xi)A}f(\xi)d\xi, \forall t \geq s,$$

i.e., $L_{\mathcal{M}}u = f$. This completes the proof of the lemma.

The following lemma will be needed in the sequel.

Lemma 2.10 *Let A be the generator of a C_0-semigroup and $\mathcal{M}$ be a closed translation invariant subspace of $AAP(\mathbf{X})$ which satisfies condition H1. Then*

$$\overline{\mathcal{D}_{\mathcal{M}} - \mathcal{A}_{\mathcal{M}}} = L_{\mathcal{M}}.$$

Proof. Let us consider the semigroup $(T^h)_{h\geq 0}$

$$T^h v(t) := e^{hA} v(t-h), v \in \mathcal{M}, h \geq 0.$$

By condition H1, clearly, $(T^h)_{h\geq 0}$ leaves $\mathcal{M}$ invariant. By Corollary 2.1, since $\mathcal{M} \subset AAP(\mathbf{X})$ this semigroup is strongly continuous which has $-L_{\mathcal{M}}$ as its generator. On the other hand, since $(T^h)_{h\geq 0}$ is the composition of two commuting and strongly continuous semigroups, by [163, p. 24] this generator is nothing but $\overline{-\mathcal{D}_{\mathcal{M}} + \mathcal{A}_{\mathcal{M}}}$.

Corollary 2.5 *Let A be the generator of a C_0-semigroup and $\mathcal{M}$ be a translation invariant closed subspace of $BUC(\mathbf{R}, \mathbf{X})$. Then the notions of admissibility, weak admissibility and mild admissibility of $\mathcal{M}$ for Eq.(2.24) are equivalent provided one of the following conditions is satisfied:*

i) $\mathcal{M}$ *satisfies condition H1 and* $\mathcal{M} \subset AAP(\mathbf{X})$;

ii) $\mathcal{M}$ *satisfies condition H2 and A is the generator of an analytic C_0-semigroup.*

Proof. i) Since the admissibility of $\mathcal{M}$ for Eq.(2.24) implies in particular that $0 \in \rho(\overline{\mathcal{D}_{\mathcal{M}} - \mathcal{A}_{\mathcal{M}}})$, and by Lemmas 2.9, 2.10

$$\overline{\mathcal{D}_{\mathcal{M}} - \mathcal{A}_{\mathcal{M}}} = \overline{\mathcal{D}_{\mathcal{M}} - \mathcal{A}_{\mathcal{M}}}^A = L_{\mathcal{M}}$$

the implication "*admissibility $\Rightarrow$ mild admissibility*" is clear. Also, the equivalence between mild admissibility and weak admissibility is obvious. *It remains only to show* "*mild admissibility $\Rightarrow$ admissibility*", i.e., if

$$0 \in \rho(L_\mathcal{M}),$$

then $\mathcal{M}$ is admissible with respect to Eq.(2.24). In fact, by assumption, for every $f \in \mathcal{M}$ there is a unique mild solution $u := L_\mathcal{M}^{-1} f$ of Eq.(2.24). It can be seen that the function $u(\tau + \cdot) \in \mathcal{M}$ is a mild solution of Eq.(2.24) with the forcing term $f(\tau + \cdot)$ for every fixed $\tau \in \mathbf{R}$. Hence, by the uniqueness, $u(\tau + \cdot) = L_\mathcal{M}^{-1} f(\tau + \cdot)$. We can rewrite this fact as

$$S(\tau) L_\mathcal{M}^{-1} f = L_\mathcal{M}^{-1} S(\tau) f, \forall f \in \mathcal{M}, \tau \in \mathbf{R}.$$

From this and the boundedness of $L_\mathcal{M}^{-1}$,

$$\lim_{\tau \downarrow 0} \frac{S(\tau) u - u}{\tau} = L_\mathcal{M}^{-1} \lim_{\tau \downarrow 0} \frac{S(\tau) f - f}{\tau}.$$

Thus, the assumption that $f \in \mathcal{M}_0$ implies that the left hand side limit exists. Thus, $u = L_\mathcal{M}^{-1} f \in \mathcal{M}_0$. As is well known, since f is differentiable $\int_s^t e^{(t-\xi)A} f(\xi) d\xi$ is differentiable (see [71, Theorem, p. 84]). Thus, by definition of mild solutions, from the differentiability of u it follows that $e^{(t-s)A} u(s)$ is differentiable with respect to $t \geq s$. Thus, $u(s) \in D(A)$ for every $s \in \mathbf{R}$. Finally, this shows that $u(\cdot)$ is a classical solution to Eq.(2.24) on $\mathbf{R}$. Hence the admissibility of $\mathcal{M}$ for Eq.(2.24) is proved.
ii) We first show the implications "admissibility $\Rightarrow$ weak admissibility" and "admissibility $\Rightarrow$ mild admissibility". In fact, since by Lemma 2.9

$$\overline{\mathcal{D}_\mathcal{M} - \mathcal{A}_\mathcal{M}} \subset \overline{\mathcal{D}_\mathcal{M} - \mathcal{A}_\mathcal{M}}^A \subset L_\mathcal{M}$$

we see that $\overline{\mathcal{D}_\mathcal{M} - \mathcal{A}_\mathcal{M}}^A, L_\mathcal{M}$ are surjective. Thus it remains to show that the mappings $\overline{\mathcal{D}_\mathcal{M} - \mathcal{A}_\mathcal{M}}^A, L_\mathcal{M}$ are injective. Actually, it suffices to show only that $L_\mathcal{M}$ is injective. Let $u \in \mathcal{M}$ such that $L_\mathcal{M} u = 0$. By definition,

$$u(t) = e^{(t-s)A} u(s), \forall t \geq s.$$

Since $(e^{tA})_{t \geq 0}$ is analytic this yields that $d^k u(t)/dt^k \in D(A), \forall t, \forall k \in \mathbf{N}$ (see Theorem 1.5). Moreover

$$u^{(j)}(t) = e^{(t-s)A} A^j u(s), \forall t \geq s, j = 1, 2.$$

Thus, by Theorem 1.5,

$$\|u^{(j)}(t)\| = \|A^j e^{1A} u(t-1)\| \leq N \|u\|, j = 1, 2,$$

where $N \geq 0$ is a constant independent of u. This shows that $u'(\cdot) \in BUC(\mathbf{R}, \mathbf{X})$. Hence, by assumption, $u'(\cdot) = Au(\cdot) \in \mathcal{M}$, i.e., u is a classical solution on $\mathbf{R}$ to Eq.(2.24). By assumption, from the admissibility it follows that $u = 0$.

Now we show the implication "weak admissibility $\Rightarrow$ mild admissibility". As $D(\overline{\mathcal{D}_\mathcal{M} - \mathcal{A}_\mathcal{M}}^A)$ contains all classical solutions on $\mathbf{R}$, the proof of this implication can be done as in that of the previous ones.
The proof of the corollary is complete if we prove the implication "mild admissibility $\Rightarrow$ admissibility". In fact, this can be shown as in the proof of i).

2.2.2. Admissibility for Abstract Ordinary Differential Equations

In this subsection we shall demonstrate some advantages of using the operator $d/dt - \mathcal{A}$ as the sum of two commuting operators to study the admissibility theory for Eq.(2.24).

Recall that by definition $\Lambda(\mathbf{X}) = \{f \in BUC(\mathbf{R}, \mathbf{X}) : sp(f) \subset \Lambda\}$, where Λ is a given closed subset of $\mathbf{R}$. Obviously, $\Lambda(\mathbf{X})$ is a translation invariant closed subspace of $BUC(\mathbf{R}, \mathbf{X})$. Moreover, it satisfies all conditions H1, H2, H3. Thus we have the following

Theorem 2.3 *Let $\mathcal{M}$ be a translation invariant subspace of $BUC(\mathbf{R}, \mathbf{X})$ satisfying condition H1, Λ be a closed subset of the real line, and A have non-empty resolvent set. Moreover let $i\Lambda \cap \sigma(A) = \emptyset$. Then for every $f \in \mathcal{M} \cap \Lambda(\mathbf{X})$ Eq.(2.24) has a unique bounded solution in $\mathcal{M} \cap \Lambda(\mathbf{X})$ provided one of the following conditions holds*

i) either Λ is compact, or

ii) the operator A is bounded on $\mathbf{X}$.

In particular, the subspace $\mathcal{M} \cap \Lambda(\mathbf{X})$ is admissible for Eq.(2.24) in both cases.

Proof. i) First of all by assumption, it is seen that the operator $\mathcal{D}_{\mathcal{M} \cap \Lambda(\mathbf{X})}$ is bounded (see e.g. [137, p. 88]). Since $\mathcal{M} \cap \Lambda(\mathbf{X})$ satisfies also condition H1, by Lemma 2.8,

$$\sigma(\mathcal{A}_{\mathcal{M} \cap \Lambda(\mathbf{X})}) \subset \sigma(A). \tag{2.29}$$

On the other hand, by Lemma 2.6 and the translation invariance and closedness of $\mathcal{M}$, we observe that for every $f \in \mathcal{M} \cap \Lambda(\mathbf{X})$ and $\lambda \in \rho(\mathcal{D}_{\Lambda(\mathbf{X})}), Re\lambda > 0$

$$R(\lambda, \mathcal{D}_{\Lambda(\mathbf{X})})f = \int_0^\infty e^{-\lambda t} S(t) f dt \in \mathcal{M} \cap \Lambda(\mathbf{X}).$$

Since $R(\mu, \mathcal{D}_{\Lambda(\mathbf{X})})f$ is continuous in $\mu \in \rho(\mathcal{D}_{\Lambda(\mathbf{X})}) = \mathbf{C} \backslash i\Lambda$, we get

$$R(\lambda, \mathcal{D}_{\Lambda(\mathbf{X})})f \in \mathcal{M} \cap \Lambda(\mathbf{X})$$

for all $\lambda \in \rho(\mathcal{D}_{\Lambda(\mathbf{X})})$. This yields that for every $\lambda \in \rho(\mathcal{D}_{\Lambda(\mathbf{X})})$ the equation

$$\lambda u - \mathcal{D}_{\mathcal{M} \cap \Lambda(\mathbf{X})} u = f$$

is uniquely solvable in $\mathcal{M} \cap \Lambda(\mathbf{X})$, i.e.,

$$\rho(\mathcal{D}_{\Lambda(\mathbf{X})}) \subset \rho(\mathcal{D}_{\mathcal{M} \cap \Lambda(\mathbf{X})}).$$

Hence by Lemma 2.6,

$$\sigma(\mathcal{D}_{\mathcal{M} \cap \Lambda(\mathbf{X})}) \subset \sigma(\mathcal{D}_{\Lambda(\mathbf{X})}) = i\Lambda. \tag{2.30}$$

Now (2.29) and (2.30) yield that

$$\sigma(\mathcal{D}_{\mathcal{M}\cap\Lambda(\mathbf{X})}) \cap \sigma(\mathcal{A}_{\mathcal{M}\cap\Lambda(\mathbf{X})}) = \varnothing.$$

Applying Theorem 4.10 to the pair of operators $\mathcal{D}_{\mathcal{M}\cap\Lambda(\mathbf{X})}$ and $\mathcal{A}_{\mathcal{M}\cap\Lambda(\mathbf{X})}$ we get the assertion of the theorem.

ii) The second case can be proved in the same manner.

Remark 2.6 The case where A generates a C_0-semigroup has been considered in [222]. In [186], the theorem has been proved for a more general form of equations in the case where the function space $\mathcal{M} = \Lambda(\mathbf{X})$. Note that, the methods used in [186] and [222] are quite different.

We now consider the case where the operators $\mathcal{A}$ and $\mathcal{D}$ satisfy condition P which we recall in the following: For $\theta \in (0, \pi), R > 0$ we denote $\Sigma(\theta, R) = \{z \in \mathbf{C} : |z| \geq R, |arg z| \leq \theta\}$.

Definition 2.8 Let A and B be commuting operators. Then

i) A is said to be of *class* $\Sigma(\theta + \pi/2, R)$ if there are positive constants θ, R such that $\theta < \pi/2$,

$$\Sigma(\theta + \pi/2, R) \subset \rho(A) \quad \text{and} \quad \sup_{\lambda \in \Sigma(\theta+\pi/2,R)} \|\lambda R(\lambda, A)\| < \infty,$$

ii) A and B are said to satisfy *condition P* if there are positive constants $\theta, \theta', R, \theta' < \theta$ such that A and B are of class $\Sigma(\theta + \pi/2, R), \Sigma(\pi/2 - \theta', R)$, respectively.

Theorem 2.4 *Let $(A + \alpha)$ be of class $\Sigma(\theta + \pi/2, R)$ for some real α and $\mathcal{M}$ be a translation - invariant subspace of $BUC(\mathbf{R}, \mathbf{X})$. Moreover, let $\sigma(A) \cap \sigma(\mathcal{D}_{\mathcal{M}}) = \varnothing$. Then the following assertions hold true:*

i) *If $\mathcal{M}$ satisfies condition H1, then $\mathcal{M}$ is weakly admissible for (2.24).*

ii) *If $\mathcal{M}$ satisfies condition H2 and A is the generator of a C_0-semigroup, then $\mathcal{M}$ is admissible, weakly admissible and mildly admissible for (2.24).*

iii) *If $\mathcal{M} \subset AAP(\mathbf{X})$ satisfies condition H1 and A is the generator of a C_0-semigroup, then $\mathcal{M}$ is admissible, weakly admissible and mildly admissible for (2.24).*

Proof. Note that under the theorem's assumption the operators $\mathcal{A} + \alpha$ and $\mathcal{D}$ satisfy condition P for some real α. In fact, we can check only that

$$\sup_{\lambda \in \Sigma(\pi/2-\varepsilon,R)} \|\lambda R(\lambda, \mathcal{D}_{\mathcal{M}})\| < \infty,$$

where $0 < \varepsilon < \pi/2$. Since $\lambda \in \Sigma(\pi/2 - \varepsilon, R)$ with $0 < \varepsilon < \pi/2$

$$\|\lambda R(\lambda, \mathcal{D}_\mathcal{M})f\| = |\lambda| \|\int_0^\infty e^{-\lambda t} f(\cdot + t)dt\|$$
$$\leq |\lambda| \int_0^\infty e^{-\mathrm{Re}\lambda t} dt \|f\|$$
$$\leq \frac{|\lambda|}{\mathrm{Re}\lambda} \|f\|$$
$$\leq M\|f\|,$$

where M is a constant independent of f. Thus, by Theorem 4.10,

$$\sigma(\overline{\mathcal{D}_\mathcal{M} - \mathcal{A}_\mathcal{M}})^A - \alpha = \sigma(\overline{\mathcal{D}_\mathcal{M} - \mathcal{A}_\mathcal{M}}^A - \alpha)$$
$$= \sigma(\overline{\mathcal{D}_\mathcal{M} - (\mathcal{A}_\mathcal{M} + \alpha)})^A$$
$$\subset \sigma(\mathcal{D}_\mathcal{M}) - \sigma(\mathcal{A}_\mathcal{M} + \alpha)$$
$$\subset \sigma(\mathcal{D}_\mathcal{M}) - \sigma(\mathcal{A}_\mathcal{M}) - \alpha.$$

Hence
$$\sigma(\overline{\mathcal{D}_\mathcal{M} - \mathcal{A}_\mathcal{M}}^A) \subset \sigma(\mathcal{D}_\mathcal{M}) - \sigma(\mathcal{A}). \tag{2.31}$$

By assumption and by Lemma 2.8 since $\sigma(\mathcal{D}_\mathcal{M}) \cap \sigma(A) = \varnothing$ we have $\sigma(\mathcal{D}_\mathcal{M}) \cap \sigma(\mathcal{A}) = \varnothing$. From (2.31) and this argument we get

$$0 \notin \sigma(\overline{\mathcal{D}_\mathcal{M} - \mathcal{A}_\mathcal{M}}^A).$$

Hence, this implies in particular the weak admissibility of the function space $\mathcal{M}$ for Eq.(2.24) proving i). Now in addition suppose that A generates a strongly continuous semigroup. Then ii) and iii) are immediate consequences of Corollary 2.5 and i).

2.2.3. Higher Order Differential Equations

Our approach can be naturally extended to higher order differential equations. We consider the admissibility of the function space $\mathcal{M} \cap \Lambda(\mathbf{X})$ where $\mathcal{M}$ is assumed to satisfy condition H1 and Λ is a closed subset of the real line for the equation

$$\frac{d^n u}{dt^n} = Au + f(t), \tag{2.32}$$

where n is a natural number. To this end, first we compute the spectrum of the operator $d^n u/dt^n := \mathcal{D}^n$ on $\mathcal{M} \cap \Lambda(\mathbf{X})$.

Proposition 2.2 *With the above notation the following assertions hold true:*

i)
$$\sigma(\mathcal{D}^n_{\mathcal{M} \cap \Lambda(\mathbf{X})}) \subset (i\Lambda)^n.$$

ii)
$$\sigma(\mathcal{D}^n_{\Lambda(\mathbf{X})}) = (i\Lambda)^n.$$

Proof. We associate with the equation

$$\frac{d^n u}{dt^n} = \mu u + f(t), f \in \mathcal{M} \cap \Lambda(\mathbf{X})$$

the following first order equation

$$\begin{cases} x_1' = x_2 \\ x_2' = x_3, \\ \cdots \\ x_n' = \mu x_1 + f(t) \end{cases}, f \in \mathcal{M} \cap \Lambda(\mathbf{X}). \tag{2.33}$$

It is easily seen that the unique solvability of these equations in $\mathcal{M} \cap \Lambda(\mathbf{X})$ are equivalent. On the other hand, by Theorem 2.3 for every $f \in \mathcal{M} \cap \Lambda(\mathbf{X})$ Eq.(2.33) has a unique (classical) solution $x(\cdot) \in \mathcal{M} \cap \Lambda(\mathbf{X}), x = (x_1, \cdots, x_n)^T$ if

$$i\Lambda \cap \sigma(I(\mu)) = \varnothing,$$

where $I(\mu)$ denotes the operator matrix associated with Eq.(2.33). A simple computation shows that $\sigma(I(\mu))$ consists of all solutions to the equation $t^n - \mu = 0$. Thus,

$$\sigma(\mathcal{D}^n_{\mathcal{M} \cap \Lambda(\mathbf{X})}) \subset \{\mu \in \mathbf{C} : \mu = (i\lambda)^n \text{for some} \lambda \in \Lambda\}.$$

Hence i) is proved. On the other hand, let $\mu \in \Lambda$. Then $g(\cdot) := xe^{i\mu \cdot} \in \Lambda(\mathbf{X})$. Obviously, $\mathcal{D}^n_{\Lambda(\mathbf{X})} g = (i\mu)^n g$ and thus, $(i\mu)^n \in \sigma(\mathcal{D}^n_{\Lambda(\mathbf{X})})$. Hence, ii) is proved.

To proceed we recall that the definition of *admissibility for the first order equations can be naturally extended to higher order equations*. Now we observe that $(i\Lambda)^n$ is compact if Λ is compact.

Theorem 2.5 *Let Λ be a compact subset of the real line and $\mathcal{M}$ be a translation invariant subspace of $BUC(\mathbf{R}, \mathbf{X})$ satisfying condition H1. Moreover, let A be any closed operator in $\mathbf{X}$ such that $\sigma(A) \cap (i\Lambda)^n = \varnothing$. Then for every $f \in \mathcal{M} \cap \Lambda(\mathbf{X})$ there exists a unique (classical) solution $u_f \in \mathcal{M} \cap \Lambda(\mathbf{X})$ of Eq.(2.32). In particular, $\mathcal{M} \cap \Lambda(\mathbf{X})$ is admissible for Eq.(2.32).*

Proof. The theorem is an immediate consequence of Theorem 2.3 and the above computation of the spectrum of $\mathcal{D}^n$.

We recall the following notion.

Definition 2.9 By a *mild solution* of Eq.(2.32) we understand a bounded uniformly continuous function $u : \mathbf{R} \to \mathbf{X}$ such that

$$\int_0^t dt_1 \int_0^{t_1} dt_2 \ldots \int_0^{t_{n-1}} u(s) ds \in D(A)$$

and

$$u(t) = x_0 + tx_1 + \ldots t^{n-1}x_{n-1} + A\int_0^t dt_1 \int_0^{t_1} dt_2 \ldots \int_0^{t_{n-1}} u(s)ds +$$
$$+ \int_0^t dt_1 \int_0^{t_1} dt_2 \ldots \int_0^{t_{n-1}} f(s)ds \quad (t \in \mathbf{R})$$

for some fixed $x_0, x_1, \ldots, x_{n-1} \in \mathbf{X}$. For $u \in BUC(\mathbf{R}, \mathbf{X})$ we say that u is a *classical solution* to Eq.(2.32) if $u(t) \in D(A), \forall t \in \mathbf{R}$ and the n-th derivative of u (denoted by $u^{(n)}$) exists as an element of $BUC(\mathbf{R}, \mathbf{X})$ such that Eq.(2.32) holds for all $t \in \mathbf{R}$.

Remark 2.7 It may be noted that a classical solution is also a mild solution. In case $n = 1$ if A generates a strongly continuous semigroup the above definition of mild solution on $\mathbf{R}$ coincides with that in Section 1 of Chapter 1. In fact we have:

Lemma 2.11 *Let A be the generator of a C_0-semigroup and u satisfy $\int_0^t u(s)ds \in D(A), \forall t$ such that*

$$u(t) = u(0) + A\int_0^t u(s)ds + \int_0^t f(s)ds. \tag{2.34}$$

Then u satisfies

$$u(t) = T(t-s)u(s) + \int_s^t T(t-\xi)f(\xi)d\xi, \forall t \geq s \tag{2.35}$$

where $T(t) = e^{tA}$. Conversely, if u satisfies Eq.(2.35), then $\int_0^t u(s)ds \in D(A), \forall t$ and u satisfies Eq.(2.34).

Proof. Suppose that u is a solution to Eq.(2.34). Then, we will show that it is also a solution to Eq.(2.35). In fact, without loss of generality we verify that u satisfies Eq.(2.35) for $s = 0$. To this purpose let us define the function

$$w(t) = T(t)u(0) + \int_0^t T(t-\xi)f(\xi)d\xi, t \geq 0.$$

We now show that w satisfies Eq.(2.34) for $t \geq 0$ as well. In fact, using the following facts from semigroup theory

$$T(t)x - x = A\int_0^t T(s)xds, \forall x \in \mathbf{X},$$

and the following which can be verified directly by definition

$$A\int_\eta^t T(s-\eta)f(\eta)ds = \lim_{h\downarrow 0}(1/h)(T(h) - I)\int_\eta^t T(s-\eta)f(\eta)ds =$$
$$= T(t-\eta)f(\eta) - f(\eta)$$

we have

$$A \int_0^t w(s)ds = T(t)u(0) - u(0) + A \int_0^t \int_0^s T(s-\eta)f(\eta)d\eta.$$

By a change of order of integrating we get

$$A \int_0^t \int_0^s T(s-\eta)f(\eta)d\eta = A \int_0^t d\eta \int_\eta^t T(s-\eta)f(\eta)ds.$$

By the above facts we have

$$\int_0^t d\eta A \int_\eta^t T(s-\eta)f(\eta)ds = \int_0^t T(t-\eta)f(\eta)d\eta - \int_0^t f(\eta)d\eta.$$

This shows that w satisfies Eq.(2.34) for all $t \geq 0$. Define $g(t) = w(t) - u(t)$. Obviously,

$$g(t) = A \int_0^t g(s)ds, \forall t \geq 0.$$

Since A generates a strongly continuous semigroup the Cauchy problem

$$x' = Ax, x(0) = 0 \in D(A)$$

has a unique solution zero. Hence, $u(t) = w(t), \forall t \geq 0$, i.e. $u(t)$ satisfies Eq.(2.35) for all $t \geq 0$.

By reversing the above argument we can easily show the converse. Hence, the lemma is proved.

Lemma 2.12 *Let A be a closed operator and u be a mild solution of Eq.(2.32) and $\phi \in L^1(\mathbf{R})$ such that its Fourier transform has compact support. Then $\phi * u$ is a classical solution to Eq.(2.32) with forcing term $\phi * f$.*

Proof. Let us define

$$U_1(t) = \int_0^t u(s)ds, \quad F_1(t) = \int_0^t f(s)ds, \ t \in \mathbf{R},$$

$$U_k(t) = \int_0^t U_{k-1}(s)ds, \quad F_k(t) = \int_0^t F_{k-1}(s)ds, \ t \in \mathbf{R}, k \in \mathbf{N}.$$

Then, by definition, we have

$$u(t) = P_n(t) + A(U_n(t)) + F_n(t), t \in \mathbf{R},$$

where P_n is a polynomial of order of $n-1$. From the closedness of A, we have

$$u * \phi(t) = P_n * \phi(t) + A(U_n * \phi(t)) + F_n * \phi(t), \ t \in \mathbf{R}.$$

Since the Fourier transform ϕ has compact support all convolutions above are infinitely differentiable. From the closedness of A we have that $(U_n * \phi)^{(k)}(t) \in D(A), t \in \mathbf{R}$ and

$$A((U_n * \phi)^{(k)}(t)) = \frac{d^k}{dt^k} A(U_n * \phi(t)), \ t \in \mathbf{R}.$$

Set $V_n(t) = U_n * \phi(t)$. Since $U_n^{(k)}(t) = U_{n-k}(t), k = 0,1,2\cdots,n$ and $U_0(t) = u(t)$ we have $V^{(k)}(t) = U_{n-k} * \phi(t), k = 0,1,2,\cdots,n$ and $V_n^{(n)}(t) = u * \phi(t)$. Hence,

$$\begin{aligned} U_n * \phi(t) &= V_n(t) \\ &= \sum_{k=0}^{n-1} \frac{t^k}{k!} U_{n-k} * \phi(0) + \frac{1}{(n-1)!} \int_0^t (t-s)^{n-1} u * \phi(s) ds. \end{aligned}$$

Since $U_n * \phi(t) \in D(A), U_{n-k} * \phi(0) \in D(A), k = 1,2,\cdots,n-1$ it follows that the integral above belongs also to $D(A)$. Furthermore, we can check that

$$\begin{aligned} u * \phi(t) &= P_n * \phi(t) + Q_n(t) + A \left(\frac{1}{(n-1)!} \int_0^t (t-s)^{n-1} u * \phi(s) ds \right) \\ &\quad + \frac{1}{(n-1)!} \int_0^t (t-s)^{n-1} f * \phi(s) ds, \end{aligned}$$

where P_n, Q_n are polynomials of order of $n-1$ which appears when one expands $A(U_n * \phi(t))$ and $F_n * \phi(t)$, respectively. Now, since all functions in the above expression are infinitely differentiable, P_n, Q_n are polynomials of order of $n-1$ and A is closed we can differentiate the expression to get

$$\frac{d^n}{dt^n}(u * \phi)(t) = A(u * \phi(t)) + f * \phi(t), \ \forall t \in \mathbf{R}.$$

This proves the lemma.

We now recall the notion of B-class[1] of functions.

Definition 2.10 *A translation invariant subspace $\mathcal{F} \subset BUC(\mathbf{R}, \mathbf{X})$ is said to be a B-class if and only if it satisfies*

i) $\mathcal{F}$ is a closed subspace of $BUC(\mathbf{R}, \mathbf{X})$;

ii) $\mathcal{F}$ contains all constant functions;

iii) $\mathcal{F}$ satisfies condition H1;

iv) $\mathcal{F}$ is invariant by multiplication by $e^{i\xi \cdot}, \forall \xi \in \mathbf{R}$.

In connection with the notion of B-classes we recall the following notion of spectrum. Let $\mathcal{F}$ be a B-class and u be in $BUC(\mathbf{R}, \mathbf{X})$. Then, by definition

$$sp_{\mathcal{F}}(u) := \{ \xi \in \mathbf{R} \ : \ \forall \varepsilon > 0 \exists f \in L^1(\mathbf{R})$$
$$\text{such that} \ \ supp \mathcal{F} f \subset (\xi - \varepsilon, \xi + \varepsilon) \ \text{and} \ f * u \notin \mathcal{F} \}.$$

---

[1] Originally, in [26, p. 60] this notion is called Λ-*class*

Lemma 2.13 *If $f \in \mathcal{F}$, where $\mathcal{F}$ is a B-class, then $\psi * f \in \mathcal{F}$, $\forall \psi \in L^1(\mathbf{R})$ such that the Fourier transform of ψ has compact support.*

Proof. For the proof we refer the reader to [26, p.60].

Hence, Theorem 2.5 yields the following:

Theorem 2.6 *Let $\mathcal{F}$ be a B-class, A be a closed linear operator with non-empty resolvent set. Then for any mild solution u to Eq.(2.32) with $f \in \mathcal{F}$,*

$$sp_{\mathcal{F}}(u) \subset \{\lambda \in \mathbf{R} : (i\lambda)^n \in \sigma(A)\}. \tag{2.36}$$

Proof. Let $\lambda_0 \in \mathbf{R}$ such that $(i\lambda_0)^n \not\in \sigma(A)$. Then, since $\sigma(A)$ is closed there is a positive number δ such that for all $\lambda \in (\lambda_0 - 2\delta, \lambda_0 + 2\delta)$ we have $(i\lambda)^n \not\in \sigma(A)$. Let us define $\Lambda := [\lambda_0 - \delta, \lambda_0 + \delta]$. Then by Theorem 2.5 for every $y \in \Lambda(\mathbf{X}) \cap \mathcal{F}$ there is a unique (classical) solution $x \in \Lambda(\mathbf{X}) \cap \mathcal{F}$. Let $\psi \in L^1(\mathbf{R})$ such that $supp \mathcal{F}\psi \subset \Lambda$. Put $v := \psi * u, g := \psi * f$. Then, by Lemma 2.13 $g \in \mathcal{F}$ and by [26, Proposition 2.5] $sp_{\mathcal{F}}(g) \subset supp \mathcal{F}\psi \cap sp_{\mathcal{F}}(f) \subset \Lambda$. Thus $g \in \Lambda(\mathbf{X}) \cap \mathcal{F}$. Since $sp_{\mathcal{F}}(v) \subset \Lambda$ by Theorem 2.5 we see that Eq.(2.32) has a unique solution in $\Lambda(\mathbf{X})$ which should be v. Moreover, applying again Theorem 2.5 we can see that the function v should belong to $\Lambda(\mathbf{X}) \cap \mathcal{F}$. We have in fact proved that $\lambda_0 \not\in sp_{\mathcal{F}}(u)$. Hence the assertion of the theorem has been proved.

In a standard manner we get the following:

Corollary 2.6 *Let $\mathcal{F}$ be a B-class, $\sigma(A) \cap (i\mathbf{R})^n$ be countable. Moreover, let u be such a mild solution to Eq.(2.32) that*

$$\lim_{t \to \infty} \frac{1}{t} \int_0^t e^{-i\lambda s} u(x+s) ds$$

exists for every $\lambda \in sp_{\mathcal{F}}(u)$ uniformly with respect to $x \in \mathbf{R}$. Then $u \in \mathcal{F}$.

Proof. The corollary is an immediate consequence of Theorem 1.20 and Theorem 2.6.

In particular, we can take $\mathcal{F} = AP(\mathbf{X}), AAP(\mathbf{X})$ and get spectral criteria for almost periodicity and asymptotic almost periodicity for solutions to the higher order equations (2.32). Other B-classes can be read in [26]. It is interesting that starting from the spectral estimate (2.36) and the above corollary in the case $n = 1$ various criteria for stability for C_0-semigroups can be established (see [26]).

We now consider the admissibility of a given translation invariant closed subspace $\mathcal{M}$ for the higher order equation (2.32). Since the geometric properties of the set $(i\mathbf{R})^n$ play an important role, we consider here only the case $n = 2$, i.e., the following equation

$$\frac{d^2 u}{dt^2} = Au + f(t). \tag{2.37}$$

It turns out that for higher order equations conditions on A are much weaker than for the first order ones. Indeed, we have

Theorem 2.7 *Let A be a linear operator on $\mathbf{X}$ such that there are positive constants R, θ and*

$$\Sigma(\theta, R) \subset \rho(A) \quad \text{and} \quad \sup_{\lambda \in \Sigma(\theta, R)} |\lambda| \|R(\lambda, A)\| < \infty.$$

Furthermore, let $\mathcal{M}$ be a translation invariant closed subspace of $BUC(\mathbf{R}, \mathbf{X})$ which satisfies condition H1 such that

$$\sigma(\mathcal{D}_{\mathcal{M}}^2) \cap \sigma(A) = \emptyset.$$

Then $\mathcal{M}$ is admissible for the second order equation (2.37).

Proof. We will apply Proposition 4.1 to the pair of linear operators $\mathcal{D}_{\mathcal{M}}^2$ and $\mathcal{A}_{\mathcal{M}}$. To this end, by Proposition 2.2 we observe that

$$\sigma(\mathcal{D}_{\mathcal{M}}^2) \subset (i\mathbf{R})^2 = (-\infty, 0].$$

On the other hand, for $0 < \varepsilon < \theta$ there is a constant M such that the following estimate holds

$$\|R(\lambda, \mathcal{D}_{\mathcal{M}}^2)\| \leq \frac{M}{|\lambda|^{1/2}}, \forall \lambda \neq 0, |arg(\lambda) - \pi| < \varepsilon.$$

In fact, this follows immediately from well known facts in [55, Chap. 2]. To make it more clear, we consider the first order equation of the form (2.33) for the case $n = 2$. For every $\lambda \in \rho(\mathcal{D}_{\mathcal{M}}^2)$ the associated equation has an exponential dichotomy and its Green function is nothing but $R(\lambda, \mathcal{D}_{\mathcal{M}}^2)$. Moreover, the norm of the Green function can be estimated via the infimum of the modulus of all the real parts of square root of λ (see [55, Chap.2, pp. 80-89]). Furthermore, since $\mathcal{M}$ is translation invariant note that $D(\mathcal{D}_{\mathcal{M}}^2)$ is dense in $\mathcal{M}$. Thus, applying Proposition 4.1 to the pair of operators $\mathcal{D}_{\mathcal{M}}^2, \mathcal{A}_{\mathcal{M}}$ we have

$$0 \in \rho(\overline{\mathcal{D}_{\mathcal{M}}^2 - \mathcal{A}_{\mathcal{M}}}).$$

It remains to show that for every $f \in \mathcal{M}_0 := D(\mathcal{D}_{\mathcal{M}}^2)$ there is a unique classical solution u on $\mathbf{R}$. In fact, denoting

$$G := (\overline{\mathcal{D}_{\mathcal{M}}^2 - \mathcal{A}_{\mathcal{M}}})^{-1},$$

we can easily see that since $\mathcal{D}_{\mathcal{M}}^2, \mathcal{A}_{\mathcal{M}}$ commute with $\mathcal{D}_{\mathcal{M}}^2$, so does G. By definition, for $\lambda \in \rho(\mathcal{D}_{\mathcal{M}}^2)$, since G is bounded on $\mathcal{M}$

$$GR(\lambda, \mathcal{D}_{\mathcal{M}}^2) = R(\lambda, \mathcal{D}_{\mathcal{M}}^2)G.$$

Hence there is $g \in \mathcal{M}$ such that $f = R(\lambda, \mathcal{D}_{\mathcal{M}}^2)g$. Thus, by the above equality $Gf = R(\lambda, \mathcal{D}_{\mathcal{M}}^2)Gg \in D(\mathcal{D}_{\mathcal{M}}^2)$. This shows the admissibility of $\mathcal{M}$ for Eq.(2.37).

2.2.4. Abstract Functional Differential Equations

This subsection will be devoted to some generalization of the method discussed in the previous ones for functional differential equations of the form

$$\frac{dx(t)}{dt} = Ax(t) + [\mathcal{B}x](t) + f(t), \forall t \in \mathbf{R}, \tag{2.38}$$

where the operator A is a linear operator on $\mathbf{X}$ and $\mathcal{B}$ is assumed to be an autonomous functional operator.

We first make precise the notion of *autonomousness* for functional operators $\mathcal{B}$:

Definition 2.11 Let $\mathcal{B}$ be an operator, everywhere defined and bounded on the function space $BUC(\mathbf{R}, \mathbf{X})$ into itself. $\mathcal{B}$ is said to be an *autonomous functional operator* if for every $\phi \in BUC(\mathbf{R}, \mathbf{X})$

$$S(\tau)\mathcal{B}\phi = \mathcal{B}S(\tau)\phi, \forall \tau \in \mathbf{R},$$

where $(S(\tau))_{\tau \in \mathbf{R}}$ is the translation group $S(\tau)x(\cdot) := x(\tau + \cdot)$ in $BUC(\mathbf{R}, \mathbf{X})$.

In connection with autonomous functional operators we will consider closed translation invariant subspaces $\mathcal{M} \subset BUC(\mathbf{R}, \mathbf{X})$ which satisfy condition H3. Recall that if $\mathcal{B}$ is an autonomous functional operator and $\mathcal{M}$ satisfies condition H3, then by definition, $\mathcal{M}$ is left invariant under $\mathcal{B}$.

Definition 2.12 Let A be the generator of a C_0-semigroup and $\mathcal{B}$ be an autonomous functional operator. A function u on $\mathbf{R}$ is said to be a *mild solution* of Eq.(2.38) on $\mathbf{R}$ if

$$u(t) = e^{(t-s)A}u(s) + \int_s^t e^{(t-\xi)A}[(Bu)(\xi) + f(\xi)]d\xi, \quad \forall t \geq s.$$

As we have defined the notion of mild solutions it is natural to extend the notion of mild admissibility for Eq.(2.38) in the case where the operator A generates a strongly continuous semigroup. It is interesting to note that in this case because of the arbitrary nature of an autonomous functional operator $\mathcal{B}$ nothing can be said on the "well posedness" of Eq.(2.38). We refer the reader to Chapter 1 for particular cases of "finite delay" and "infinite delay" in which Eq.(2.38) is well posed. However, as shown below we can extend our approach to this case. Now we formulate the main result for this subsection.

Theorem 2.8 *Let A be the infinitesimal generator of an analytic strongly continuous semigroup, $\mathcal{B}$ be an autonomous functional operator on the function space $BUC(\mathbf{R}, \mathbf{X})$ and $\mathcal{M}$ be a closed translation invariant subspace of $AAP(\mathbf{X})$ which satisfies condition H3. Moreover, assume that*

$$\sigma(\mathcal{D}_\mathcal{M}) \cap \sigma(\mathcal{A} + \mathcal{B}) = \varnothing.$$

Then $\mathcal{M}$ is mildly admissible for Eq. (2.38), i.e., for every $f \in \mathcal{M}$ there is a unique mild solution $u_f \in \mathcal{M}$ of Eq.(2.38).

Proof. Since $\mathcal{M}$ satisfies condition H3, for every $f \in \mathcal{M}$ we have $\mathcal{B}f \in \mathcal{M}$. Thus,

$$\begin{aligned} D((\mathcal{A}+\mathcal{B})_\mathcal{M}) &= \{f \in \mathcal{M} : Af(\cdot) + \mathcal{B}f \in \mathcal{M}\} \\ &= \{f \in \mathcal{M} : Af(\cdot) \in \mathcal{M}\} \\ &= D(\mathcal{A}_\mathcal{M}). \end{aligned}$$

Hence

$$(\mathcal{A}+\mathcal{B})_\mathcal{M} = \mathcal{A}_\mathcal{M} + \mathcal{B}_\mathcal{M}.$$

As $\mathcal{M}$ satisfies condition H3 it satisfies condition H1 as well. Thus, by Lemma 2.8,

$$\sigma(\mathcal{A}_\mathcal{M}) \subset \sigma(A) \subset \sigma(A)$$

and

$$\|R(\lambda, \mathcal{A}_\mathcal{M})\| \le \|R(\lambda, A)\|, \forall \lambda \in \rho(A).$$

Since $\mathcal{B}$ is bounded $\mathcal{D}_\mathcal{M}$ and $(\mathcal{A}+\mathcal{B})_\mathcal{M} = \mathcal{A}_\mathcal{M} + \mathcal{B}_\mathcal{M}$ satisfy condition P. From [167, Lemma 2 and the remarks follows] it may be seen that $\mathcal{A}_\mathcal{M}$ is the infinitesimal generator of the strongly continuous semigroup $(T(t))_{t \ge 0}$

$$T(t)f(\xi) := e^{tA}f(\xi), \forall f \in \mathcal{M}, \xi \in \mathbf{R}.$$

Hence $D((\mathcal{A}+\mathcal{B})_\mathcal{M}) = D(\mathcal{A}_\mathcal{M})$ is dense everywhere in $\mathcal{M}$. It may be noted that $R(\lambda, \mathcal{A}+\mathcal{B})$ commutes with the translation group. Since $\mathcal{M}$ satisfies condition H3 we can easily show that

$$\sigma((\mathcal{A}+\mathcal{B})_\mathcal{M}) \subset \sigma(\mathcal{A}+\mathcal{B}).$$

Applying Theorem 4.10 we get

$$\sigma(\overline{\mathcal{D}_\mathcal{M} - (\mathcal{A}+\mathcal{B})_\mathcal{M}}) \subset \sigma(\mathcal{D}_\mathcal{M}) - \sigma((\mathcal{A}+\mathcal{B})_\mathcal{M}).$$

Hence

$$0 \in \rho(\overline{\mathcal{D}_\mathcal{M} - (\mathcal{A}+\mathcal{B})_\mathcal{M}}).$$

On the other hand, since $\mathcal{B}_\mathcal{M}$ is bounded on $\mathcal{M}$

$$\begin{aligned} \overline{\mathcal{D}_\mathcal{M} - (\mathcal{A}+\mathcal{B})_\mathcal{M}} &= \overline{\mathcal{D}_\mathcal{M} - \mathcal{A}_\mathcal{M}} - \mathcal{B}_\mathcal{M} \\ &= L_\mathcal{M} - \mathcal{B}_\mathcal{M} \end{aligned}$$

we have

$$0 \in \rho(L_\mathcal{M} - \mathcal{B}_\mathcal{M}). \tag{2.39}$$

If $u, f \in \mathcal{M}$ such that $(L_\mathcal{M} - \mathcal{B}_\mathcal{M})u = f$, then

$$L_\mathcal{M} u = \mathcal{B}_\mathcal{M} u + f.$$

By definition of the operator $L_\mathcal{M}$, this is equivalent to the following

$$u(t) = e^{(t-s)A}u(s) + \int_s^t e^{(t-\xi)A}[(\mathcal{B}_\mathcal{M} u)(\xi) + f(\xi)]d\xi, \forall t \ge s,$$

i.e., u is a mild solution to Eq.(2.38). Thus (2.39) shows that $\mathcal{M}$ is mildly admissible for Eq.(2.38).

Remark 2.8 Sometime it is convenient to re-state Theorem 2.8 in other forms than that made above. In fact, in practice we may encounter difficulty in computing the spectrum $\sigma(\mathcal{A} + \mathcal{B})$. Hence, alternatively, we may consider $\mathcal{D} - \mathcal{A} - \mathcal{B}$ as a sum of two commuting operators $\mathcal{D} - \mathcal{B}$ and $\mathcal{A}$ if $\mathcal{B}$ commutes with $\mathcal{A}$. In subsection 3.4 we again consider this situation.

We formulate here the analogs of Theorems 2.5, 2.7 for higher order functional differential equations

$$\frac{d^n x(t)}{dt^n} = Ax(t) + [\mathcal{B}x](t) + f(t). \tag{2.40}$$

Theorem 2.9 *Let $\mathcal{M}$ be a closed translation invariant subspace of the function space $BUC(\mathbf{R}, \mathbf{X})$ which satisfies condition H3, A be a closed linear operator on $\mathbf{X}$ with nonempty resolvent set, $\mathcal{B}$ be an autonomous functional operator on $BUC(\mathbf{R}, \mathbf{X})$ and Λ be a closed subset of the real line. Moreover, let*

$$(i\Lambda)^n \cap \sigma(\mathcal{A} + \mathcal{B}) = \varnothing.$$

Then for every $f \in \mathcal{M} \cap \Lambda(\mathbf{X})$ there exists a unique (classical) solution u_f to Eq.(2.39) provided one of the following conditions is satisfied:

i) Either Λ is compact or

ii) A is bounded on $\mathbf{X}$.

In particular, in both cases $\mathcal{M} \cap \Lambda(\mathbf{X})$ is admissible for Eq.(2.39).

Proof. The proof can be done in the same way as that of Theorem 2.5. So the details are omitted.

In applications we frequently meet the operator $\mathcal{B}$ in the integral form. This implies the commutativeness of $\mathcal{B}$ with the convolution, i.e.,

$$\mathcal{B}(u * v) = u * (\mathcal{B}v), \forall u \in L^1(\mathbf{R}), v \in BUC(\mathbf{R}, \mathbf{X}).$$

Hence, as a consequence of Theorem 2.9 we have

Corollary 2.7 *Let A be the generator of a C_0-semigroup and $\mathcal{B}$ be an autonomous functional operator on $BUC(\mathbf{R}, \mathbf{X})$ which commutes with the convolution. Moreover, let u be a bounded uniformly continuous mild solution of Eq.(2.38) with almost periodic f. Then,*

$$sp_{AP}(u) \subset i\mathbf{R} \cap \sigma(\mathcal{A} + \mathcal{B}).$$

Proof. The proof can be done identically as that of Theorem 2.6. So the details are omitted.

Theorem 2.10 *Let A be a linear operator on $\mathbf{X}$ such that there are positive constants R, θ and*

$$\Sigma(\theta, R) \subset \rho(A) \, and \sup_{\lambda \in \Sigma(\theta, R)} |\lambda| \|R(\lambda, A)\| < \infty,$$

and $\mathcal{B}$ be an autonomous functional operator. Furthermore, let $\mathcal{M}$ be a translation invariant closed subspace of $BUC(\mathbf{R}, \mathbf{X})$ which satisfies condition H3 such that

$$\sigma(\mathcal{D}_\mathcal{M}^2) \cap \sigma(\mathcal{A} + \mathcal{B}) = \varnothing.$$

Then, $\mathcal{M}$ is admissible for the following equation:

$$\frac{d^2 x(t)}{dt^2} = Ax(t) + [\mathcal{B}x](t) + f(t).$$

Proof. The proof can be done as in that of Theorem 2.7. So the details are omitted.

We now study the mild admissibility of a function space $\mathcal{M}$ for the nonlinearly perturbed equation

$$\frac{dx(t)}{dt} = Ax(t) + [\mathcal{B}x](t) + [Fx](t), \forall t \in \mathbf{R}, \qquad (2.41)$$

where F is not necessarily an autonomous functional operator. Note that the notion of mild solutions to Eq.(2.40) in the case where A is the generator of a C_0-semigroup can be extended to Eq.(2.41).

Theorem 2.11 *Let A be the generator of a C_0-semigroup and $\mathcal{B}$ be any autonomous functional operator on $BUC(\mathbf{R}, \mathbf{X})$, and $\mathcal{M}$ be a closed translation invariant subspace of $BUC(\mathbf{R}, \mathbf{X})$ which satisfies condition H3 and is mildly admissible for Eq.(2.40). Moreover, let F be a (possibly nonlinear) operator defined on $\mathcal{M}$ which satisfies the Lipschitz condition*

$$\|F(u) - F(v)\| \leq \delta \|u - v\|, \forall u, v \in \mathcal{M}.$$

Then, for sufficiently small δ, Eq.(2.41) has a unique mild solution $u_F \in \mathcal{M}$.

Proof. Under the assumptions of the theorem the closed linear operator $L_\mathcal{M} - \mathcal{B}_\mathcal{M}$ is invertible. Thus if we define the normed space $\mathbf{B}$ to be the set $D(L_\mathcal{M} - \mathcal{B}_\mathcal{M})$ with graph norm $\|u\|_\mathbf{B} := \|(L_\mathcal{M} - \mathcal{B}_\mathcal{M})u\| + \|u\|$, for every $u \in D(L_\mathcal{M} - \mathcal{B}_\mathcal{M})$, then $\mathbf{B}$ becomes a Banach space. Moreover, $L_\mathcal{M} - \mathcal{B}_\mathcal{M}$ is an isomorphism from $\mathbf{B}$ onto $\mathcal{M}$. Thus, by the Inverse Lipschitz Continuous Mapping Theorem, for sufficiently small δ there exists the inverse function to $L_\mathcal{M} - \mathcal{B}_\mathcal{M} - F$ which is Lipschitz continuous. This proves the theorem.

Remark 2.9 In the case where $\mathcal{B} = 0$ we can weaken considerably conditions on the function space $\mathcal{M}$ (see the previous section). Here the translation invariance and condition H3 are needed to use the differential operator $L_\mathcal{M} - \mathcal{B}_\mathcal{M}$.

2.2.5. Examples and Applications

In this subsection we will present several examples and applications and discuss the relation between our results and the previous ones.

As typical examples of the function spaces $\Lambda(\mathbf{X})$, where Λ is a closed subset of the real line we will take the following ones:

Example 2.3

The space of all $\mathbf{X}$ valued continuous τ-periodic functions $\mathcal{P}(\tau)$. In this case $\Lambda = \{2k\pi/\tau, k \in \mathbf{Z}\}$.

Example 2.4

Let Λ be a discrete subset of $\mathbf{R}$. Then $\Lambda(\mathbf{X})$ will consists of almost periodic functions (see [12]).

Example 2.5

Let Λ be a countable subset of $\mathbf{R}$. Then $\Lambda(\mathbf{X})$ will consists of almost periodic functions if in addition one assumes that $\mathbf{X}$ does not contain any subspace which is isomorphic to the space c_0 (see [137]).

Below we will revisit one of the main results of [134] to show how our method fits in the problem considered in [134]. Moreover, our method can be easily extended to the infinite dimensional case.

Example 2.6

(cf. [134]) Consider the following ordinary functional differential equation

$$x'(t) = \int_0^\infty [dE(s)]x(t-s) + f(t), x \in \mathbf{C}^n, t \in \mathbf{R}, \qquad (2.42)$$

where E is an $n \times n$ matrix function with elements in $\mathbf{C}$, f is a $\mathbf{C}^n$-valued almost periodic function. In addition, we assume that E is continuous from the left and of bounded total variation on $[0, \infty)$, i.e.

$$0 < \gamma = \int_0^\infty |dE(s)| < \infty.$$

As is well known for every $f \in AP(\mathbf{C}^n)$ there is a corresponding Fourier series

$$\sum_{k=0}^\infty a_k e^{i\lambda_k t}.$$

We define

$$\mathcal{A}_q^0 := \{f \in AP(\mathbf{C}^n) : a_0 = 0, |\lambda_k| \geq q, k = 1, 2, \cdots\}$$

and $\mathcal{A}_q = \mathcal{A}_q^0 + V_c$, where V_c is the set of all $\mathbf{C}^n$-valued constant functions. Now we define our operator

$$\mathcal{B}u(t) := \int_0^\infty [dE(s)]u(t-s), t \in \mathbf{R}, u \in AP(\mathbf{C}^n).$$

Obviously, $\mathcal{B}$ is an autonomous functional operator with $\|\mathcal{B}\| \leq \gamma$. If we define $\Lambda := \{\eta \in \mathbf{R} : |\eta| \geq q\}$, then $\mathcal{A}_q^0 = AP(\mathbf{C}^n) \cap \Lambda(\mathbf{C}^n)$. We are now in a position to apply Theorem 2.9, ii).

Assertion 1 *Under the above notations and assumptions Eq.(2.42) has a unique almost periodic solution $x_f \in \mathcal{A}_q^0$ for every $f \in \mathcal{A}_q^0$ if $\gamma < q$.*

Proof. In fact, by assumption it is obvious that the spectral radius $r_\sigma(\mathcal{B}) < \gamma$. Hence, $i\Lambda \cap \sigma(\mathcal{B}) = \emptyset$.

If in addition we assume that

$$M := \int_0^\infty dE(s) \qquad (2.43)$$

is a nonsigular matrix, then Assertion 1 implies the following:

Assertion 2 *Under Assertion 1's assumptions and the nonsingularity of the matrix (2.43) there exists a unique solution $x_f \in \mathcal{A}_q$ to Eq.(2.42) for every $f \in \mathcal{A}_q$.*

Proof. In this case the operator $d/dt - \mathcal{B}$ is a direct sum of two invertible operators in $\mathcal{A}_q^0$ and V_c.

Remark 2.10 In [134, Section 3] Assertion 2 has been proved with a little stronger assumption, namely, $\gamma\delta < q$, where $\delta \geq 1$ is an "absolute constant" (in terminology of [134, p.401]). The condition $\gamma < q$ of Assertion 2 becomes also necessary in many cases. To show this, we consider the case

$$\mathcal{B}u(t) = Bu(t + \tau), \forall t \in \mathbf{R}, u \in BUC(\mathbf{R}, \mathbf{X}),$$

where τ is a given constant, B is a matrix. Now suppose that there exists a unique solution $x_f \in \mathcal{A}_q$ to Eq.(2.42) for every $f \in \mathcal{A}_q$. Denoting $Gf := x_f$ we see that G is a bounded linear operator on $\mathcal{A}_q$. Moreover, since $\mathcal{B}$ commutes with translation group so does G , i.e., $\mathcal{D}Gf = G\mathcal{D}f, \forall f \in D(\mathcal{D})$. Taking $f := e^{i\lambda t}y$ we have $\mathcal{D}x_f = \mathcal{D}Gf = G\mathcal{D}f = \lambda Gf = \lambda x_f$. Hence, $x_f(t) = e^{i\lambda t}x$ for some x. Substituting this into Eq.(2.42) we get the assertion that given $|\lambda| \geq q$ for every $y \in \mathbf{C}^n$ there exists a unique $x \in \mathbf{C}^n$ such that

$$i\lambda x - e^{i\tau\lambda}Bx = y.$$

This shows that $i\lambda \in \rho(e^{i\tau\lambda}B) = e^{i\tau\lambda}\rho(B)$ and yields $\gamma < q$.

In the following example we will revisit a problem discussed in [211] with an unbounded A.

Example 2.7

Let us consider the equation

$$\frac{dx(t)}{dt} = Ax(t) + \sum_{k=1}^{N} B_k x(t + \tau_k) + f(t), t \in \mathbf{R}, \tag{2.44}$$

where A is the infinitesimal generator of an analytic C_0-semigroup, $B_k, k = 1, \cdots, N$ are bounded linear operators on $\mathbf{X}$ which are commutative with each other and A, $\tau_k, k = 1, \cdots, N$ are given reals and f is a bounded uniformly continuous function. Let us denote $\Lambda = sp(f)$. Then

Assertion 3 *Let Λ be bounded. Then if*

$$\sigma(A) \cap \cup_{\lambda \in \Lambda} \sigma(i\lambda - \sum_{k=1}^{N} B_k)e^{i\tau_k \lambda}) = \varnothing,$$

Eq.(2.44) has a unique classical solution in $\Lambda(\mathbf{X})$.

Proof. If Λ is bounded, then $\mathcal{D}_{\Lambda(\mathbf{X})}$ is bounded. Hence, if

$$\sigma(\mathcal{D}_{\Lambda(\mathbf{X})} - \sum_{k=1}^{N} B_k e^{\tau_k \mathcal{D}_{\Lambda(\mathbf{X})}}) \cap \sigma(A) = \varnothing$$

Eq.(2.44) has a unique classical solution in $\Lambda(\mathbf{X})$. In turn, using the estimates of spectra as in [211] we get

$$\sigma(\mathcal{D}_{\Lambda(\mathbf{X})} - \sum_{k=1}^{N} B_k e^{\tau_k \mathcal{D}_{\Lambda(\mathbf{X})}}) \subset \cup_{\lambda \in \Lambda} \sigma(i\lambda - \sum_{k=1}^{N} B_k e^{i\tau_k \lambda}).$$

Assertion 4 *Let f be almost periodic and $B_k, k = 1, \cdots, N$ be commutative with each other and A and*

$$i\Lambda \cap (\sigma(A) + \cup_{\lambda_k \in \overline{e^{i\tau_k \Lambda}}} \sigma(\sum_{k=1}^{N} B_k \lambda_k)) = \varnothing.$$

Then Eq.(2.44) has a unique almost periodic mild solution in $\Lambda(\mathbf{X})$.

Proof. First using the Weak Spectral Mapping Theorem (see Theorem 1.8) we have

$$\sigma(S(\tau_k)) = \overline{e^{i\tau_k \Lambda}}.$$

CHAPTER 2. SPECTRAL CRITERIA

In view of [211, Theorem 1], denoting the multiplication operator by B_k by also B_k for the sake of simplicity, we have

$$\sigma(\sum_{k=1}^{N} B_k S(\tau_k)) \subset \cup_{\lambda_k \in \overline{e^{i\tau_k \Lambda}}} \sigma(\sum_{k=1}^{N} B_k \lambda_k).$$

By the commutativeness assumption applying Theorem 2.8 and then Theorem 4.10 we get the conclusion of the assertion.

In case A is the generator of a C_0-semigroup which is not necessarily analytic we can still apply Theorem 2.2 and the commutativeness of the operators $\mathcal{A}, \mathcal{B}$ as shown in the following example:

Example 2.8

Let A be the infinitesimal generator of a strongly continuous semigroup of linear operators on $\mathbf{X}$, $\mathcal{B}$ be an autonomous functional operator on $BUC(\mathbf{R}, \mathbf{X})$ and $\mathcal{M}$ be a translation invariant subspace of $AAP(\mathbf{X})$. Moreover, we assume that $\mathcal{B}$ and $\mathcal{A}$ commute. The only difference between this example and the previous one is that the semigroup generated by A may not be analytic. However, we can find conditions for the admissibility of $\mathcal{M}$ by using evolution semigroup associated with A as in Theorem 2.2. In fact, in $\mathcal{M}$, since $\mathcal{B}_\mathcal{M}$ is bounded it generates the norm continuous semigroup $(B^h)_{h \geq 0}$. Hence, $\overline{-\mathcal{D}_\mathcal{M} + \mathcal{A}_\mathcal{M}} + \mathcal{B}_\mathcal{M}$ generates a strongly continuous semigroup $(T^h B^h)_{h \geq 0}$. Thus, in view of the spectral inclusion of strongly continuous semigroups, this generator is invertible if $1 \notin \sigma(T^1 B^1)$. Using the commutativeness of the operators under consideration and the Weak Spectral Mapping Theorem for the translation group on $\mathcal{M}$ we have

$$\sigma(T^1 B^1) \subset \sigma(T^1).\sigma(B^1) \subset \overline{e^{-\mathcal{D}_\mathcal{M}}} \sigma(e^A).\sigma(B^1). \tag{2.45}$$

Hence the following is obvious:

Assertion 5 *If*
$$1 \notin \overline{e^{-\mathcal{D}_\mathcal{M}}} \sigma(e^A).\sigma(B^1),$$

then
$$\frac{dx(t)}{dt} = Ax(t) + [\mathcal{B}x](t) + f(t), \tag{2.46}$$

has a unique mild solution in $\mathcal{M}$ for every given $f \in \mathcal{M}$.

As an application suppose that we are given an almost periodic function f. Let $\mathcal{M} \subset AP(\mathbf{X})$ consisting of all functions g such that $sp(g) \subset sp(f)$. Then the above condition can be written as

$$1 \notin \overline{e^{-isp(f)}} \sigma(e^A).e^{\sigma(\mathcal{B}_\mathcal{M})} \tag{2.47}$$

which implies the existence of an almost periodic mild solution to Eq.(2.45). To illustrate the usefulness of (2.47) we consider the following case of Eq.(2.46)

$$\frac{dx(t)}{dt} = Ax(t) + bx(t+1) + f(t), \tag{2.48}$$

where $b \in \mathbf{R}$ and f is 1-periodic and continuous. In this case, $\mathcal{B} = bS(1)$. Hence, $sp(f) = 2\pi \mathbf{Z}$ and condition (2.47) can be written as

$$1 \notin \sigma(e^A)e^b. \tag{2.49}$$

Hence, if condition (2.49) holds true, then Eq.(2.48) has a unique 1-periodic mild solution.

In the following example we will demonstrate another way than Theorem 2.8 to use the evolution semigroups and sums of commuting operators method to study the admissibility of function spaces. In fact, sometime it is convenient to apply Remark 2.8. We refer the reader to [186, Theorem 2] for related results in the case of sufficiently small variation.

Example 2.9

We consider now in this example the following equation:

$$\dot{x}(t) = Ax(t) + \int_0^\infty x(t-s)db(s), \tag{2.50}$$

where A is the infinitesimal generator of an analytic strongly continuous semigroup of linear operators, and $b: R^+ \mapsto \mathbf{C}$ is a function of bounded variation satisfying

$$\exists \gamma > 0 : \int_0^\infty e^{\gamma s} d|b(s)| < \infty.$$

To this end we consider the equation

$$\dot{x}(t) = \int_0^\infty x(t-s)db(s), \tag{2.51}$$

and the associated operator $\mathcal{B}$ defined by

$$[\mathcal{B}(\phi)](t) = \int_0^\infty \phi(t-s)db(s), \qquad \phi \in BUC(\mathbf{R}, \mathbf{C}), \ t \in R.$$

Let $\mathcal{N}$ be any translation invariant subspace of $AAP(\mathbf{X})$ which satisfies condition H3. To study Eq.(2.50) below we will use the following esimate

$$\sigma(\overline{\mathcal{D}_\mathcal{N} - \mathcal{B}_\mathcal{N} - \mathcal{A}_\mathcal{N}}) \subset \sigma(\mathcal{D}_\mathcal{N} - \mathcal{B}_\mathcal{N}) - \sigma(\mathcal{A}_\mathcal{N})$$

the validity of which is easily established under the above-mentioned assumptions.

The main result we are going to prove in this example is the following which will be then applied to study Eq.(2.50):

Theorem 2.12 Let Λ be a subset of $\mathbf{R}$ and let $M = \Lambda(\mathbf{C}) = \{\phi \in BUC(\mathbf{R}, \mathbf{C}) : \sigma(\phi) \subset \Lambda\}$. Then

$$\sigma(\mathcal{D}_M - \mathcal{B}_M) = \{i\lambda - \int_0^\infty e^{-i\lambda s} db(s) : \lambda \in \Lambda\} \ (=: (i\tilde{\Lambda})).$$

Proof. If $\mu = i\lambda - \int_0^\infty e^{-i\lambda s} db(s)$ for some $\lambda \in \Lambda$, then $\phi := \exp(i\lambda\cdot)$ belongs to M, and

$$\begin{aligned}(\mathcal{D}_M - \mathcal{B}_M)\phi &= i\lambda\phi - \phi \int_0^\infty e^{-i\lambda s} db(s) \\ &= \mu\phi,\end{aligned}$$

and hence $\mu \in P_\sigma(\mathcal{D}_M - \mathcal{B}_M)$. Thus $(i\tilde{\Lambda}) \subset \sigma(\mathcal{D}_M - \mathcal{B}_M)$.

Next we shall show that $(i\tilde{\Lambda}) \supset \sigma(\mathcal{D}_M - \mathcal{B}_M)$. To do this, it is sufficient to prove the claim:

Assertion 6 If $i\lambda \neq \int_0^\infty e^{-i\lambda s} db(s) + k \ (\forall \lambda \in \Lambda)$, then $k \in \rho(\mathcal{D}_M - \mathcal{B}_M)$.

To establish the claim, we will show that for each $f \in M$, the equation

$$\dot{x}(t) = \int_0^\infty x(t-s)db(s) + kx(t) + f(t), \quad t \in \mathbf{R} \tag{2.52}$$

possesses a unique solution $x_f \in M$ and that the map $f \in M \mapsto x_f \in M$ is continuous. We first treat the homogeneous functional differential equation (FDE)

$$\dot{x}(t) = \int_0^\infty x(t-s)db(s) + kx(t), \tag{2.53}$$

which may be considered as a FDE on the uniform fading memory space $C_\gamma = \{\phi \in C((-\infty, 0]; \mathbf{C}) : \sup_{\theta \leq 0} |\phi(\theta)|e^{\gamma\theta} < \infty\}$ which is equipped with norm $\|\phi\|_{C_\gamma} = \sup_{\theta \leq 0} |\phi(\theta)|e^{\gamma\theta}$. Let us consider the solution semigroup $T(t) : C_\gamma \mapsto C_\gamma, t \geq 0$, of (2.53) which is defined as

$$T(t)\phi = x_t(\phi), \quad \phi \in C_\gamma,$$

where $x(\cdot, \phi)$ denotes the solution of (2.53) through $(0, \phi)$ and x_t is an element in C_γ defined as $x_t(\theta) = x(t+\theta), \theta \leq 0$. Let G be the infinitesimal generator of the solution semigroup $T(t)$. We assert that

$$i\mathbf{R} \cap \sigma(G) = \{i\lambda \in i\mathbf{R} : i\lambda = \int_0^\infty e^{-i\lambda s} db(s) + k\}.$$

Indeed, if $i\lambda \in i\mathbf{R} \cap \sigma(G)$, then it follows from [107, p. 155, Th. 4.4] that $i\lambda \in P_\sigma(G)$ because of $\text{Re}(i\lambda) = 0$, and consequently from [107, p.135, Th. 2.1] we get that

$i\lambda - (\int_0^\infty e^{-i\lambda\theta}db(s) + k) = 0$, which shows that $i\lambda$ belongs to the set of the right hand side in the assertion. Conversely, if $i\lambda$ is an element of the set of the right hand side in the assertion, then the function w defined by $w(\theta) = \exp(i\lambda\theta), \theta \leq 0$, together with the derivative $\dot{w}$ belong to the space C_γ, and satisfy the relation

$$\dot{w}(0) = \lambda i = \int_0^\infty e^{-i\lambda s}db(s) + k$$
$$= \int_0^\infty w(-s)db(s) + kw(0),$$

and hence $w \in D(G)$ and $Gw = \dot{w} = i\lambda w$ by [107, p.150, Th. 4.1]. Thus $i\lambda \in \sigma(G) \cap i\mathbf{R}$, and the assertion is proved.

Now consider the sets $\Sigma_C := \{\lambda \in \sigma(G) : \text{Re}\lambda = 0\}$ and $\Sigma_U := \{\lambda \in \sigma(G) : \text{Re}\lambda > 0\}$. Then the sets $\Sigma = \Sigma_C \cup \Sigma_U$ is a finite set [107, p. 144, Prop. 3.2]. In regard of the set Σ, we get the decomposition of the space C_γ:

$$C_\gamma = S \oplus C \oplus U,$$

where S, C, U are invariant under $T(t)$, the restriction $T(t)|_U$ can be extended to a group, and there exist positive constants c_1 and α such that

$$\|T(t)|_S\| \leq c_1 e^{-\alpha t} \quad (t \geq 0), \quad \|T(t)|_U\| \leq c_1 e^{\alpha t} \quad (t \leq 0)$$

([107, p.145, Ths.3.1, 3.3]). Let Φ be a basis vector in C, and let Ψ be the basis vector associated with Φ. From [107, p. 149, Cor. 3.8] we know that the C-component $u(t)$ of the segment x_t for each solution $x(\cdot)$ of (2.52) is given by the relation $u(t) = \langle \Psi, \Pi_C x_t \rangle$ (where Π_C denotes the projection from C_γ onto C which corresponds to the decomposition of the space C_γ), and $u(t)$ satisfies the ordinary differential equation

$$\dot{u}(t) = Lu(t) - \hat{\Psi}(0^-)f(t), \tag{2.54}$$

where L is a matrix such that $\sigma(L) = \sigma(G) \cap i\mathbf{R}$ and the relation $T(t)\Phi = \Phi e^{tL}$ holds. Moreover, $\hat{\Psi}$ is the one associated with the Riesz representation of Ψ. Indeed, $\hat{\Psi}$ is a normalized vector-valued function which is of locally bounded variation on $(-\infty, 0]$ satisfying $\langle \Psi, \phi \rangle = \int_{-\infty}^0 \phi(\theta)d\hat{\Psi}(\theta)$ for any $\phi \in C_\gamma$ with compact support. Observe that $\Sigma_C \subset i\mathbf{R} \setminus i\Lambda$. Indeed, if $\mu \in \Sigma_C$, then $\mu = i\lambda = \int_0^\infty e^{-i\lambda s}db(s) + k$ with some $\lambda \in \mathbf{R}$ by the preceding assertion. Hence we get $\lambda \notin \Lambda$ by the assumption of the claim, and $\mu \in i\mathbf{R} \setminus i\Lambda$, as required. This observation leads to $\sigma(L) \cap i\Lambda = \emptyset$. Since $\hat{\Psi}(0^-)f \in M$, the ordinary differential equation (2.54) has a unique solution $u \in M$ with $\|u\| \leq c_2\|\hat{\Psi}(0^-)f\| \leq c_3\|f\|$ for some constants c_2 and c_3. Consider a function $\xi : \mathbf{R} \mapsto C_\gamma$ defined by

$$\xi(t) = \int_{*-\infty}^t T^{**}(t-s)\Pi_S^{**}\Gamma f(s)ds + \Phi u(t) + \int_{*t}^\infty T^{**}(t-s)\Pi_U^{**}\Gamma f(s)ds,$$

where Γ is the one defined in [107, p. 118] and $\int_*$ denotes the weak-star integration (cf. [107, p. 116]). If $t \geq 0$, then

$$T(t)\xi(\sigma) + \int_{*\sigma}^{t+\sigma} T^{**}(t+\sigma-s)\Gamma f(s)ds =$$
$$= T(t)[\int_{*-\infty}^{\sigma} T^{**}(\sigma-s)\Pi_S^{**}\Gamma f(s)ds +$$
$$+ \Phi u(\sigma) + \int_{*\sigma}^{\infty} T^{**}(\sigma-s)\Pi_U^{**}\Gamma f(s)ds] +$$
$$+ \int_{*\sigma}^{t+\sigma} T^{**}(t+\sigma-s)\Gamma f(s)ds$$
$$= \int_{*-\infty}^{\sigma} T^{**}(t+\sigma-s)\Pi_S^{**}\Gamma f(s)ds + \Phi e^{tL}u(\sigma)$$
$$+ \int_{*\sigma}^{\infty} T^{**}(t+\sigma-s)\Pi_U^{**}\Gamma f(s)ds +$$
$$+ \int_{*\sigma}^{t+\sigma} T^{**}(t+\sigma-s)(\Pi_S^{**}+\Pi_C^{**}+\Pi_U^{**})\Gamma f(s)ds$$
$$= \int_{*-\infty}^{t+\sigma} T^{**}(t+\sigma-s)\Pi_S^{**}\Gamma f(s)ds + \Phi[e^{tL}u(\sigma) +$$
$$+ \int_{\sigma}^{t+\sigma} e^{(t+\sigma-s)L}(-\hat{\Psi}(0^-)f(s))ds] + \int_{*t+\sigma}^{\infty} T^{**}(t+\sigma-s)\Pi_U^{**}\Gamma f(s)ds$$
$$= \int_{*-\infty}^{t+\sigma} T^{**}(t+\sigma-s)\Pi_S^{**}\Gamma f(s)ds + \Phi u(t+\sigma) +$$
$$+ \int_{*t+\sigma}^{\infty} T^{**}(t+\sigma-s)\Pi_U^{**}\Gamma f(s)ds$$
$$= \xi(t+\sigma),$$

where we used the relation $T^{**}(t)\Pi_C^{**}\Gamma = T^{**}(t)\Phi\langle\Psi,\Gamma\rangle = \Phi e^{tL}(-\hat{\Psi}(0^-))$. Then [107, Theorem 2.9, p.121] yields that $x(t) := [\xi(t)](0)$ is a solution of (2.52). Define a $\psi \in C_\gamma^*$ by $\langle\psi,\phi\rangle = \phi(0), \phi \in C_\gamma$. Then

$$x(t) - \Phi(0)u(t) = \langle\psi, \xi(t) - \Phi u(t)\rangle$$
$$= \langle\psi, \int_{*-\infty}^{t} T^{**}(t-s)\Pi_S^{**}\Gamma f(s)ds +$$
$$+ \int_{*t}^{\infty} T^{**}(t-s)\Pi_U^{**}\Gamma f(s)ds\rangle$$
$$= \int_{-\infty}^{t} \langle\psi, T^{**}(t-s)\Pi_S^{**}\Gamma\rangle f(s)ds +$$
$$+ \int_{t}^{\infty} \langle\psi, T^{**}(t-s)\Pi_U^{**}\Gamma\rangle f(s)ds$$
$$= \int_{-\infty}^{\infty} K(t-s)f(s)ds = K*f(t),$$

where $K(t) = \langle \psi, T^{**}\Pi_S^{**}\Gamma\rangle\chi_{(-\infty,0]} + \langle \psi, T^{**}\Pi_U^{**}\Gamma\rangle\chi_{[0,\infty)}$ and it is an integrable function on $\mathbf{R}$. Then $\sigma(x - \Phi(0)u) \subset \sigma(f)$, and hence $x - \Phi(0)u \in M$. Thus we get $x \in M$ because of $u \in M$. Moreover, the map $f \in M \mapsto x \in M$ is continuous.

Finally, we will prove the uniqueness of solutions of (2.52) in M. Let x be any solution of (2.52) which belongs to M. By [107, Th. 2.8, p. 120] the C_γ-valued function $\Pi_S x_t$ satisfies the relation

$$\Pi_S x_t = T(t - \sigma)\Pi_S x_\sigma + \int_{*\sigma}^{t} T^{**}(t-s)\Pi_S^{**}\Gamma f(s)ds$$

for all $t \geq \sigma > -\infty$. Note that $\sup_{\sigma \in \mathbf{R}} \|x_\sigma\|_{C_\gamma} < \infty$. Therefore, letting $\sigma \to -\infty$ we get

$$\Pi_S x_t = \int_{*-\infty}^{t} T^{**}(t-s)\Pi_S^{**}\Gamma f(s)ds,$$

because

$$\lim_{\sigma \to \infty} \int_{*\sigma}^{t} T^{**}(t-s)\Pi_S^{**}\Gamma f(s)ds = \int_{*-\infty}^{t} T^{**}(t-s)\Pi_S^{**}\Gamma f(s)ds$$

converges. Similarly, one gets

$$\Pi_U x_\sigma = -\int_{\sigma}^{\infty} T^{**}(t-s)\Pi_U^{**}\Gamma f(s)ds.$$

Also, since $\langle \Psi, x_t\rangle$ satisfies the ordinary differential equation (2.54), $\Pi_C u_t = \Phi\langle\Psi, x_t\rangle = \Phi u(t)$ for all $t \in \mathbf{R}$ by the uniqueness of the solution of (2.54) in M. Consequently, we have $x_t \equiv \xi(t)$ or $x(t) \equiv [\xi(t)](0)$, which shows the uniqueness of the solution of (2.52) in M.

Corollary 2.8 *Suppose that $i\lambda \neq \int_0^\infty e^{-i\lambda s}db(s)$ for all $\lambda \in \Lambda$. Then (2.51) is admissible for $M = \Lambda(\mathbf{C})$; that is, for any $f \in M$ the equation*

$$\dot{x}(t) = \int_0^\infty x(t-s)db(s) + f(t), \quad t \in \mathbf{R},$$

is uniquely solvable.

Proof. The corollary is a direct consequence of Theorem 2.12, since $0 \notin \sigma(\mathcal{D}_M - \mathcal{B}_M)$.

Corollary 2.9 *Let Λ be a closed subset of $\mathbf{R}$. If $k \neq i\lambda - \int_0^\infty e^{-i\lambda s}db(s)$ for all $\lambda \in \Lambda$, then there exists an $F \in L^1(\mathbf{C})$ such that*

$$1/(i\lambda - \int_0^\infty e^{-i\lambda s}db(s) - k) = \mathcal{F}F(\lambda) := \int_{-\infty}^{\infty} F(t)e^{-i\lambda t}dt \quad (\forall \lambda \in \Lambda).$$

Proof. As seen in the proof of Theorem 2.12, there exists an integrable function K such that $(\mathcal{D}_M - \mathcal{B}_M - k)^{-1}f - \Phi(0)u(t) = K * f$ for all $f \in M$. We first claim that there exists an integrable function F_1 such that $u = F_1 * f$ is a unique solution of (2.54) in $M \cap AP(\mathbf{C})$ for each $f \in M \cap AP(\mathbf{C})$. In fact, by considering a linear transform which changes the matrix L to the Jordan form, this claim is reduced to the one for the case that (2.54) is a scalar equation. Therefore, we will restrict our consideration to the equation

$$\dot{u}(t) = i\lambda u(t) + g(t), \tag{2.55}$$

where $g \in \Lambda(\mathbf{C}) \cap AP(\mathbf{C})$ and $\lambda \notin \Lambda$, and find an integrable function H such that $H * g$ is a solution of (2.55). Take an interval $I = [\lambda - \alpha, \lambda + \alpha]$ such that $I \cap \Lambda = \varnothing$. In the same way as in [137, pp.89-90], we choose a complex function H on $\mathbf{R}$ such that it is integrable on $\mathbf{R}$, continuous everywhere except zero, $H(+0) - H(-0) = 1$, and it has the Fourier transform $\mathcal{F}H(\xi) = 1/i(\xi - \lambda)$ for $\xi \notin I$. For $g \in \mathcal{M}$ put

$$u(t) = g * H(t) = \int_{-\infty}^{\infty} g(t-s)H(s)ds.$$

Then $u \in AP(\mathbf{C})$ and $sp(u) \subset sp(g) \cap supp\mathcal{F}H \subset sp(g) \subset \Lambda$. Hence $u \in M$. To establish the claim, we must prove that u satisfies (2.55). First, let g be a trigonometric polynomial with spectrum outside I:

$$g = \sum a_m e^{i\lambda_m t}, \lambda_m \notin I.$$

Then

$$\begin{aligned} u(t) &= \sum a_m e^{i\lambda_m t} \int_{-\infty}^{\infty} e^{-i\lambda_m s} H(s) ds \\ &= \sum a_m e^{i\lambda_m t} \mathcal{F}H(\lambda_m) \\ &= \sum \frac{a_m}{i(\lambda_m - \lambda)} e^{i\lambda_m t}, \end{aligned}$$

which implies $d/dt[e^{-i\lambda t}u(t)] = e^{-i\lambda t}g(t)$. For a general almost periodic function $g \in M$, we can choose a sequence of trigonometric functions

$$g_n(t) = \sum a_{n,m} e^{i\lambda_{n,m} t},$$

such that $\lambda_{n,m}$ are all in the Bohr spectrum of g and that $g_n \to g$ uniformly on $\mathbf{R}$. Hence $sp(g_n) \subset \Lambda$, and $sp(g_n)$ is outside I. Put $u_n = g_n * H$. Then $d/dt[e^{-i\lambda t}u_n(t)] = e^{-i\lambda t}g_n(t)$. Since $g_n \to g$ uniformly on $\mathbf{R}$, $u_n \to u = g * H$ uniformly on $\mathbf{R}$, so that $d/dt[e^{-i\lambda t}u(t)] = e^{-i\lambda t}g(t)$. Consequently, $u' - i\lambda u = g$, as required.

We now set $F = K + \Phi(0)F_1$. Then F is an integrable function on $\mathbf{R}$, and $F * f$ is a unique solution of (2.52) in $M \cap AP(\mathbf{C})$ for each $f \in M \cap AP(\mathbf{C})$. Let $\lambda \in \Lambda$, and set $x(t) = F(t) * e^{i\lambda t}$. Since $sp(x) \subset sp(e^{i\lambda t}) = \{\lambda\}$, we must get $sp(x) = \{\lambda\}$ because of $x \neq 0$. Note that $x \in AP(\mathbf{C})$. Then the limit $\lim_{T \to \infty} \frac{1}{2T} \int_{s-T}^{s+T} x(t) e^{-i\lambda t} dt (=: a \neq 0)$ exists uniformly for $s \in \mathbf{R}$. Note that

$$a = \mathcal{F}F(\lambda) = \frac{1}{2T}\int_{s-T}^{s+T} x(t)e^{-i\lambda t}dt$$

for all $s \in \mathbf{R}$. Indeed, we get

$$\int_{s-T}^{s+T} x(t)e^{-i\lambda t}dt = \int_{s-T}^{s+T}(\int_{-\infty}^{\infty} F(\tau)e^{i\lambda(t-\tau)}d\tau)e^{-i\lambda t}dt$$
$$= \int_{s-T}^{s+T}\int_{-\infty}^{\infty} F(\tau)e^{-i\lambda\tau}d\tau dt = 2T\mathcal{F}F(\lambda).$$

Then

$$\frac{1}{2T}(x(T)e^{-i\lambda T} - x(-T)e^{i\lambda T}) = \frac{1}{2T}\int_{-T}^{T}\{-i\lambda x(t) + \dot{x}(t)\}e^{-i\lambda t}dt$$
$$= \frac{1}{2T}\int_{-T}^{T}(-i\lambda x(t) + \int_{-\infty}^{\infty} x(t-s)d\bar{b}(s) + kx(t) + e^{i\lambda t})e^{-i\lambda t}dt$$
$$= (k - i\lambda)\mathcal{F}F(\lambda) + 1 + \frac{1}{2T}\int_{-\infty}^{\infty}(\int_{-T-s}^{T-s} x(\tau)e^{-i\lambda\tau}d\tau)e^{-i\lambda s}d\bar{b}(s)$$
$$= (k - i\lambda + \int_{0}^{\infty} e^{-i\lambda s}db(s))\mathcal{F}F(\lambda) + 1,$$

where $\bar{b}(t) = b(t)$ if $t \geq 0$, and $\bar{b}(t) = b(0)$ if $t < 0$. Letting $T \to \infty$ in the above, then we get $0 = (k - i\lambda + \int_{0}^{\infty} e^{-i\lambda s}db(s))\mathcal{F}F(\lambda) + 1$, or $\mathcal{F}F(\lambda) = 1/(i\lambda - \int_{0}^{\infty} e^{-i\lambda s}db(s) - k)$, as required.

We next consider Eq.(2.50). To this purpose, we need

Lemma 2.14 Let $M(\mathbf{C}) = \Lambda(\mathbf{C}) \cap AP(\mathbf{C})$ and $M(\mathbf{X}) = \Lambda(\mathbf{X}) \cap AP(\mathbf{X})$. Then $\rho(\mathcal{D}_{M(\mathbf{C})} - \mathcal{B}_{M(\mathbf{C})}) \subset \rho(\mathcal{D}_{M(\mathbf{X})} - \mathcal{B}_{M(\mathbf{X})})$.

Proof. Let $k \in \rho(\mathcal{D}_{M(\mathbf{C})} - \mathcal{B}_{M(\mathbf{C})})$. Then Theorem 2.12 and Corollary 2.9 imply that there exists an integrable function F such that $\mathcal{F}F(\lambda) = 1/(i\lambda - \int_{0}^{\infty} e^{-i\lambda s}db(s) - k)$ for all $\lambda \in \Lambda$. Then, by almost the same argument as in the first paragraph of the proof of Corollary 2.9, we see that for any $f \in M(\mathbf{X})$ the equation

$$\dot{x}(t) = kx(t) + \int_{0}^{\infty} x(t-s)db(s) + f(t)$$

has a unique solution $u(t) = F * f(t)$ in $M(\mathbf{X})$. Thus $k \in \rho(\mathcal{D}_{M(\mathbf{X})} - \mathcal{B}_{M(\mathcal{X})})$.

Applying this lemma, we get a condition under which $M(\mathbf{X})$ is admissible for Eq.(2.50)

Corollary 2.10 Assume that

$$[i\lambda - \int_{0}^{\infty} e^{-i\lambda s}db(s) - A]^{-1} \in L(\mathbf{X}) \text{ for all } \lambda \in \Lambda. \tag{2.56}$$

Then $\mathcal{M} := \Lambda(\mathbf{X}) \cap AP(\mathbf{X})$ is admissible for Eq.(2.50).

Proof. In fact, since A is the generator of an analytic strongly continuous semigroup and $\mathcal{B}$ is bounded one sees that $\mathcal{D}_\mathcal{M} - \mathcal{B}_\mathcal{M}$ and $\mathcal{A}_\mathcal{M}$ satisfy all conditions of Theorem 4.10. Moreover, in view of Lemma 2.10 they are densely defined. Thus,

$$\sigma(\overline{\mathcal{D}_\mathcal{M} - \mathcal{B}_\mathcal{M} - \mathcal{A}_\mathcal{M}}) \subset \sigma(\mathcal{D}_\mathcal{M} - \mathcal{B}_\mathcal{M}) - \sigma(\mathcal{A}_\mathcal{M}).$$

Hence, condition (2.56) means that $0 \in \sigma(\overline{\mathcal{D}_\mathcal{M} - \mathcal{B}_\mathcal{M} - \mathcal{A}_\mathcal{M}})$ which, by Theorem 2.8 (or more precisely its proof and Remark 2.8) yields the existence and uniqueness of mild solution in $\mathcal{M}$ to Eq.(2.50).

2.3. DECOMPOSITION THEOREM AND PERIODIC, ALMOST PERIODIC SOLUTIONS

We consider in this section the following linear inhomogeneous integral equation

$$x(t) = U(t,s)x(s) + \int_s^t U(t,\xi)g(\xi)d\xi, \forall t \geq s; t, s \in \mathbf{R}, \tag{2.57}$$

where f is continuous, $x(t) \in \mathbf{X}$, $\mathbf{X}$ is a Banach space, $(U(t,s))_{t \geq s}$ is assumed to be a 1-periodic evolutionary process on $\mathbf{X}$. As is known, continuous solutions of Eq.(2.57) correspond to the *mild solutions* of evolution equations

$$\frac{dx}{dt} = A(t)x + f(t), t \in \mathbf{R}, x \in \mathbf{X}, \tag{2.58}$$

where $A(t)$ is a (in general, unbounded) linear operator for every fixed t and is 1-periodic in t, and $(U(t,s))_{t \geq s}$ is generated by Eq.(2.58).

In Sections 1 and 2 we have studied conditions for the existence and uniqueness (in some classes of function spaces) of almost periodic solutions of Eq.(2.57). In fact we have shown that if the following *nonresonant condition* holds

$$(\sigma(P) \cap S^1) \cap \overline{e^{isp(f)}} = \varnothing, \tag{2.59}$$

where $P := U(1,0)$, S^1 denotes the unit circle of the complex plane, and f is almost periodic, then there exists an almost periodic solution x_f to Eq.(2.57) which is unique if one requires

$$\overline{e^{isp(x_f)}} \subset \overline{e^{isp(f)}}.$$

We may ask a question as what happens in *the resonant case* where condition (2.59) fails. Historically, this question goes back to a classical result of ordinary differential equations saying that supposing the finite dimension of the phase space $\mathbf{X}$ and the 1-periodicity of f Eq.(2.57) has a 1-periodic solution if and only if it has a bounded solution (see e.g. [4, Theorem 20.3, p. 278]). It is the purpose of this section to give an answer to the general problem as mentioned above (*Massera type problem*): Let Eq.(2.57) have a bounded (uniformly continuous) solution x_f with given almost *periodic forcing term f. Then, when does Eq.(2.57) have an almost periodic solution w (which may be different from x_f) such that*

$$\overline{e^{isp(w)}} \subset \overline{e^{isp(f)}} \quad ?$$

In connection with this problem we note that various conditions are found on the bounded solution, itself, and the countability of the part of spectrum $\sigma(P) \cap S^1$ so that the bounded solution itself is almost periodic, or more generally, together with f belongs to a given function space $\mathcal{F}$ (see Corollary 2.6). Here we note that this philosophy in general does not apply to the Massera type problem for almost periodic solutions. In fact, it is not difficult to give a simple example in which f is 1-periodic and a bounded (uniformly continuous) solution to Eq.(2.57) exists, but this bounded solution itself is not 1-periodic.

Our method is to employ the evolution semigroup associated with the process $(U(t,s))_{t \geq s}$ to study the harmonic analysis of bounded solutions to Eq.(2.57). As a result we will prove a spectral decomposition theorem for bounded solutions (Theorem 2.14 and Theorem 2.15) which seems to be useful in dealing with the above Massera type problem. In fact, we will apply the spectral decomposition theorem to find new spectral criteria for the existence of almost periodic solutions and will consider particular cases to show the usefulness of this spectral decomposition technique. More concretely, even in the case where condition (2.59) fails we can still prove the existence of a bounded uniformly continuous solution w to Eq.(2.57) such that $\overline{e^{isp(w)}} = \overline{e^{isp(f)}}$ provided that $(\sigma(P) \cap S^1) \backslash \overline{e^{isp(f)}}$ is closed, and that Eq.(2.57) has a bounded uniformly continuous solution u (Corollary 2.11). Since w is a "spectral component" of u in case u is almost periodic the Fourier series of w is part of that of u (Corollary 2.12). Our Corollary 2.13 will deal with a particular autonomous case in which Corollary 2.11 fails to give a spectral criterion for the existence of quasi-periodic mild solutions.

For the sake of simplicity of notations we will use throughout the section the following notation: $\sigma(g) := \overline{e^{isp(g)}}$ for every bounded uniformly continuous function g. Throughout the section we will denote by $\sigma_\Gamma(P) = \sigma(P) \cap S^1$. Throughout this section $(U(t,s))_{t \geq s}$ will be assumed to be a 1-periodic strongly continuous evolutionary process. The operator $U(1,0)$ will be called the *monodromy operator* of the evolutionary process $(U(t,s))_{t \geq s}$ and will be denoted by P throughout this section. Note that the period of the evolutionary processes is assumed to be 1 merely for the sake of simplicity. Recall that

Definition 2.13 Let $(U(t,s))_{t \geq s}, (t,s \in \mathbf{R})$ be a 1-periodic strongly continuous evolutionary process and $\mathbf{F}$ be a closed subspace of $BUC(\mathbf{R}, \mathbf{X})$ such that for every fixed $h > 0, g \in \mathbf{F}$ the map $t \mapsto U(t, t-h)g(t-h)$ belongs to $\mathbf{F}$. Then the semigroup of operators $(T^h)_{h \geq 0}$ on $\mathbf{F}$, defined by the formula

$$T^h g(t) = U(t, t-h)g(t-h), \forall t \in \mathbf{R}, h \geq 0, g \in \mathbf{F},$$

is called *evolution semigroup* associated with the process $(U(t,s))_{t \geq s}$.

We refer the reader to the previous section for further information on this semigroup. In the case where the evolution semigroup $(T^h)_{h\geq 0}$ is strongly continuous on $\mathbf{F}$ the explicit formula for the generator $\mathcal{A}$ of $(T^h)_{h\geq 0}$ is as follows:

Lemma 2.15 *Let $(T^h)_{h\geq 0}$ be strongly continuous on $\mathbf{F}$, a closed subspace of $BUC(\mathbf{R}, \mathbf{X})$. Then its generator $\mathcal{A}$ is the operator on $\mathbf{F}$ with $D(\mathcal{A})$ consisting of all $g \in \mathbf{F}$ such that g is a solution to Eq.(2.57) with some $f \in \mathbf{F}$ (in this case such a function f is unique), by definition, $\mathcal{A}g = -f$.*

Proof. For the proof of the lemma we need only a minor modification of that of Lemma 2.1, so details are omitted.

2.3.1. Spectral Decomposition

Let us consider the subspace $\mathcal{M} \subset BUC(\mathbf{R}, \mathbf{X})$ consisting of all functions $v \in BUC(\mathbf{R}, \mathbf{X})$ such that
$$\overline{e^{isp(v)}} := \sigma(v) \subset S_1 \cup S_2 , \tag{2.60}$$
where $S_1, S_2 \subset \mathbf{S}^1$ are disjoint closed subsets of the unit circle. We denote by $\mathcal{M}_v = \overline{span\{S(t)v, t \in \mathbf{R}\}}$, where $(S(t))_{t\in\mathbf{R}}$ is the translation group on $BUC(\mathbf{R}, \mathbf{X})$, i.e. $S(t)v(s) = v(t+s), \forall t, s \in \mathbf{R}$.

Theorem 2.13 *Under the above notation and assumptions the function space $\mathcal{M}$ can be split into a direct sum $\mathcal{M} = \mathcal{M}_1 \oplus \mathcal{M}_2$ such that $v \in \mathcal{M}_i$ if and only if $\sigma(v) \subset S_i$ for $i = 1, 2$.*

Proof. Let $v \in \mathcal{M}$. Then, as is known (see Theorem 1.16)
$$isp(v) = \sigma(\mathcal{D}_{\mathcal{M}_v}). \tag{2.61}$$
Thus, by the Weak Spectral Mapping Theorem (see Theorem 1.8)
$$\sigma(S(1)|_{\mathcal{M}_v}) = \overline{e^{\sigma(\mathcal{D}_{\mathcal{M}_v})}} = \sigma(v) \subset S_1 \cup S_2. \tag{2.62}$$
Hence there is a spectral projection in $\mathcal{M}_v$ (note that in general we do not claim that this projection is defined on the whole space $\mathcal{M}$)
$$P_v^1 := \frac{1}{2i\pi} \int_\gamma R(\lambda, S(1)|_{\mathcal{M}_v}) d\lambda ,$$
where γ is a contour enclosing S_1 and disjoint from S_2, (or in general a union of fintely many such countours) by which we have
$$\sigma(P_v^1 S(1) P_v^1) \subset S_1. \tag{2.63}$$
On the other hand, denote $\Lambda_i \subset BUC(\mathbf{R}, \mathbf{X})$ consisting of all functions u such that $\sigma(u) \subset S_i$ for $i = 1, 2$. Then obviously, $\Lambda_i \subset \mathcal{M}$. Moreover, they are closed subspaces of $\mathcal{M}$, $\Lambda_1 \cap \Lambda_2 = \{0\}$. Now we show that if $v \in \mathcal{M}$, then $P_v^1 v \in \Lambda_1$ and $v - v_1 := v_2 \in \Lambda_2$. If this is true, then it yields that

$$\mathcal{M} = \Lambda_1 \oplus \Lambda_2.$$

To this end, we will prove
$$\sigma(v_j) \subset S_j, \forall j = 1, 2. \tag{2.64}$$

In fact we show that $\mathcal{M}_{v_1} = ImP_v^1$. Obviously, in view of the invariance of ImP_v^1 under translations we have $\mathcal{M}_{v_1} \subset ImP_v^1$. We now show the inverse. To this end, let $y \in ImP_v^1 \subset \mathcal{M}_v$. Then, by definition, we have
$$y = \lim_{n \to \infty} x_n,$$
where x_n can be represented in the form
$$x_n = \sum_{k=1}^{N(n)} \alpha_{k,n} S(t_{k,n}) v, \alpha_{k,n} \in \mathbf{C}, t_{k,n} \in \mathbf{R} \, \forall n.$$

Hence, since $y, x_n \in \mathcal{M}_v$
$$\begin{aligned} y = P_v^1 y &= \lim_{n \to \infty} \sum_{k=1}^{N(n)} \alpha_{k,n} S(t_{k,n}) P_v^1 v \\ &= \lim_{n \to \infty} \sum_{k=1}^{N(n)} \alpha_{k,n} S(t_{k,n}) v_1 \, . \end{aligned} \tag{2.65}$$

This shows that $y \in \mathcal{M}_{v_1}$. Thus, by the Weak Spectral Mapping Theorem and (2.63),
$$\overline{e^{isp(v_1)}} = \sigma(S(1)|_{\mathcal{M}_{v_1}}) = \sigma(S(1)|_{ImP_v^1}) \subset S_1.$$

By definition, $v_1 \in \Lambda_1$ and similarly, $v_2 \in \Lambda_2$. Thus the theorem is proved.

Remark 2.11 Below for every $v \in \mathcal{M}$ we will call $v_j, j = 1, 2$, as defined in the proof of Theorem 2.13, *spectral components* of the functions v. It is easily seen that if in the proof of Theorem 2.13, v is assumed to be almost periodic, then both spectral components v_j are almost periodic.

We will need the following lemma in the sequel.

Lemma 2.16 *Let f be in $BUC(\mathbf{R}, \mathbf{X})$ and $(U(t,s))_{t \geq s}$ be a 1-periodic strongly continuous evolutionary process. Then the following assertions hold:*

i) If $T : \mathbf{R} \to L(\mathbf{X})$ be 1-periodic and strongly continuous, then
$$\sigma(T(\cdot)f(\cdot)) \subset \sigma(f).$$

ii) If f is of precompact range, then
$$\sigma\left(\int_t^{t+1} U(t+1, \xi) f(\xi) d\xi\right) \subset \sigma(f).$$

Proof. (i) Let T_n be the nth Cesaro mean of the Fourier series of T, so T_n is 1-periodic trigonometric polynomial with value in $L(\mathbf{X})$ and $\|T_n(s)\| \leq \sup_{0 \leq t \leq 1} \|T(t)\|$ and $T_n(s)x \to T(s)x$ uniformly in s for fixed $x \in \mathbf{X}$. For every n it is easily seen that $\sigma(T_n(\cdot)f(\cdot)) \subset \sigma(f)$. Set $\Lambda := \{\lambda \in \mathbf{R} : e^{i\lambda} \in \sigma(f)\}$. Obviously, Λ is closed and $sp(T_n(\cdot)f(\cdot)) \subset \Lambda$. Thus, if $\phi \in L^1(\mathbf{R})$ and $(supp \mathcal{F}\phi) \cap \Lambda = \varnothing$, then

$$0 = \int_{-\infty}^{\infty} \phi(t-s)T_n(s)f(s)ds \to \int_{-\infty}^{\infty} \phi(t-s)T(s)f(s)ds$$

as $n \to \infty$, by Dominanated Convergence Theorem. Thus

$$\int_{-\infty}^{\infty} \phi(t-s)T(s)f(s)ds = 0$$

for all such ϕ. This proves (i).

(ii) Since f is of precompact range the evolution semigroup $(T^h)_{h \geq 0}$ associated with the process $(U(t,s))_{t \geq s}$ is strongly continuous at f. Thus, in view of (i)

$$\sigma\left(\int_0^h T^\xi f d\xi\right) \subset (f), \forall h \geq 0.$$

On the other hand

$$\int_0^1 T^\xi f d\xi(t+1) = \int_0^1 U(t+1, t+1-\xi)f(t+1-\xi)d\xi = \int_t^{t+1} U(t+1, \eta)f(\eta)d\eta.$$

This proves (ii).

Lemma 2.17 *Let u be a bounded uniformly continuous solution to (2.57) and f be of precompact range. Then the following assertions hold true:*

i)
$$\sigma(u) \subset \sigma_\Gamma(P) \cup \sigma(f) , \qquad (2.66)$$

ii)
$$\sigma(u) \supset \sigma(f) . \qquad (2.67)$$

Proof. (i) Set $P(t) := U(t, t-1)$, $\forall t \in \mathbf{R}$, $G := \{\lambda \in \mathbf{C} : e^\lambda \in \rho(P)\}$.

$$g(t) := \int_t^{t+1} U(t+1, \xi)f(\xi)d\xi, t \in \mathbf{R}.$$

By Lemma 2.16 $\sigma(g) \subset \sigma(f)$. By the definition of Carleman spectrum,

$$\hat{u}(\lambda) = \begin{cases} \int_0^\infty e^{-\lambda t} u(t) dt, & (Re\lambda > 0); \\ -\int_0^\infty e^{\lambda t} u(-t) dt, & (Re\lambda < 0). \end{cases}$$

Hence, for $Re\lambda > 0$ and $\lambda \in G$ we have

$$\begin{aligned}
\hat{u}(\lambda) &= \int_0^\infty e^{-\lambda t} u(t) dt, \\
&= \int_0^\infty e^{-\lambda t} R(e^\lambda, P(t))(e^\lambda - P(t)) u(t) dt \\
&= \int_0^1 e^{-\lambda t} e^\lambda R(e^\lambda, P(t)) u(t) dt + \\
&\quad + \int_0^\infty e^{-\lambda t} R(e^\lambda, P(t))(u(t+1) - P(t) u(t)) dt \\
&= H(\lambda) + \int_0^\infty e^{-\lambda t} R(e^\lambda, P(t)) g(t) dt, \quad\quad (2.68)
\end{aligned}$$

where

$$H(\lambda) := \int_0^1 e^{-\lambda t} e^\lambda R(e^\lambda, P(t)) u(t) dt.$$

Obviously, $H(\lambda)$ is analytic in G. On the other hand, since $R(e^\lambda, P(t))$ is 1-periodic strongly continuous (see Lemma 2.2), by Lemma 2.16 the function $g_1(t) := R(e^\lambda, P(t)) g(t)$ has the property that $\sigma(g_1) \subset \sigma(f)$. Thus from (2.68), for $Re\lambda > 0$, $\lambda \in G$,

$$\hat{u}(\lambda) = H(\lambda) + \hat{g}_1(\lambda). \quad\quad (2.69)$$

Finally, if $\zeta_0 \in \mathbf{R} : e^{\zeta_0} \notin \sigma_\Gamma(P) \cup \sigma(f)$, then $\hat{u}$ has an analytic continuation at ζ_0. This completes the proof of (i).

(ii) Under the assumptions it may be seen that the evolution semigroup $(T^h)_{h\geq 0}$ associated with $(U(t,s))_{t\geq s}$ is strongly continuous at f and $u \in BUC(\mathbf{R}, \mathbf{X})$ (this can be checked directly using Eq.(2.57) as in Lemma 2.19 below). Hence, by Lemma 2.15

$$\lim_{h\downarrow 0} \frac{T^h u - u}{h} = \mathcal{A} u = -f . \quad\quad (2.70)$$

Hence, to prove (2.67) it suffices to show that $\sigma(T^h u) \subset \sigma(u)$. In turn, this is clear in view of Lemma 2.16.

We are now in a position to state the main result of this section.

Theorem 2.14 *(Spectral Decomposition Theorem) Let u be a bounded, uniformly continuous solution to Eq.(2.57). Moreover, let f have precompact range and the sets $\sigma(f)$ and $\sigma(P) \cap S^1$ be contained in a disjoint union of the closed subsets $S_1, \cdots, S_k$ of the unit circle. Then the solution u can be decomposed into a sum of k spectral components $u_j, j = 1, \cdots, k$ such that each $u_j, j = 1, \cdots, k$ is a solution to Eq.(2.57) with $f = f_j, j = 1, \cdots, k$, respectively, where $f = \sum_{j=1}^k f_j$ is the decomposition of f into the sum of spectral components as described in Theorem 2.13, i.e. $u = \sum_{j=1}^k u_j, \sigma(u_j), \sigma(f_j) \subset S_j, j = 1, \cdots, k$ and $u_j \in BUC(\mathbf{R}, \mathbf{X})$ is a solution to Eq.(2.57) with $f := f_j$ for $j = 1, \cdots, k$.*

Proof. Let us denote by $\mathcal{N}$ the subspace of $BUC(\mathbf{R}, \mathbf{X})$ consisting of all functions u such that $\sigma(u) \subset \cup_{j=1}^k S_j$. Then, by assumptions and Theorem 2.13 there are corresponding spectral projections $P_1, \cdots, P_k$ on $\mathcal{N}$ with properties that

i) $P_j P_n = 0$ if $j \neq n$,

ii) $\Sigma_{j=1}^k P_j = I$,

iii) If $u \in Im P_j$, then $\sigma(P_j u) \subset S_j$ for all $j = 1, \cdots, k$.

Note that by Lemma 2.17 for every positive h and $j = 1, \cdots, k$ the operator T^h leaves $Im P_j$ invariant. Hence, $\mathcal{N}$ and $Im P_1, \cdots, Im P_k$ are invariant under the semigroup $(T^h)_{h \geq 0}$. Consequently, since u is a solution to Eq.(2.57) and f has precompact range the evolution semigroup $(T^h)_{h \geq 0}$ is strongly continuous at u and f. Using the explicit formula for the generator of $(T^h)_{h \geq 0}$ as described in Lemma 2.15 we have

$$P_j f = P_j \lim_{h \downarrow 0} \frac{T^h u - u}{h} = P_j \lim_{h \downarrow 0} \Sigma_{n=1}^k P_n \frac{T^h u - u}{h}$$
$$= \lim_{h \downarrow 0} \frac{T^h P_j u - P_j u}{h}. \qquad (2.71)$$

This yields that $P_j u$ is a solution to Eq.(2.57) with corresponding $f_j = P_j f$.

Remark 2.12 If in Theorem 2.14 we assume furthermore that f and u are both almost periodic, then the spectral components $u_j, j = 1, \cdots, k$ are all almost periodic. This is not the case if neither u, nor f is almost periodic. However, if we have some additional information on the spectral sets S_j, e.g., their countability and the phase space **X** does not contain c_0, then the almost periodicity of u_j are guaranteed.

Now we are going to focus our special attention on autonomous equations of the form
$$dx/dt = Ax + f(t), \qquad (2.72)$$
where A is the generator of a C_0-semigroup $(T(t))_{t \geq 0}$, and $f \in BUC(\mathbf{R}, \mathbf{X})$ has precompact range. Below we will use the following notation: $\sigma_i(A) = \{\lambda \in \mathbf{R} : i\lambda \in (\sigma(A) \cap i\mathbf{R})\}$. By mild solutions of Eq.(2.72) we will understand in a standard way that they are solutions to Eq.(2.57) with $U(t,s) := T(t-s), \forall t \geq s$. As shown below, in this case we can refine the spectral decomposition technique to get stronger assertions which usefulness will be shown in the next subsection when we deal with quasi-periodic solutions. To this purpose, we now prove the following lemma.

Lemma 2.18 Let Eq.(2.72) satisfy the above conditions, i.e., A generates a C_0-semigroup and $f \in BUC(\mathbf{R}, \mathbf{X})$ has precompact range. Moreover, let u be a bounded uniformly continuous mild solution to Eq.(2.72). Then the following assertions hold:

i)
$$sp(u) \subset \sigma_i(A) \cup sp(f) , \qquad (2.73)$$

ii)
$$sp(u) \supset sp(f). \qquad (2.74)$$

Proof. (i) For (2.73) we compute the Carleman transform of u. Since u is a mild solution, by Lemma 2.11 $\int_0^t u(\xi)d\xi \in D(A), \forall t \in \mathbf{R}$ and

$$u(t) - u(0) = A\int_0^t u(\xi)d\xi + \int_0^t f(\xi)d\xi. \tag{2.75}$$

Hence, taking the Laplace transform of u we have

$$\hat{u}(\lambda) + \frac{1}{\lambda}u(0) = \frac{1}{\lambda}A\hat{u}(\lambda) + \frac{1}{\lambda}\int_0^\infty e^{-\lambda t}f(t)dt$$
$$= \frac{1}{\lambda}A\hat{u}(\lambda) + \frac{1}{\lambda}\hat{f}(\lambda), \tag{2.76}$$

and hence, for $\Re\lambda \neq 0$, $(\lambda - A)\hat{u}(\lambda) = -u(0) + \hat{f}(\lambda)$. Obviously, for $\xi \notin sp(f)$ and $i\xi \notin \sigma(A) \cap i\mathbf{R}$, one has that $\hat{u}(\lambda)$ has a holomorphic extension around $i\xi$, i.e., $\xi \notin sp(u)$.

(ii) Note that since for every $h > 0$ the operator T^h is a multiplication by a bounded operator $T(h)$ we have $sp(T^h u) \subset sp(u)$. Thus, using the argument of the proof of Lemma 2.17 (ii) we have

$$sp(f) = sp(-f) = sp\left(\lim_{h\downarrow 0}\frac{T^h u - u}{h}\right)$$
$$\subset sp(u) .$$

This completes the proof of (ii).

The main result for the autonomous case is the following:

Theorem 2.15 *Let A generate a C_0-semigroup and $f \in BUC(\mathbf{R}, \mathbf{X})$ have precompact range. Moreover, let u be a bounded uniformly continuous mild solution to Eq.(2.72). Then the following assertions hold true:*

i) If

$$\overline{e^{i\sigma_i(A)}}\backslash\sigma(f)$$

is closed, Eq.(2.72) has a bounded uniformly continuous mild solution w such that $\sigma(w) = \sigma(f)$,

ii) If $\sigma_i(A)$ is bounded and

$$\sigma_i(A)\backslash sp(f) \tag{2.77}$$

is closed, then Eq.(2.72) has a bounded uniformly continuous mild solution w such that $sp(w) = sp(f)$.

Proof. (i) Note that in this case together with (2.73) the proof of Theorem 2.14 applies.
(ii) Under the assumptions there exists a continuous function ψ which belongs to the Schwartz space of all C^∞-functions on $\mathbf{R}$ with each of its derivatives decaying

faster than any polynomial such that its Fourier transform $\tilde\psi$ has $\sigma_i(A)\backslash sp(f)$ as its support (which is compact in view of the assumptions). Hence, every bounded uniformly continuous function g such that $sp(g) \subset \sigma_i(A)\cup sp(f)$ can be decomposed into the sum of two spectral components as follows:

$$g = g_1 + g_2 = \psi * g + (g - \psi * g),$$

where $g_1 = \psi * g, g_2 = (g - \psi * g)$. Moreover, this decomposition is continuous in the following sense: If $g^{(n)}, n = 1, 2, \cdots$ is a sequence in $BUC(\mathbf{R}, \mathbf{X})$ with $sp(g^{(n)}) \subset \sigma_i(A) \cup sp(f)$ such that $\lim_n g^{(n)} = g$ in $BUC(\mathbf{R}, \mathbf{X})$, then $\lim_n g_1^{(n)} = g_1, \lim_n g_2^{(n)} = g_2$. Hence we have in fact proved a version of Theorem 2.13 which allows us to employ the proof of Theorem 2.14 for this assertion (ii).

Remark 2.13 i) In view of the failure of the Spectral Mapping Theorem for general C_0-semigroups the condition in the assertion (i) is a little more general than that formulated in terms of $\sigma(T(1))$.

ii) If we know beforehand that u is almost periodic, then in the statement of Theorem 2.15 we can claim that the spectral component w is almost periodic.

2.3.2. Spectral Criteria For Almost Periodic Solutions

This subsection will be devoted to some applications of the spectral decomposition theorem to prove the existence of almost periodic solutions with specific spectral properties. In particular, we will revisit the classical result by Massera on the existence of periodic solutions as well as its extensions. To this end, the following notion will play the key role.

Definition 2.14 Let $\sigma(f)$ and $\sigma_\Gamma(P)$ be defined as above. We say that the set $\sigma(f)$ and $\sigma_\Gamma(P)$ satisfy *the spectral separation condition* if the set $\sigma_\Gamma(P)\backslash\sigma(f)$ is closed.

Corollary 2.11 Let f be almost periodic, $\sigma(f)$ and $\sigma_\Gamma(P)$ satisfy the spectral separation condition. Moreover, let $\sigma(f)$ be countable and $\mathbf{X}$ not contain any subspace which is isomorphic to c_0. Then if there exists a bounded uniformly continuous solution u to Eq.(2.57), there exists an almost periodic solution w to Eq.(2.57) such that $\sigma(w) = \sigma(f)$.

Proof. We define in this case $S_1 := \sigma(f)$, $S_2 := \sigma_\Gamma(P)\backslash\sigma(f)$. Then, by Theorem 2.14 there exists a solution w to Eq.(2.57) such that $\sigma(w) \subset \sigma(f)$. Using the estimate (2.67) we have $\sigma(w) = \sigma(f)$. In particular, since $\sigma(w)$ is countable and $\mathbf{X}$ does not contain c_0, w is almost periodic.

Remark 2.14 i) If $\sigma(f)$ is finite, then $sp(w)$ is discrete. Thus, the condition that $\mathbf{X}$ does not contain any subspace isomorphic to c_0 can be dropped.

ii) In the case where $\sigma_\Gamma(P)$ is countable it is known that with additional ergodic conditions on u the solution u has "similar spectral properties" as f (see Corollary 2.6). However, in many cases it is not expected that the solution u itself has similar spectral properties as f as in the Massera type problem (see [147], [45], [206], [168] e.g.).

iii) In the case where P is compact (or merely $\sigma_\Gamma(P)$ is finite) the spectral separation condition is always satisfied. Hence, we have a natural extension of a classical result for almost periodic solutions. In this case see also Corollary 2.12 below.

iv) We emphasize that the solution w in the statement of Corollary 2.11 is a "$\sigma(f)$-spectral component" of the bounded solution u. This will be helpful to find the Fourier coefficients of w as part of those of u.

v) In view of estimate (2.67) w may be seen as a "minimal" solution in some sense.

Corollary 2.12 *Let all assumptions of Corollary 2.11 be satisfied. Moreover, let $\sigma_\Gamma(P)$ be countable. Then if there exists a bounded uniformly continuous solution u to Eq.(2.57), it is almost periodic. Moreover, the following part of the Fourier series of u*

$$\Sigma b_\lambda e^{i\lambda t} \quad , b_\lambda = \lim_{T\to\infty} \frac{1}{2T} \int_{-T}^{T} e^{-i\lambda \xi} u(\xi) d\xi, \tag{2.78}$$

where $e^{i\lambda} \in \sigma(f)$, is again the Fourier series of another almost periodic solution to Eq.(2.57).

Proof. The assertion that u is almost periodic is standard in view of (2.66) (see Chapter 1). It may be noted that in the case u is almost periodic, the spectral decomposition can be carried out in the function space $AP(\mathbf{X})$ instead of the larger space $BUC(\mathbf{R}, \mathbf{X})$. Hence, we can decompose the solution u into the sum of two almost periodic solutions with spectral properties described in Theorem 2.14. Using the definition of Fourier series of almost periodic functions we arrive at the next assertion of the corollary.

The next corollary will show the advantage of Theorem 2.15 which allows us to take into account the structure of $sp(f)$ rather than that of $\sigma(f)$. To this end, we introduce the following terminology. A set of reals S is said to have an *integer and finite basis* if there is a finite subset $T \subset S$ such that any element $s \in S$ can be represented in the form $s = n_1 b_1 + \cdots + n_m b_m$, where $n_j \in \mathbf{Z}, j = 1, \cdots, m$, $b_j \in T, j = 1, \cdots, m$. If f is quasi-periodic and the set of its Fourier-Bohr exponents is discrete (which coincides with $sp(f)$ in this case), then the spectrum $sp(f)$ has an integer and finite basis (see [137, p.48]). Conversely, if f is almost periodic and $sp(f)$ has an integer and finite basis, then f is quasi-periodic. We refer the reader to [137, pp. 42-48] more information on the relation between quasi-periodicity and spectrum, Fourier-Bohr exponents of almost periodic functions.

Corollary 2.13 *Let all assumptions of the second assertion of Theorem 2.15 be satisfied. Moreover, assume that* $\mathbf{X}$ *does not contain* c_0. *Then if* $sp(f)$ *has an integer and finite basis, Eq.(2.72) has a quasi-periodic mild solution* w *with* $sp(w) = sp(f)$.

Proof. Under the corollary's assumptions the spectrum $sp(w)$ of the solution w, as described in Theorem 2.15, is in particular countable. Hence w is almost periodic. Since $sp(w) = sp(f)$, $sp(w)$ has an integer and finite basis. Thus w is quasi-periodic.

Below we will consider some particular cases

Example 2.10 *Periodic solutions.*

If $\sigma(f) = \{1\}$ we are actually concerned with the existence of periodic solutions. Hence, Corollary 2.11 extends the classical result to a large class of evolution equations which has 1 as an isolated point of $\sigma_\Gamma(P)$. Moreover, Corollary 2.12 provides a way to approximate the periodic solution. In particular, suppose that $\sigma_\Gamma(P)$ has finitely many elements, then we have the following:

Corollary 2.14 *Let* $\sigma_\Gamma(P)$ *have finitely many elements* $\{\mu_1, \cdots, \mu_N\}$ *and* $u(\cdot)$ *be a bounded uniformly continuous solution to Eq.(2.57). Then it is of the form*

$$u(t) = u_0(t) + \sum_{k=1}^{N} e^{i\lambda_k t} u_k(t), \qquad (2.79)$$

where u_0 *is a bounded uniformly continuous mild 1-periodic solution to the inhomogeneous equation (2.57),* $u_k, k = 1, \cdots, N$, *are 1-periodic solutions to Eq.(2.57) with* $f = -i\lambda_k u_k$, *respectively,* $v(t) = \sum_{k=1}^{N} e^{i\lambda_k t} u_k(t)$ *is a quasi periodic solution to the corresponding homogeneous equation of Eq.(2.57) and* $\lambda_1, \cdots, \lambda_N$ *are such that* $0 < \lambda_1, \cdots, < \lambda_N < 2\pi$ *and* $e^{i\lambda_j} = \mu_j, j = 1, \cdots, N$.

Example 2.11 *Anti-periodic solutions.*

An anti-periodic (continuous) function f is defined to be a continuous one which satisfies $f(t+\omega) = -f(t), \forall t \in \mathbf{R}$ and here $\omega > 0$ is given. Thus, f is $2-\omega$-periodic. It is known that, the space of anti-periodic functions f with antiperiod ω, which is denoted by $\mathcal{AP}(\omega)$, is a subspace of $BUC(\mathbf{R}, \mathbf{X})$ with spectrum

$$sp(f) \subset \{\frac{2k+1}{\omega}, k \in \mathbf{Z}\}.$$

Without loss of generality we can assume that $\omega = 1$. Obviously, $\sigma(f) = \{-1\}, \forall f \in \mathcal{AP}(\omega)$. In this case the spectral separation condition is nothing but the condition that $\{-1\}$ is an isolated point of $\sigma_\Gamma(P)$.

Example 2.12

Let u be a bounded uniformly continuous solution to Eq.(2.57) with f 2-periodic. Let us define

$$F(t) = \frac{f(t) - f(t+1)}{2}, G(t) = \frac{f(t) + f(t+1)}{2}, \forall t \in \mathbf{R}.$$

Then, it is seen that F is 1-anti-periodic and G is 1-periodic. Applying Theorem 2.14 we see that there exist two solutions to Eq.(2.57) as two components of u which are 1-antiperiodic and 1-periodic with forcing terms F, G, respectively. In particular, the sum of these solutions is a 2-periodic solution of Eq.(2.57) with forcing term f.

Example 2.13

Let A be a sectorial operator in a Banach space $\mathbf{X}$ and the map $t \mapsto B(t) \in L(\mathbf{X}^\alpha, \mathbf{X})$ be Hölder continuous and 1-periodic. Then, as shown in [90, Theorem 7.1.3] the equation

$$\frac{dx}{dt} = (-A + B(t))x, \qquad (2.80)$$

generates an 1-periodic strongly continuous evolutionary process $(U(t,s))_{t\geq s}$. If, furthermore, A has compact resolvent, then the monodromy operator P of the process is compact. Hence, for every almost periodic function f the sets $\sigma(f)$ and $\sigma_\Gamma(P)$ always satisfy spectral separation condition. In Section 1 we have shown that if $\sigma_\Gamma(P) \cap \sigma(f) = \emptyset$, then there is a unique almost periodic solution x_f to the inhomogeneous equation

$$\frac{dx}{dt} = (-A + B(t))x + f(t) \qquad (2.81)$$

with property that $\sigma(x_f) \subset \sigma(f)$. Now suppose that $\sigma_\Gamma(P) \cap \sigma(f) \neq \emptyset$. By Corollary 2.11, if u is any bounded solution (the uniform continuity follows from the boundedness of such a solution to Eq.(2.81)), then there exists an almost periodic solution w such that $\sigma(w) = \sigma(f)$. We refer the reader to [90] and [179] for examples from parabolic differential equations which can be included into the abstract equation (2.81).

Example 2.14

Consider the heat equation in materials

$$\begin{cases} v_t(t,x) = \Delta v(t,x) + f(t,x), t \in \mathbf{R}, x \in \Omega \\ v(t,x) = 0, t \in \mathbf{R}, x \in \partial\Omega, \end{cases} \qquad (2.82)$$

where $\Omega \subset \mathbf{R}^n$ denotes a bounded domain with smooth boundary $\partial\Omega$. Let $\mathbf{X} = L^2(\Omega)$, $A = \Delta$ with $D(A) = W^{2,2}(\Omega) \cap W_0^{1,2}(\Omega)$. Then A is selfadjoint and negative definite (see e.g. [179]). Hence $\sigma(A) \subset (-\infty, 0)$. In particular $\sigma_i(A) = \emptyset$. Eq.(22) now becomes

$$\frac{dv}{dt} = Av + f. \qquad (2.83)$$

We assume further that $f(t,x) = a(t)g(x)$ where a is a bounded uniformly continuous real function with $sp(a) = \mathbf{Z} \cup \pi\mathbf{Z}$, $g \in L^2(\Omega), g \neq 0$. It may be seen that $\sigma(f) = S^1$ and $sp(f)$ has an integer and finite basis. Hence, Theorem 2.14 does not give any information on the existence of a solution w with specific spectral properties. However, in this case Theorem 2.15 applies, namely, if Eq.(2.82) has a bounded solution, then it has a quasi periodic solution with the same spectrum as f.

Example 2.15

We consider the case of Eq.(2.57) having exponential dichotomy. In this case, as is well known from Section 1, for every almost periodic f there exists a unique almost periodic solution x_f to Eq.(2.57). From the results above we see that $\sigma(x_f) = \sigma(f)$.

2.3.3. When Does Boundedness Yield Uniform Continuity ?

It turns out that in many cases the uniform continuity follows readily from the boundedness of the solutions under consideration. Below we will discuss some particular cases frequently met in applications in which boundedness implies already uniform continuity.

Definition 2.15 A τ-periodic strongly continuous evolutionary process $(U(t,s))_{t \geq s}$ is said to satisfy *condition C* if the (monodromy) operators $P(t) := U(t, t - \tau)$ is norm continuous with respect to t.

Consider the nonautonomous perturbation of autonomous equations. Namely, let $U(t,s) = T(t-s)$, where $(T(t))_{t \geq 0}$ is a compact C_0-semigroup. It can be checked easily that the perturbed equation

$$x(t) = T(t-s)x(s) + \int_s^t T(t-\xi)B(\xi)x(\xi)d\xi, \ \forall t \geq s, \qquad (2.84)$$

where $t \mapsto B(t) \in L(\mathbf{X})$ is τ-periodic and continuous, determines a process satisfying condition C (see also Proposition 2.4 and the next subsection for parabolic equations which also satisfies condition C).

Proposition 2.3 *Let $(U(t,s))_{t \geq s}$ be a τ-periodic strongly continuous evolutionary process satisfying condition C and for every t the operator $P(t) := U(t, t - \tau)$ be compact. Then if u is a bounded mild solution to Eq.(2.57) on the whole line, it is uniformly continuous.*

Proof. For the sake of simplicity we assume that $\tau = 1$. We consider the difference

$$\begin{aligned} u(t') - u(t) &= U(t', t-1)u(t-1) - U(t, t-1)u(t-1) + \\ &\quad + \int_{t-1}^{t'} U(t', \xi)f(\xi)d\xi - \int_{t-1}^{t} U(t, \xi)f(\xi)d\xi. \end{aligned} \qquad (2.85)$$

We now show that
$$\lim_{|t'-t|\to 0} \|U(t', t-1) - U(t, t-1)\| = 0. \tag{2.86}$$

To this end, let us denote by B the unit ball of $\mathbf{X}$. Then the set $K = \{P(t)x, t \in \mathbf{X}, x \in B\}$ is precompact. In fact, let $\varepsilon > 0$. Then by condition C and the τ-periodicity of $P(t)$ there are $0 < t_1 < \cdots < t_n < 1$ such that
$$\|P(t+t_i) - P(t+\xi)\| < \varepsilon/2, \forall \xi \in [t+t_{i-1}, t+t_i]. \tag{2.87}$$

Since for every t_i the operator $P(t_i)B$ is precompact, there exists
$$\{x_1^i, x_2^i, \cdots, x_{k(i)}^i\} \subset B$$
such that if $x \in B$ then
$$\|P(t_i)x_j^i - P(t_i)x\| < \varepsilon/2, \tag{2.88}$$
for some x_j^i. Now let $y = P(\eta)x$ for some $\eta \in \mathbf{R}, x \in B$. Then by the τ-periodicity of $P(t)$ we can assume $\eta \in [0,1]$ and $\eta \in [t_{i-1}, t_i]$ for some i. From (2.87), (2.88) it follows that y is contained in the ball centered at $P(t_i)x_j^i$ with radius ε. This shows that K is precompact. Thus, (2.86) follows from the following
$$\lim_{t'-t\downarrow 0} \sup_{x \in K} \|U(t'-t, 0)x - x\| = 0. \tag{2.89}$$

In turn, (2.89) follows from the precompactness of K and the strong continuity of the evolutionary process $(U(t,s))_{t \geq s}$. On the other hand, we have
$$\int_{t-1}^{t'} U(t', \xi)f(\xi)d\xi - \int_{t-1}^{t} U(t, \xi)f(\xi)d\xi =$$
$$= [U(t', t)\int_{t-1}^{t} U(t, \xi)f(\xi)d\xi - \int_{t-1}^{t} U(t, \xi)f(\xi)d\xi] +$$
$$+ \int_{t}^{t'} U(t', \xi)f(\xi)d\xi. \tag{2.90}$$

By the 1-periodicity of $(U(t,s))_{t \geq s}$ and f one sees that $\int_{t-1}^{t} U(t,\xi)f(\xi)d\xi$ is 1 periodic with respect to t. Hence its range is precompact. As above,
$$\lim_{t'-t\downarrow 0}[U(t',t)\int_{t-1}^{t} U(t,\xi)f(\xi)d\xi - \int_{t-1}^{t} U(t,\xi)f(\xi)d\xi] = 0$$
uniformly in t. This and the fact that
$$\lim_{t'-t\downarrow 0} \int_{t}^{t'} U(t',\xi)f(\xi)d\xi = 0,$$
implies that
$$\lim_{t'-t\downarrow 0} \|u(t') - u(t)\| = 0.$$

Remark 2.15 It is interesting to study the question as how many τ-periodic solutions Eq.(2.57) may have. This depends on the space of τ-periodic mild solutions of the homogeneous equation. A moment of reflection shows that if $v(\cdot)$ is a τ-periodic solution to the corresponding homogeneous equation of Eq.(2.57), then $v(0)$ is a solution to the equation
$$y - Py = 0, y \in \mathbf{X},$$
where P is the monodromy operator of the process $(U(t,s))_{t\geq s}$. Hence, if P is compact, then $v(0)$ belongs to a finite dimensional subspace of $\mathbf{X}$. That is the possible τ-periodic mild solutions to Eq.(2.57) forms a finite dimensional subspace. Note that in Section 1 we have shown that it is necessary and sufficient for Eq.(2.57) (without the compactness assumption) to have a unique τ-periodic solution that the unity belongs to the resolvent set of the monodromy operator P.

2.3.4. Periodic Solutions of Partial Functional Differential Equations

In this subsection we will prove some analogs of the results of the previous subsection for periodic solutions of the following abstract functional differential equation

$$\frac{dx}{dt} = Ax + F(t)x_t + f(t), t \in \mathbf{R}, \tag{2.91}$$

where A is the generator of a compact C_0-semigroup, $F(t) \in L(C, \mathbf{X})$ is τ-periodic and continuous with respect to t, f is continuous and τ-periodic. As usual, we denote $C := C([-r, 0], \mathbf{X})$ where $r > 0$ is a given real number. If $u : [s, s+\alpha) \mapsto \mathbf{X}, \alpha > r$ we denote $x_t(\theta) = x(t+\theta), \theta \in [-r, 0], t \in [s+r, s+\alpha-r)$.

We say that u is a *mild solution* to Eq.(2.91) on $[s, +\infty)$ if there is $\phi \in C$ such that

$$u(t) = T(t-s)\phi(0) + \int_s^t T(t-\xi)[F(\xi)u_\xi + f(\xi)]d\xi, \forall t \geq s. \tag{2.92}$$

u is said to be a *mild solution on the whole line* if instead of arbitrary ϕ in Eq.(2.92) one takes u_s for every $s \in \mathbf{R}$. We now recall some facts on equations of the form (2.91) which the reader can find in Chapter 1. Under the above made assumption for every $s \in \mathbf{R}, \phi \in C$ there exists a unique mild solution $u(t)$ to Eq.(2.92) on $[0, \infty)$. And, by definition, the map taking $\phi \in C$ into $u_t \in C$ (denoted by $V(t,s)\phi$) in the case $f = 0$ is called *solution operator* of the homogeneous equation associated with Eq.(2.92). It may be noted that $(V(t,s))_{t\geq s}$ is a τ-periodic strongly continuous evolutionary process on C. For the sake of simplicity we assume that $\tau > r$, otherwise we can use the fact that the monodromy operator has finitely many elements on the unit circle of the complex plane. Thus the monodromy operator $V(\tau, 0)$ is compact (see Theorem 1.13). This will suggests us to generalize the result obtained in the previous section to Eq.(2.91).

Now suppose that we have the following variation-of-constants formula.

$$\begin{cases} u(t) = [V(t,s)\phi](0) + \int_s^t [V(t,\xi)X_0 f(\xi)](0)d\xi, \\ u_0 = \phi, \end{cases} \quad (2.93)$$

for Eq.(2.91) where $X_0 : [-r, 0] \mapsto L(\mathbf{X})$ is given by $X_0(\theta) = 0 \ -r \leq \theta < 0$ and $X_0(0) = I$.

Hence from the τ-periodicity of the process $(V(t,s))_{t \geq s}$ and f we can write

$$u_t = V(t, t-\tau)u_{t-\tau} + g_t, \quad (2.94)$$

where g_t is τ-periodic and independent of u. Unfortunately, the validity of the variation-constants formula (2.93) is not clear[2]. However, we can show that the representation (2.94) holds true. In fact, we now show that

$$u(t+\theta) - [V(t, t-\tau)u_{t-\tau}](\theta) =$$
$$= \int_{t-\tau}^{t+\theta} T(t+\theta-\xi)F(\xi)[u_\xi - V(\xi, t-\tau)u_{t-\tau}]d\xi +$$
$$+ \int_{t-\tau}^{t+\theta} T(t+\theta-\xi)f(\xi)d\xi, \ \theta \in [-r, 0] \quad (2.95)$$

defines a τ-periodic function in C which is nothing but g_t. In turn, this follows immediately from the existence and uniqueness of solutions and the τ-periodicity of F, f. Hence in (2.94) g_t is a τ-periodic function which depends only on $(T(t))_{t \geq 0}, F(\cdot), f(\cdot)$. Actually, g_t is the solution of (2.92) with initial value 0.

Hence, by the same argument as in Lemma 2.17 we obtain

Corollary 2.15 *Let u_t be the solution of Eq.(2.94). Then, as a function taking value in C,*

$$e^{i\tau sp(u)} \subset \{1\} \cup (\sigma(P) \cap S^1).$$

Thus by Theorem 2.13 in the phase space C

$$u_t = u_0(t) + \sum_{k=1}^M e^{i\lambda_k t} u_k(t),$$

where $u_j : \mathbf{R} \mapsto C$ are τ-periodic and continous for all $j = 0, 1, \cdots, M$, and M is the number of elements of the spectrum of the monodromy operator of the process $(V(t,s))_{t \geq s}$ on the unit circle. In particular, from this u_t is almost periodic and so is $u(\cdot)$. Hence

$$u(t) = u_t(0) = u_0(t)(0) + \sum_{k=1}^M e^{i\lambda_k t} u_k(t)(0), \forall t.$$

Finally we arrive at

---

[2]In general, $(V(t))_{t \geq 0}$ is not defined at discontinuous functions. If one extends its domain as done in [150] or [226, p. 115], then this semigroup is not strongly continuous even in the simpleast case. So, the integral in (2.119) seems to be undefined. The authors owe this remark to S. Murakami for which we thank him.

$$u(t) = u^0(t) + \sum_{k=1}^{M} e^{i\lambda_k t} u^k(t), \forall t,$$

where $u^k(t) := u_k(t)(0) \forall k = 0, 1, \cdots, M$. We now check that u^0 and $\sum_{k=1}^{M} e^{i\lambda_k t} u^k$ are a τ-periodic mild solution to the inhomogeneous equation (2.91) and a quasi periodic mild solution to its homogeneous equation, respectively. In fact, to prove it we again use Lemma 2.1 to show that

$$Lu^0 = \mathcal{F}u^0 + f,$$

where $\mathcal{F}w(t) = F(t)w_t$.

Theorem 2.16 *Let u be a bounded mild solution to Eq.(2.91) and $p \in \mathbf{N}$ such that $(p-1)\tau \leq r < p\tau$. Then u can be represented in the form*

$$u(t) = u^0(t) + \sum_{k=1}^{M} e^{i\lambda_k t} u^k(t), \forall t, \qquad (2.96)$$

where u^0 is a $p\tau$-periodic mild solution to Eq.(2.91) and $\sum_{k=1}^{M} e^{i\lambda_k t} u^k(t)$ a quasi periodic solution to its homogeneous equation.

Proof. The uniform continuity of u is guaranteed by Proposition 2.4 below. Hence, we can proceed as in the proof of the Spectral Decomposition Theorem (Theorem 2.14). In fact, with the above notation, since u is a mild solution of Eq.(2.91) and is almost periodic,

$$Lu = \mathcal{F}u + f,$$

where the operator L is defined as in Lemma 2.1, i.e., $Lw = g$ if and only if w is bounded, uniformly continuous such that

$$w(t) = T(t-s)w(s) + \int_s^t T(t-\xi)f(\xi)d\xi, \ \forall t \geq s; t, s \in \mathbf{R}.$$

From here we can proceed identically as in the proof of the Spectral Decomposition Theorem by using Theorem 2.13.

Corollary 2.16 *Eq.(2.91) has a τ-periodic mild solution if and only if it has a bounded mild solution. Moreover, if u is a bounded mild solution to Eq.(2.91), then the following two-sided sequence*

$$a_k = \lim_{T \to \infty} \frac{1}{2T} \int_{-T}^{T} e^{-(2ik\pi/\tau)\xi} u(\xi) d\xi, k \in \mathbf{Z} \qquad (2.97)$$

determines the Fourier coefficients of a τ periodic mild solution to Eq.(2.91).

As in Proposition 2.3, for Eq.(2.91) we will show that the uniform continuity follows readily from the boundedness of the mild solution u. In fact we have

Proposition 2.4 Let $(T(t))_{t\geq 0}$ be a compact C_0-semigroup and the maps $t \mapsto F(t) \in L(\mathbf{X}), t \mapsto f(t) \in \mathbf{X}$ be τ-periodic and continuous. Moreover, let u be a bounded mild solution to Eq.(2.91) on the whole line. Then it is uniformly continuous.

Proof. We have to show that $\lim_{|t'-t|\to 0} \|u(t') - u(t)\| = 0$. Without loss of generality we assume that $t' \geq t$. By the compactness of $(T(t))_{t\geq 0}$, the map $T(t)$ is norm continuous for $t > 0$. Hence, we have

$$\lim_{t'-t\downarrow 0} \|u(t') - u(t)\| \leq \lim_{t'-t\downarrow 0} \|T(t'-t+\tau) - T(\tau)\| \|u(t-\tau)\| +$$

$$+ \lim_{t'-t\downarrow 0} \|\int_{t-\tau}^{t'} T(t'-\xi)[F(\xi)u_\xi + f(\xi)]d\xi -$$

$$- \int_{t-\tau}^{t'} T(t'-\xi)[F(\xi)u_\xi + f(\xi)]d\xi\| \leq$$

$$\leq \lim_{t'-t\downarrow 0} \|\int_{t-\tau}^{t'} T(t'-\xi)[F(\xi)u_\xi + f(\xi)]d\xi$$

$$- \int_{t-\tau}^{t'} T(t'-\xi)[F(\xi)u_\xi + f(\xi)]d\xi\|. \tag{2.98}$$

On the other hand, by the exponential boundedness of the semigroup $(T(t))_{t\geq 0}$ and the boundedness of u, F, f

$$\lim_{t'-t\downarrow 0} \|\int_{t-\tau}^{t'} T(t'-\xi)[F(\xi)u_\xi + f(\xi)]d\xi -$$

$$- \int_{t-\tau}^{t} T(t-\xi)[F(\xi)u_\xi + f(\xi)]d\xi\| =$$

$$\leq \lim_{t'-t\downarrow 0} \|T(t'-t)\int_{t-\tau}^{t} T(t-\xi)[F(\xi)u_\xi + f(\xi)]d\xi -$$

$$- \int_{t-\tau}^{t} T(t-\xi)[F(\xi)u_\xi + f(\xi)]d\xi\| +$$

$$+ \lim_{t'-t\downarrow 0} \|\int_{t}^{t'} T(t'-\xi)[F(\xi) + f(\xi)]d\xi\| \leq$$

$$\leq \lim_{t'-t\downarrow 0} \|T(t'-t)\int_{t-\tau}^{t} T(t-\xi)[F(\xi)u_\xi + f(\xi)]d\xi -$$

$$- \int_{t-\tau}^{t} T(t-\xi)[F(\xi)u_\xi + f(\xi)]d\xi\|. \tag{2.99}$$

As shown in [216, Lemma 2.5] the set $\int_c^\tau T(\eta)F(t+\eta)u_{t+\eta}d\eta$ is precompact for every fixed positive c. From this follows easily the precompactness of the set $\int_0^\tau T(\eta)F(t+\eta)u_{t+\eta}d\eta$ and then that of the set $\int_{t-\tau}^t T(t-\xi)[F(\xi)u_\xi + f(\xi)]d\xi$. This and the strong continuity of $(T(t))_{t\geq 0}$ imply that

$$\lim_{t'-t\downarrow 0} \|T(t'-t)\int_{t-\tau}^{t} T(t-\xi)[F(\xi)u_\xi + f(\xi)]d\xi -$$
$$-\int_{t-\tau}^{t} T(t-\xi)[F(\xi)u_\xi + f(\xi)]d\xi\| = 0. \quad (2.100)$$

Thus $\lim_{t'-t\downarrow 0} \|u(t') - u(t)\| = 0$.

2.3.5. Almost Periodic Solutions of Partial Functional Differential Equations

In this subsection we are concerned with necessary and sufficient conditions for the following abstract autonomous functional differential equation to have almost periodic solutions with the same structure of spectrum as f

$$\frac{dx(t)}{dt} = Ax(t) + Fx_t + f(t), \quad x \in \mathbf{X}, t \in \mathbf{R}, \quad (2.101)$$

where A is the infinitesimal generator of a strongly continuous semigroup, $x_t \in C([-r,0], \mathbf{X}), x_t(\theta) := x(t+\theta), r > 0$ is a given positive real number, $F\varphi := \int_{-r}^{0} d\eta(s)\varphi(s), \forall \varphi \in C([-r,0], \mathbf{X}), \eta : [-r, 0] \to L(\mathbf{X})$ is of bounded variation and f is a $\mathbf{X}$-valued almost periodic function.

Recall that a continuous function $x(\cdot)$ on $\mathbf{R}$ is said to be a mild solution on $\mathbf{R}$ of Eq.(2.101) if for all $t \geq s$

$$x(t) = T(t-s)x(s) + \int_{s}^{t} T(t-\xi)[Fx_\xi + f(\xi)]d\xi. \quad (2.102)$$

Below we will denote by $\mathcal{F}$ the operator acting on $BUC(\mathbf{R}, \mathbf{X})$ defined by the formula
$$\mathcal{F}u(\xi) := Fu_\xi, \quad \forall u \in BUC(\mathbf{R}, \mathbf{X}).$$

Recall that by *autonomous operator* in $BUC(\mathbf{R}, \mathbf{X})$ we mean a bounded linear operator $\mathcal{K}$ acting on $BUC(\mathbf{R}, \mathbf{X})$ such that it commutes with the translation group, i.e.,
$$\mathcal{K}S(\tau) = S(\tau)\mathcal{K}, \quad \forall \tau \in \mathbf{R}.$$

An example of autonomous operator is the above-defined operator $\mathcal{F}$. Let $(T(t))_{t\geq s}$ be a C_0-semigroup. Then, we denote by $\mathcal{S}$ the space of all elements of $BUC(\mathbf{R}, \bar{\mathbf{X}})$ at which $(T^h)_{h\geq 0}$ is strongly continuous. In the same way as in the previous sections we define the operator $\mathcal{L}$ acting on $\mathcal{S}$ as follows: $u \in D(\mathcal{L})$ if and only if there exists a function $g \in \mathcal{S}$ such that

$$u(t) = T(t-s)u(s) + \int_{s}^{t} T(t-\xi)g(\xi)d\xi, \quad \forall t \geq s,$$

and, by definition, $\mathcal{L}u := g$. Hence, $\mathcal{L}$ is well-defined as a singled-valued operator. Moreover, it is closed. For bounded uniformly continuous mild solutions $x(\cdot)$ the following characterization will be used:

Theorem 2.17 $x(\cdot)$ *is a bounded uniformly continuous mild solution of Eq.(2.101) if and only if* $\mathcal{L}x(\cdot) = \mathcal{F}x(\cdot) + f$.

Lemma 2.19 *Let* $(T^h)_{h\geq 0}$ *be the evolution semigroup associated with a given strongly continuous semigroup* $(T(t))_{t\geq s}$. *Then the following assertions hold true:*

  i) *Every mild solution* $u \in BUC(\mathbf{R}, \mathbf{X})$ *of Eq.(2.101) is an element of* $\mathcal{S}$,

  ii) $AP(\mathbf{X}) \subset \mathcal{S}$,

  iii) *For the infinitesimal generator* $\mathcal{G}$ *of* $(T^h)_{h\geq 0}$ *in the space* $\mathcal{S}$ *one has the relation:* $\mathcal{G}g = -\mathcal{L}g$ *if* $g \in D(\mathcal{G})$.

Proof. (i) By the definition of mild solutions (2.102) we have

$$\begin{aligned} \|u(t) - T(h)u(t-h)\| &\leq \int_{t-h}^{t} \|T(t-\xi)\|(\|\mathcal{F}\|\|u\| + \|f\|)d\xi \\ &\leq hNe^{\omega h}, \end{aligned} \quad (2.103)$$

where N is a positive constant independent of h, t. Hence

$$\lim_{h\downarrow 0} \|T^h u - u\| = \lim_{h\downarrow 0} \sup_{t\in\mathbf{R}} \|T(h)u(t-h) - u(t)\| = 0,$$

i.e., the evolution semigroup $(T^h)_{h\geq 0}$ is strongly continuous at u.
(ii) The second assertion is a particular case of Lemma 2.1. (iii) The relation between the infinitesimal generator $\mathcal{G}$ of $(T(t))_{t\geq 0}$ and the operator $\mathcal{L}$ can be proved similarly as in Lemma 2.1.

Spectrum of a mild solution of Eq.(2.101)

We will denote
$$\Delta(\lambda) := \lambda - A - B_\lambda, \ \forall \lambda \in \mathbf{C}, \quad (2.104)$$

where $B_\lambda = \int_{-r}^{0} d\eta(s) e^{\lambda s}$, and

$$\rho(A, \eta) := \{\lambda \in \mathbf{C} : \exists \Delta^{-1}(\lambda) \in L(\mathbf{X})\}. \quad (2.105)$$

Lemma 2.20 $\rho(A, \eta)$ *is open in* $\mathbf{C}$, *and* $\Delta^{-1}(\lambda)$ *is analytic in* $\rho(A, \eta)$.

Proof. The proof of the lemma can be taken from that of [75, Lemma 3.1, pp.207-208].

Below we will assume that $u \in BUC(\mathbf{R}, \mathbf{X})$ is any mild solution of Eq.(2.101). Since u is a mild solution of Eq.(2.101), by Lemma 2.11 $\int_0^t u(\xi)d\xi \in D(A), \forall t \in \mathbf{R}$ and

$$u(t) - u(0) = A\int_0^t u(\xi)d\xi + \int_0^t g(\xi)d\xi, \quad (2.106)$$

where $g(\xi) := Fu_\xi + f(\xi)$. Hence, taking the Laplace transform of u we have

$$\hat{u}(\lambda) + \frac{1}{\lambda}u(0) = \frac{1}{\lambda}A\hat{u}(\lambda) + \frac{1}{\lambda}\int_0^\infty e^{-\lambda t}g(t)dt$$
$$= \frac{1}{\lambda}A\hat{u}(\lambda) + \frac{1}{\lambda}\hat{f}(\lambda) + \int_0^\infty e^{-\lambda t}\int_{-r}^0 d\eta(s)u(s+t)dt$$
$$= \frac{1}{\lambda}A\hat{u}(\lambda) + \frac{1}{\lambda}\hat{f}(\lambda) + \int_{-r}^0 d\eta(s)\int_0^\infty e^{-\lambda t}u(s+t)dt$$
$$= \frac{1}{\lambda}A\hat{u}(\lambda) + \frac{1}{\lambda}\hat{f}(\lambda) + \int_{-r}^0 d\eta(s)e^{\lambda s}\int_s^\infty e^{-\lambda \xi}u(\xi)dt.$$

By setting
$$\psi(\lambda) := u(0) + \int_{-r}^0 d\eta(s)e^{\lambda s}\int_s^0 e^{-\lambda \xi}u(\xi)d\xi$$

we have
$$\hat{u}(\lambda)(\lambda - A - B_\lambda) = \hat{f}(\lambda) + \psi(\lambda). \tag{2.107}$$

Obviously, $\psi(\lambda)$ has a holomorphic extension on the whole complex plane. Thus, for $\xi \notin sp(f), i\xi \in \rho(A,\eta)$, since
$$\hat{u}(\lambda) = (\lambda - A - B_\lambda)^{-1}\hat{f}(\lambda) + \psi(\lambda)$$

and by Lemma 2.20 $\hat{u}(\lambda)$ has a holomorphic extension around $i\xi$, i.e., $\xi \notin sp(u)$. So, we have in fact proved the following

Lemma 2.21

$$sp(u) \subset \{\xi \in \mathbf{R} : \nexists \Delta^{-1}(i\xi) \text{ in } L(\mathbf{X})\} \cup sp(f), \tag{2.108}$$

where $\Delta(\lambda) = \lambda I - A - B_\lambda$.

Proof. The proof is clear from the above computation.

Below for the sake of simplicity we will denote
$$\sigma_i(\Delta) := \{\xi \in \mathbf{R} : \nexists \Delta^{-1}(i\xi) \text{ in } L(\mathbf{X})\}$$

We will show that the behavior of solutions of Eq.(2.101) depends heavily on the structure of this part of spectrum.

Decomposition Theorem and its consequences

Let us consider the subspace $\mathcal{M} \subset BUC(\mathbf{R},\mathbf{X})$ consisting of all functions $v \in BUC(\mathbf{R},\mathbf{X})$ such that
$$\sigma(v) := \overline{e^{isp(v)}} \subset S_1 \cup S_2, \tag{2.109}$$

where $S_1, S_2 \subset \mathbf{S}^1$ are disjoint closed subsets of the unit circle. We denote by $\mathcal{M}_v = \overline{span\{S(t)v, t \in \mathbf{R}\}}$, where $(S(t))_{t\in\mathbf{R}}$ is the translation group on $BUC(\mathbf{R},\mathbf{X})$, i.e. $S(t)v(s) = v(t+s), \forall t, s \in \mathbf{R}$.

Lemma 2.22 *Under the above notations and assumptions the function space $\mathcal{M}$ can be split into a direct sum $\mathcal{M} = \mathcal{M}_1 \oplus \mathcal{M}_2$ such that $v \in \mathcal{M}_i$ if and only if $\sigma(v) \subset S_i$ for $i = 1, 2$. Moreover, any autonomous bounded linear operator in $BUC(\mathbf{R}, \mathbf{X})$ leaves invariant $\mathcal{M}$ as well as $\mathcal{M}_j$, $j = 1, 2$.*

Proof. The first claim has been proved in Theorem 2.13. It remains to show that if B is an autonomous bounded operator in $BUC(\mathbf{R}, \mathbf{X})$, then $sp(Bw) \subset sp(w)$ for each $w \in BUC(\mathbf{R}, \mathbf{X})$. In fact, consider

$$\begin{aligned} R(\lambda, \mathcal{D}) Bw &= \int_0^\infty e^{-\lambda t} S(t) Bw dt \ (\forall \Re \lambda > 0) \\ &= B \int_0^\infty e^{-\lambda t} S(t) w dt \\ &= B R(\lambda, \mathcal{D}) w. \end{aligned} \quad (2.110)$$

Hence, if $\xi \notin sp(w)$, then since $i\xi \in \rho(\mathcal{D}_w)$ the integral

$$\int_0^\infty e^{-\lambda t}(S(t)|_{\mathcal{M}_w}) dt = R(\lambda, \mathcal{D}_w)$$

has an analytic extension in a neighborhood of $i\xi$. This yields that

$$R(\lambda, \mathcal{D}_w) w = R(\lambda, \mathcal{D}) w, \ \forall \Re \lambda > 0$$

has an analytic extension in a neighborhood of $i\xi$. So does the function (in λ) $BR(\lambda, \mathcal{D}) w = R(\lambda, \mathcal{D}) Bw$. By Lemma 1.5 this shows that $i\xi \in \rho(\mathcal{D}_w)$, and hence by Lemma 1.16 $\xi \notin sp(Bw)$, i.e., $sp(Bw) \subset sp(w)$. In particular, this implies that B leaves invariant $\mathcal{M}$ as well as $\mathcal{M}_j, j = 1, 2$.

Remark 2.16 Similarly we can carry out the decomposition of an almost periodic function into spectral components as done in the above lemma.

Lemma 2.23 *Let $u \in BUC(\mathbf{R}, \mathbf{X})$ be a mild solution of Eq.(2.101) with $f \in AP(\mathbf{X})$. Then*

$$sp(u) \supset sp(f). \quad (2.111)$$

Proof. In the proof of Lemma 2.22 we have shown that if B is an autonomous operator acting on $BUC(\mathbf{R}, \mathbf{X})$ and $u \in BUC(\mathbf{R}, \mathbf{X})$, then $sp(Bu) \subset sp(u)$. Hence, for every $h > 0$, by Lemma 2.19 we have

$$\begin{aligned} sp(f) &= sp(-f) = sp\left(\lim_{h \downarrow 0} \frac{T^h u - u}{h} - \mathcal{F} u \right) \\ &\subset sp(u). \end{aligned} \quad (2.112)$$

Theorem 2.18 *Let the following conditions be fulfilled*

$$\overline{e^{i\sigma_i(\Delta)}} \setminus \overline{e^{isp(f)}} \ be\ closed, \quad (2.113)$$

and Eq.(2.101) have a bounded uniformly continuous mild solution on the whole line. Then there exists a bounded uniformly continuous mild solution w of Eq.(2.101) such that
$$\overline{e^{isp(w)}} \subset \overline{e^{isp(f)}}. \tag{2.114}$$

Moreover, if
$$\overline{e^{i\sigma_i(\Delta)}} \cap \overline{e^{isp(f)}} = \emptyset, \tag{2.115}$$

then such a solution w is unique in the sense that if there exists a mild solution v to Eq.(2.101) such that $e^{isp(v)} \subset \overline{e^{isp(f)}}$, then $v = w$.

Proof. By Lemma 2.21
$$sp(u) \subset \sigma_i(\Delta) \cup sp(f). \tag{2.116}$$

Let us denote by Λ the set $\overline{e^{i(\sigma_i(\Delta)}} \cup \overline{e^{isp(f))}}$, S_1 the set $\overline{e^{isp(f)}}$ and S_2 the set $\overline{e^{i\sigma_i(\Delta)}} \setminus \overline{e^{isp(f)}}$, respectively. Thus, by Lemma 2.22 there exists a projection P from $\mathcal{M}$ onto $\mathcal{M}_1$ which is commutative with $\mathcal{F}$ and T^h. Hence,

$$\begin{aligned} -P \lim_{h \downarrow 0} \frac{T^h u - u}{h} &= -\lim_{h \downarrow 0} P \frac{T^h u - u}{h} \\ &= \lim_{h \downarrow 0} \frac{T^h P u - P u}{h} \\ &= \mathcal{L} P u. \end{aligned} \tag{2.117}$$

On the other hand, since u is a mild solution, by Theorem 2.17 $\mathcal{L}u = \mathcal{F}u + f$. Since $Pf = f$ and P commutes with $\mathcal{F}$,

$$\begin{aligned} \mathcal{F}Pu + f &= P\mathcal{F}u + f \\ &= P\mathcal{F}u + Pf \\ &= P\mathcal{L}u \\ &= -P \lim_{h \downarrow 0} \frac{T^h u - u}{h} \\ &= \mathcal{L}Pu. \end{aligned} \tag{2.118}$$

By Theorem 2.17 we have that Pu is a mild solution of Eq.(2.101). Now we prove the next assertion on the uniqueness. In fact, suppose that there is another mild solution $v \in BUC(\mathbf{R}, \mathbf{X})$ to Eq.(2.101) such that $e^{i\ sp(v)} \subset \overline{e^{i\ sp(f)}}$, then it is seen that $w - v$ is a mild solution of the homogeneous equation corresponding to Eq.(2.101). Hence, $sp(w-v) \subset \sigma_i(\Delta)$. This shows that
$$e^{isp(w-v)} \subset \overline{e^{i\sigma_i(\Delta)}} \cap \overline{e^{isp(f)}} = \emptyset.$$

So, $w - v = 0$. This completes the proof of the theorem.

Remark 2.17 By Lemma 2.23, the mild solution mentioned in Theorem 2.18 is "minimal" in the sense that its spectrum is minimal. In the above theorem we have proved that under the assumption (2.115) if there is a mild solution u to Eq.(2.101) in $BUC(\mathbf{R}, \mathbf{X})$, then there is a unique mild solution w to Eq.(2.101) such that $e^{isp(w)} \subset \overline{e^{isp(f)}}$. The assumption on the existence of a mild solution u is unremovable, even in the case of equations without delay. In fact, this is due to the failure of the spectral mapping theorem in the infinite dimensional systems (for more details see e.g. [65], [163], [179]). Hence, in addition to the condition (2.115) it is necessary to impose further conditions to guarantee the existence and uniqueness of such a mild solution w. In the next subsection we will examine conditions for the existence of a bounded mild solution to Eq.(2.101).

Theorem 2.19 *Let the assumption (2.113) of Theorem 2.18 be fulfilled. Moreover, let the space $\mathbf{X}$ not contain c_0 and $\overline{e^{isp(f)}}$ be countable. Then there exists an almost periodic mild solution w to Eq.(2.101) such that $e^{isp(w)} \subset \overline{e^{isp(f)}}$. Furthermore, if (2.115) holds, then such a solution w is unique.*

Proof. The proof is obvious in view of Theorem 1.20, Remark 1.3 and Theorem 2.18.

Remark 2.18 As we have seen, the almost periodic mild solution w is a component of the mild solution u which existence is assumed. Hence, if we assume further that $\sigma_i(\Delta)$ is countable, then the solution u is also almost periodic. Thus, the Bohr-Fourier coefficients of solution w can be computed as follows:

$$a(\lambda) = \lim_{T \to \infty} \frac{1}{2T} \int_{-T}^{T} e^{-i\lambda t} u(t) dt, \ \forall e^{i\lambda} \in \overline{e^{isp(f)}}.$$

Quasi-periodic solutions

As in the above subsection the following lemma is obvious.

Lemma 2.24 *Let Λ_1, Λ_2 be disjoint closed subsets of the real line and $\Lambda := \Lambda_1 \cup \Lambda_2$. Moreover let Λ_1 be compact. Then the space $\Lambda(\mathbf{X}) = \Lambda_1(\mathbf{X}) \oplus \Lambda_2(\mathbf{X})$ and $\Lambda_1(\mathbf{X})$ and $\Lambda_2(\mathbf{X})$ are left invariant by any autonomous functional operators.*

Proof. Since the proof can be done in the similar manner as in the previous subsections we omit the details.

Remark 2.19 We can prove a similar decomposition in the function space $AP(\mathbf{X})$.

Theorem 2.20 *Let $sp(f)$ have an integer and finite basis and $\mathbf{X}$ do not contain c_0. Moreover, let $\sigma_i(\Delta)$ be bounded and $\sigma_i(\Delta) \backslash sp(f)$ be closed. Then if Eq.(2.101) has a mild solution $u \in BUC(\mathbf{R}, \mathbf{X})$, it has a quasi-periodic mild solution w such that $sp(w) = sp(f)$. If $\sigma_i(\Delta) \cap sp(f) = \emptyset$, then such a solution w is unique.*

Proof. As the proof of this theorem is analogous to that of Theorem 2.18 we omit the details.

Remark 2.20 If $\sigma_i(\Delta)$ is bounded, by the same argument as in this subsection it is more convenient to replace the condition on the closedness of $\overline{e^{\sigma_i(\Delta)}} \setminus \overline{e^{isp(f)}}$ of Theorem 2.18 by the weaker condition that $\sigma_i(\Delta)\setminus sp(f)$ is closed. A sufficient condition for the boundedness of $\sigma_i(\Delta)$ will be given in the next section.

Existence and uniqueness of almost periodic solutions of Eq.(2.101)

Recall that to the corresponding homogeneous equation of Eq.(2.101) one can associate a strongly continuous solution semigroup $(V(t))_{t\geq 0}$ on the space $C := C([-r, 0]\mathbf{X})$. Our main interest in this subsection is to prove the existence of an almost periodic mild solution to Eq.(2.101) under the condition that $\overline{e^{isp(f)}} \cap \sigma(V(1)) = \emptyset$. For the sake of simplicity, we always assume in this section that $r < 1$. Having proved this, Theorem 2.2 can be extended to almost periodic solutions of Eq.(2.101) by using Theorem 2.18. To this end, we first recall the *variation-of-constants formula* for Eq.(33) (see e.g. [226, p.115-116])

$$\begin{cases} u(t) = [V(t-s)\phi](0) + \int_s^t [V(t-\xi)X_0 f(\xi)](0)d\xi, \\ u_0 = \phi, \end{cases} \quad (2.119)$$

where $X_0 : [-r, 0] \mapsto L(\mathbf{X})$ is given by $X_0(\theta) = 0$ for $-r \leq \theta < 0$ and $X_0(0) = I$ and $(V(t)_{t\geq 0})$ is the solution semigroup generated by Eq.(2.101) in C. Although this formula seems to be ambiguous it suggests some insights to prove the existence of a bounded mild solution. In fact, let $u \in BUC(\mathbf{R}, \mathbf{X})$ be a mild solution of Eq.(2.101). Then we will examine the spectrum of the function

$$w : \mathbf{R} \ni t \mapsto w(t) := u_t - V(1)u_{t-1} \in C([-r,0], \mathbf{X}),$$

which may be "defined by the formula"

$$w(t)(\xi) \text{ "} = \text{"} \int_s^{t+\xi} [V(t-\xi)X_0 f(\xi)](0)d\xi, \quad \forall \xi \in [-r, 0]. \quad (2.120)$$

Hence, $w(t)$ may be defined independent of $u(\cdot)$. Moreover, if this is the case, we can use the equation $u_t = V(1)u_{t-1} + w(t)$ to solve u, and so to prove the existence of a bounded mild solution to Eq.(2.101). It turns out that all these can be done without using the variation-of-constants (2.119). In fact, we begin with another definition of the function $w(t)$. For every fixed $t \in \mathbf{R}$, let us consider the Cauchy problem

$$\begin{cases} y(\xi) = \int_{t-1}^{\xi} T(\xi - \eta)[Fy_\eta + f(\eta)]d\eta, & \xi \in [t-1, t], \\ y_{t-1} = 0 \in C. \end{cases} \quad (2.121)$$

It is easy to see that if there exists a bounded mild solution $u(\cdot)$ to Eq.(2.101), then $w(t) := u_t - V(1)u_{t-1}$ satisfies Eq.(2.121). In what follows we will consider the function $v : \mathbf{R} \ni t \mapsto y_t \in C$, where y_t is defined by (2.121).

Lemma 2.25 *The operator* $L : BUC(\mathbf{R}, \mathbf{X}) \ni f \mapsto v \in BUC(\mathbf{R}, C)$ *is well-defined as a continuous linear operator. Moreover,* $S_t L f = L S(t) f, \forall t \in \mathbf{R}$, *where* $S_t, t \in \mathbf{R}$ *is the translation group in* $BUC(\mathbf{R}, C)$.

Proof. First we show that if $f \in BUC(\mathbf{R}, \mathbf{X})$, then $v(\cdot)$ is uniformly continuous. In fact, for every $\varepsilon > 0$, there is a $\delta > 0$ such that $\sup_{t \in \mathbf{R}} \|f(\xi + h) - f(\xi)\| < \varepsilon$, $\forall |h| < \delta$. For the function $v(t + h)$ we consider the following Cauchy problem

$$\begin{cases} x(t+h+\theta) = \int_{t+h-1}^{t+h+\theta} T(t+h+\theta-\zeta)[Fx_\zeta + f(\zeta)]d\zeta, \\ \qquad\qquad\qquad\qquad\qquad \forall \theta \in [t+h-1, t+h], \\ x_{t+h-1} = 0. \end{cases} \qquad (2.122)$$

By denoting $z(\delta) := x(\delta + h)$ we can see that $z(\cdot)$ is the solution of the equation

$$\begin{cases} z(t+\theta) = \int_{t-1}^{t+\theta} T(t+\theta-\zeta)[Fz_\zeta + f(h+\zeta)]d\zeta, \\ \qquad\qquad\qquad\qquad \forall \theta \in [t-1, t], \\ z_{t-1} = 0. \end{cases} \qquad (2.123)$$

Hence, taking into account (2.121) and (2.123), by the Gronwall inequality

$$\begin{aligned}
\sup_{t \in \mathbf{R}} \|v(t+h) - v(t)\| &= \sup_{t \in \mathbf{R}} \sup_{\theta \in [-r, 0]} \|z(t+\theta) - y(t+\theta)\| \leq \\
&\leq \sup_{t \in \mathbf{R}} \sup_{\xi \in [t-1, t]} \|z(\xi) - y(\xi)\| \\
&\leq \delta K, \qquad\qquad\qquad\qquad\qquad\qquad (2.124)
\end{aligned}$$

where K depends only on $(T(t))_{t \geq 0}, \|F\|$. This shows that $v \in BUC(\mathbf{R}, C)$. From (2.122) and (2.123) it follows immediately the relation $S_t Lf = LS(t)f$. The boundedness of the operator L is an easy estimate in which the Gronwall inequality is used.

Corollary 2.17 *Let f be almost periodic. Under the above notation, the following assertions hold true:*

i)
$$sp(v) \subset sp(f). \qquad (2.125)$$

ii) The function v is almost periodic.

Proof. To show the first assertion we can use the same argument as in the proof of Lemma 2.22. The second one is a consequence of the first one. In fact, since f is almost periodic, it can be appoximated by a sequence of trigonometric polynomials. On the other hand, from the first assertion, this yields that if P_n is a trigonometric polynomial, then so is $\mathcal{L}P_n$. Hence, $\mathcal{L}f = v$ can be approximated by a sequence of trigonometric polynomials, i.e., v is almost periodic.

We are in a position to prove the main result of this section.

Theorem 2.21 *Let*
$$\overline{e^{isp(f)}} \cap \sigma(V(1)) = \varnothing \qquad (2.126)$$

hold. Then Eq.(2.101) has a unique almost periodic mild solution x_f such that

$$e^{isp(x_f)} \subset \overline{e^{isp(f)}}.$$

Proof. Let us consider the equation

$$u(t) = V(1)u(t-1) + v(t), \qquad (2.127)$$

where $v(t)$ is defined by (2.121). It is easy to see that the spectrum of the multiplication operator $K : v \mapsto V(1)v$, where $v \in \Lambda(C([-r,0]), \Lambda := sp(f)$ has the property that $\sigma(K) \subset \sigma(V(1))$ (see e.g. [167]). In the space $\Lambda(C([-r,0])$ the spectrum of the translation $S_{-1} : u_t \mapsto u_{t-1}$ can be estimated as follows in view of the Weak Spectral Mapping Theorem (see Theorem 1.8)

$$\sigma(S_{-1}) = \overline{e^{-d/dt|_{\Lambda(C)}}} = \overline{e^{-i\Lambda}}.$$

Let us denote $W := K \cdot S_{-1}$. It may be noted that W is the composition of two commutative bounded linear operators. Thus (see e.g. [193, Theorem 11.23, p.280]),

$$\begin{aligned}\sigma(W) &\subset \sigma(K)\sigma(S_{-1}) \\ &\subset \sigma(V(1))\overline{e^{-i\Lambda}}.\end{aligned} \qquad (2.128)$$

Obviously, (2.126) and (2.128) show that $1 \notin \sigma(W)$. Hence, the equation (2.127) has a unique solution u. We are now in a position to construct a bounded mild solution of Eq.(2.101). To this end, we will establish this solution in every segment $[n, n+1)$. Then, we show that these segments give a solution on the whole real line. We consider the sequence $(u_n)_{n \in \mathbf{Z}}$. In every interval $[n, n+1)$ we consider the Cauchy problem

$$\begin{cases} x(\xi) = T(\xi-n)[u(n)](0) + \int_n^\xi T(\xi-\eta)[Fx_\eta + f(\eta)]d\eta, & \forall \xi \in [n, \infty), \\ x_n = u(n). \end{cases} \qquad (2.129)$$

Obviously, this solution is defined in $[n, +\infty)$. On the other hand, by the definition of $V(1)u(n)$ and $v(n+1)$ we have $V(1)u(n) = a_{n+1}, v(n+1) = b_{n+1}$, where

$$\begin{aligned} a(\xi) &= T(\xi-n)u(n)(0) + \int_n^\xi T(\xi-\eta)Fa_\eta d\eta, \quad \forall \xi > n, a_n = u(n) \\ b(\xi) &= \int_n^\xi T(\xi-\eta)[Fb_\eta + f(\eta)]d\eta, \quad \forall \xi \in [n, n+1], b_n = 0. \end{aligned}$$

Thus, $a(\xi) + b(\xi) = x(\xi)$. This yields that

$$x_{n+1} = a_{n+1} + b_{n+1} = V(1)u(n) + v(n+1) = u(n+1).$$

By this process we can establish the existence of a bounded continuous mild solution $x(\cdot)$ of Eq.(2.101) on the whole line. Moreover, we will prove that $x(\cdot)$ is almost periodic. As $u(\cdot)$ and f are almost periodic, so is the function $g : \mathbf{R} \ni t \mapsto (u(t), f(t)) \in C \times \mathbf{X}$. As is known, the sequence $\{g(n)\} = \{(u(n), f(n))\}$ is almost periodic. Hence, for every positive ε the following set is relatively dense (see [67, p. 163-164])

$$T := \mathbf{Z} \cap T(g,\varepsilon), \qquad (2.130)$$

where $T(g,\varepsilon) := \{\tau \in \mathbf{R} : \sup_{t\in\mathbf{R}} \|g(t+\tau) - g(t)\| < \varepsilon\}$, i.e., the set of ε periods of g. Hence, for every $m \in T$ we have

$$\|f(t+m) - f(t)\| < \varepsilon, \forall t \in \mathbf{R}, \qquad (2.131)$$
$$\|u(n+m) - u(n)\| < \varepsilon, \forall n \in \mathbf{Z}. \qquad (2.132)$$

Since x is a solution to Eq.(2.102), for $0 \le s < 1$ and all $n \in \mathbf{N}$, we have

$$\|x(n+m+s) - x(n+s)\| \le \|T(s)\| \cdot \|u(n+m) - u(n)\|$$
$$+ \int_0^s T(s-\xi)\big[\|F\| \cdot \|x_{n+m+\xi} - x_{n+\xi}\|$$
$$+ \|f(n+m+\xi) - f(n+\xi)\|\big]d\xi$$
$$\le Ne^\omega \|u(n+m) - u(n)\| + Ne^\omega \int_0^s \big[\|F\|$$
$$\cdot \|x_{n+m+\xi} - x_{n+\xi}\| + \|f(n+m+\xi) - f(n+\xi)\|\big]d\xi$$

Hence

$$\|x_{n+m+s} - x_{n+s}\| \le Ne^\omega \|u(n+m) - u(n)\|$$
$$+ Ne^\omega \int_0^s \big[\|F\| \cdot \|x_{n+m+\xi} - x_{n+\xi}\| + \|f(n+m+\xi) - f(n+\xi)\|\big]d\xi.$$

Using the Gronwall inequality we can show that

$$\|x_{n+m+s} - x_{n+s}\| \le \|x_{n+m+s} - x_{n+s}\| \le \varepsilon M, \qquad (2.133)$$

where M is a constant which depends only on $\|F\|, N, \omega$. This shows that m is a εM-period of the function $x(\cdot)$. Finally, since T is relatively dense for every ε, we see that $x(\cdot)$ is an almost periodic mild solution of Eq.(2.101). Now we are in a position to apply Lemma 2.22, Remark 2.16 and Theorem 2.18 to see that Eq.(2.101) has a unique almost periodic solution such that (2.126) holds. This completes the proof of the theorem.

We state below a version of Theorem 2.21 for the case in which the semigroup $(T(t))_{t\ge 0}$ is compact.

Corollary 2.18 *Let the semigroup $(T(t))_{t\ge 0}$ be compact and $e^{i\sigma_i(\Delta)} \cap \overline{e^{isp(f)}} = \emptyset$. Then Eq.(2.101) has a unique almost periodic mild solution x_f with $e^{isp(x_f)} \subset \overline{e^{isp(f)}}$.*

Proof. Since the solution operator $V(t)$ associated with Eq.(2.101) is compact for sufficiently large t, e.g. for $t > r$, the Spectral Mapping Theorem holds true for this semigroup (see [65] or [163]). Now by applying Theorem 2.18 and Theorem 2.21 we get the corollary.

CHAPTER 2. SPECTRAL CRITERIA

Corollary 2.19 *Let the semigroup $(T(t))_{t\geq 0}$ be compact, $sp(f)$ have an integer and finite basis. Moreover, let $\sigma_i(\Delta)$ be bounded and $e^{i\sigma_i(\Delta)} \cap \overline{e^{isp(f)}} = \varnothing$. Then there exists a unique quasi-periodic mild solution w of Eq.(2.101) such that $sp(w) \subset sp(f)$.*

Proof. By Corollary 2.18 there exists an almost periodic mild solution x_f of Eq.(2.101). Note that from the condition $e^{i\sigma_i(\Delta)} \cap \overline{e^{isp(f)}} = \varnothing$ follows $\sigma_i(\Delta) \cap sp(f) = \varnothing$. Now we can decompose the almost periodic solution x_f as done in Lemma 2.24 to get a minimal almost periodic mild solution w such that $sp(w) \subset sp(f)$.

We now consider necessary conditions for the existence and uniqueness of bounded mild solutions to Eq.(2.101) and their consequences. To this end, for a given closed subset $\Lambda \subset \mathbf{R}$ we will denote by $\Lambda_{AP}(\mathbf{X})$ the subspace of $\Lambda(\mathbf{X})$ consisting of all functions f such that $f \in AP(\mathbf{X})$.

Lemma 2.26 *For every $f \in \Lambda_{AP}(\mathbf{X})$ let Eq.(2.101) have a unique mild solution u_f bounded on the whole line. Then, u_f is almost periodic and*

$$sp(u_f) \subset sp(f). \tag{2.134}$$

In particular, $\sigma_i(\Delta) \cap \Lambda = \varnothing$.

Proof. Let us denote by $\mathcal{L}_\Lambda$ the linear operator with the domain $D(\mathcal{L}_\Lambda)$ consisting of all functions $u \in BC(\mathbf{R}, \mathbf{X})$ which are mild solutions of Eq.(2.101) with certain $f \in \Lambda_{AP}(\mathbf{X})$. For $u \in D(\mathcal{L}_\Lambda)$ we define $\mathcal{L}_\Lambda u = f$. We now show that $\mathcal{L}_\Lambda$ is well defined, i.e., for a given $u \in D(\mathcal{L}_\Lambda)$ there exists exactly one $f \in \Lambda_{AP}(\mathbf{X})$ such that u is a mild solution of Eq.(2.102). Suppose that there exists another $g \in \Lambda_{AP}(\mathbf{X})$ such that

$$u(t) = T(t-s)u(s) + \int_s^t T(t-\xi)[Fu_\xi + g(\xi)]d\xi, \quad \forall t \geq s. \tag{2.135}$$

Then,

$$0 = \int_s^t T(t-\xi)[f(\xi) - g(\xi)]d\xi, \quad \forall t \geq s. \tag{2.136}$$

Hence

$$0 = \frac{1}{t-s}\int_s^t T(t-\xi)[f(\xi) - g(\xi)]d\xi, \quad \forall t > s.$$

From the strong continuity of the semigroup $(T(t))_{t\geq 0}$ and by letting $s \to t$ it follows that $f(t) = g(t)$. From the arbitrary nature of t, this yields that $f = g$. Next, we show that the operator $\mathcal{L}_\Lambda$ is closed, i.e., if there are $u^n \in D(\mathcal{L}_\Lambda)$, $n = 1, 2, ...$ such that $\mathcal{L}_\Lambda u^n = f^n$, $n = 1, 2, ...$ and $u^n \to u \in BC(\mathbf{R}, \mathbf{X})$, $f^n \to f \in \Lambda_{AP}(\mathbf{X})$, then $u \in D(\mathcal{L}_\Lambda)$ and $\mathcal{L}_\Lambda u = f$. In fact, by definition

$$u^n(t) = T(t-s)u^n(s) + \int_s^t T(t-\xi)[Fu_\xi^n + f^n(\xi)]d\xi, \quad \forall t \geq s, \quad \forall n = 1, 2, ... \tag{2.137}$$

For every fixed $t \geq s$, letting n tend to infinity one has

$$u(t) = T(t-s)u(s) + \int_s^t T(t-\xi)[Fu_\xi + f(\xi)]d\xi, \quad \forall t \geq s, \qquad (2.138)$$

proving the closedness of the operator $\mathcal{L}_\Lambda$. Now with the new norm $\|u\|_1 := \|u\| + \|\mathcal{L}_\Lambda u\|$ the space $D(\mathcal{L}_\Lambda)$ becomes a Banach space. Hence, from the assumption the linear operator $\mathcal{L}_\Lambda$ is a bijective from the Banach space $(D(\mathcal{L}_\Lambda), \|\cdot\|_1)$ onto $\Lambda_{AP}(\mathbf{X})$. By the Banach Open Mapping Theorem the inverse of $\mathcal{L}_\Lambda^{-1}$ is continuous. Now suppose $f \in \Lambda_{AP}(\mathbf{X})$. It may be noted that for every $\tau \in \mathbf{R}$, $S(\tau)f\Lambda_{AP}(\mathbf{X})$. Thus, the function $x_f(\cdot + \tau)$ should be the unique mild solution in $BC(\mathbf{R}, \mathbf{X})$ to Eq.(2.102). So, if f is periodic with period, say, ω. Then, since $S(\omega)f = f$, one has $x_f(\cdot + \omega) = x_f(\cdot)$, i.e., x_f is ω-periodic. In the general case, by the spectral theory of almost periodic functions (see e.g. [137, Chap. 2]), f can be approximated by a sequence of trigonometric polynomials

$$P_n(t) = \sum_{k=1}^{N(n)} a_{n,k} e^{i\lambda_{n,k}t}, \quad a_{n,k} \in \mathbf{X}, \lambda_{n,k} \in \sigma_b(f) \subset \Lambda, \quad n = 1, 2, \ldots$$

By the above argument, for every n, $Q_n := \mathcal{L}_\Lambda^{-1} P_n$ is also a trigonometric polynomial. Moreover, since $\mathcal{L}_\Lambda^{-1}$ is continuous, Q_n tends to $\mathcal{L}_\Lambda^{-1} f = x_f$. This shows that x_f is almost periodic and $sp(x_f) \subset sp(f) \subset \Lambda$. Now let f be of the following form $f(t) = ae^{i\lambda t}, t \in \mathbf{R}$, where $0 \neq a \in \mathbf{X}$, $\lambda \in \Lambda$. Then, as shown above, since $sp(x_f) \subset sp(f) = \{\lambda\}$, $x_f(t) = be^{i\lambda t}$ for a unique $b \in \mathbf{X}$. If we denote by e_λ the function in $C[-r, 0]$ defined as follows: $e_\lambda(\theta) := e^{i\lambda\theta}$, $\theta \in [-r, 0]$, then $e_t^{i\lambda\cdot} = e^{i\lambda t} e_\lambda$. With this notation, one has

$$be^{i\lambda t} = T(t-s)be^{i\lambda s} + \int_s^t T(t-\xi)[e^{i\lambda\xi} F be_\lambda + ae^{i\lambda\xi}]d\xi, \quad \forall t \geq s. \qquad (2.139)$$

Since $be^{i\lambda t}$ and $\int_s^t T(t-\xi)[e^{i\lambda\xi} F be_\lambda + ae^{i\lambda\xi}]d\xi$ are differentiable with respect to $t \geq s$, so is $T(t-s)be^{i\lambda s}$. This yields $b \in D(A)$. Consequently, $be^{\lambda t}$ is a classical solution of Eq.(2.101), i.e.,

$$\frac{be^{i\lambda t}}{dt} = Abe^{i\lambda t} + e^{i\lambda t} F(be_\lambda) + ae^{i\lambda t}, \quad \forall t. \qquad (2.140)$$

In particular, this yields that for every $a \in \mathbf{X}$ there exists a unique $b \in \mathbf{X}$ such that

$$(i\lambda - A - B_\lambda)b = a, \qquad (2.141)$$

i.e., by definition $\lambda \notin \sigma_i(\Delta)$, finishing the proof of the lemma

This necessary condition has another application to the study of the asymptotic behavior of solutions as shown in the next corollary. To this end, we first recall the notion of exponential dichotomy of the corresponding homogeneous equation of Eq.(2.101). This homogeneous equation is said to have an exponential dichotomy if the C_0-semigroup of solution operators, associated with it, has an exponential dichotomy. As is known from Section 1, for a C_0-semigroup $(U(t))_{t\geq 0}$ to have an exponential dichotomy it is necessary and sufficient that $\sigma(U(1)) \cap S^1 = \emptyset$.

Corollary 2.20 *Let $(T(t))_{t\geq 0}$ be a strongly continuous semigroup of compact linear operators. Then, a necessary and sufficient condition for the corresponding homogeneous equation of Eq.(2.101) to have an exponential dichotomy is that Eq.(2.101) has a unique bounded mild solution for every given almost periodic function f.*

Proof. Necessity: Since the solution semigroup $(V(t))_{t\geq 0}$ associated with the corresponding homogeneous equation of Eq.(2.101) is a strongly continuous semigroup of compact linear operators, the Spectral Mapping Theorem holds true with respect to this semigroup. On the other hand, by Lemma 2.26 one has $\sigma_i(\Delta) \cap \mathbf{R} = \emptyset$. This yields that $\sigma(V(1)) \cap S^1 = \emptyset$, and hence, by Theorem 2.1 the solution semigroup $(V(t))_{t\geq 0}$ has an exponential dichotomy.
Sufficiency: If the corresponding homogeneous equation of Eq.(2.101) has an exponential dichotomy, then $\sigma(V(1)) \cap S^1 = \emptyset$. Hence, the sufficiency follows from Theorem 2.21.

A condition for the boundedness of $\sigma_i(\Delta)$

As shown in the previous subsection the boundedness of $\sigma_i(\Delta)$ is important for the decomposition of a bounded solution into spectral components which yields the existence of almost periodic and quasi-periodic solutions. We now show that in many frequently met situations this boundedness is available.

Proposition 2.5 *If A is the infinitesimal generator of a strongly continuous analytic semigroup of linear operators, then $\sigma_i(\Delta)$ is bounded.*

Proof. Let us consider the operator $\mathcal{A} + \mathcal{F}$ in $AP(\mathbf{X})$, where $\mathcal{A}$ is the operator of multiplication by A, i.e., $u \in D(\mathcal{A}) \subset AP(\mathbf{X})$ if and only if $u(t) \in D(A)$ $\forall t$ and $Au(\cdot) \in AP(\mathbf{X})$. As shown in the previous section, this operator is sectorial. Hence, $\sigma(\mathcal{A} + \mathcal{F}) \cap i\mathbf{R}$ is bounded. For every

$$\mu \in i\mathbf{R} \backslash \sigma(\mathcal{A} + \mathcal{F})$$

the conditions of Theorem 2.8 are satisfied with the function space $\mathcal{M}$ consisting of all functions in $t \in \mathbf{R}$ of the form $e^{i\mu t}x, x \in \mathbf{X}$. Since Eq.(2.101) has a unique mild solution in $\mathcal{M}$, by Lemma 2.26 it is easily seen that this assertion is nothing but $\mu \notin \sigma_i(\Delta)$. Hence, the proposition is proved.

Examples

To illustrate the above abstract results we will give below several examples in which our conditions can be easily verified.

Example 2.16

We consider the following evolution equation

$$\frac{du(t)}{dt} = -Au(t) + Bu_t + f(t), \qquad (2.142)$$

where A is a sectorial operator in $\mathbf{X}$ and B is a bounded linear operator from $C([-r,0], \mathbf{X}) \to \mathbf{X}$ and u_t is defined as usual, f is an almost periodic function. Moreover, let us assume that the operator A have compact resolvent. Then, $-A$ generates a compact strongly continuous analytic semigroup of linear bounded operators in $\mathbf{X}$ (see e.g. [90], [179]). Hence, for this class of equations all assertions of this subsection are applicable. Note that an important class of parabolic partial differential equations can be included into the evolution equation (2.142) (see e.g. [216], [226]).

Example 2.17

Consider the equation
$$\begin{cases} w_t(x,t) = w_{xx}(x,t) - aw(x,t) - bw(x,t-r) + f(x,t), \\ \qquad\qquad\qquad\qquad\qquad 0 \le s \le \pi,\ t \ge 0, \\ w(0,t) = w(\pi,t) = 0,\ \forall t > 0, \end{cases} \quad (2.143)$$
where $w(x,t), f(x,t)$ are scalar-valued functions. We define the space $\mathbf{X} := L^2[0,\pi]$ and $A_T : \mathbf{X} \to \mathbf{X}$ by the formula
$$\begin{cases} A_T = y'', \\ D(A_T) = \{y \in \mathbf{X} :\ y,\ y'\text{ are absolutely continuous},\ y'' \in \mathbf{X}, \\ \qquad\qquad\qquad y(0) = y(\pi) = 0\}. \end{cases} \quad (2.144)$$
We define $F : C \to \mathbf{X}$ by the formula $F(\varphi) = -a\varphi(0) - b\varphi(-r)$. The evolution equation we are concerned with in this case is the following
$$\frac{dx(t)}{dt} = A_T x(t) + F x_t + f(t),\ x(t) \in \mathbf{X}, \quad (2.145)$$
where A_T is the infinitesimal generator of a compact semigroup $(T(t))_{t \ge 0}$ in $\mathbf{X}$ (see [216, p. 414]). Moreover, the eigenvalues of A_T are $-n^2, n = 1, 2, ...$ and the set $\sigma_i(\Delta)$ is determined from the set of imaginary solutions of the equations
$$\lambda + a + be^{-\lambda r} = -n^2,\ n = 1, 2, ...\ . \quad (2.146)$$
We consider the existence of almost periodic mild solutions of Eq.(2.143) through those of Eq.(2.144). Now if the equation (2.146) have no imaginary solutions, then for every almost periodic f Eq.(2.143) has a unique almost periodic solution. This corresponds to the case of exponential dichotomy which was discussed in [226]. For instance, this happens when we put $a = 0, b = r = 1$.

Let us consider the case where $a = -1, b = \pi/2, r = 1$. It is easy to see that in this case the equation (2.146) have only imaginary solutions $\lambda = i\pi/2, -i\pi/2$. So, our system has no exponential dichotomy. However, applying our theory we can find almost periodic solutions as follows: if $\pi/2, -\pi/2 \notin sp(f)$, then Eq.(2.144) has a unique almost periodic mild solution. We may let $\pi/2, -\pi/2$ be in $sp(f)$, but as isolated points. Then, if there is a bounded mild solution u to Eq.(2.144), then it has a bounded mild solution w such that $sp(w) \subset sp(f)$. Note that in this case, the uniform continuity is automatically fulfilled (see [168]) and $\mathbf{X} = L^2[0,\pi]$ does not contain c_0, so if $sp(f)$ is countable, then w is almost periodic.

2.4. FIXED POINT THEOREMS AND FREDHOLM OPERATORS

In this section we will demonstrate how a classical technique can be applied to the infinite dimensional evolution equations with delay, cf. [206] and [209].

2.4.1. Fixed Point Theorems

Let $\mathbf{X}$ be a Banach space and T a bounded linear affine map defined by $Tx := Sx + z$ for $x \in \mathbf{X}$, where S is a bounded linear operator on $\mathbf{X}$ and $z \in \mathbf{X}$ is fixed. For the existence and uniqueness of a fixed point of T the following result is obvious.

Theorem 2.22 *If $1 \in \rho(S)$ (the resolvent set of S), then T has a unique fixed point in $\mathbf{X}$.*

Now we recall the following two fixed point theorems for linear affine maps by Chow and Hale [45] whose proofs are based on Schauder's fixed point theorem and the Hahn-Banach theorem.

Theorem 2.23 *If there is an $x^0 \in \mathbf{X}$ such that $\{x^0, Tx^0, T^2x^0, \cdots\}$ is relatively compact in $\mathbf{X}$, then T has a fixed point in $\mathbf{X}$.*

Proof. Set $\mathcal{D} = \{T^n x^0 : n = 0, 1, 2, \cdots\}$. Then $T\mathcal{D} \subset \mathcal{D}$. Since T is a linear affine map, and since T is continuous,

$$T(\overline{\operatorname{co}\mathcal{D}}) \subset \overline{\operatorname{co}\mathcal{D}},$$

where $\overline{\operatorname{co}\mathcal{D}}$ denotes the closure of the convex hull of $\mathcal{D}$. Since $\mathcal{D}$ is relatively compact, $\overline{\operatorname{co}\mathcal{D}}$ is a compact, convex set. From Schauder's fixed point theorem T has a fixed point in $\overline{\operatorname{co}\mathcal{D}}$ as desired.

Theorem 2.24 *If the range $\mathcal{R}(I - S)$, I being the identity, is closed and if there is an $x^0 \in \mathbf{X}$ such that $\{x^0, Tx^0, T^2x^0, \cdots\}$ is bounded in $\mathbf{X}$, then T has a fixed point in $\mathbf{X}$.*

Proof. Obviously, T has a fixed point if and only if $z \in \mathcal{R}(I - S)$. Suppose that $z \notin \mathcal{R}(I - S)$. Since $\mathcal{R}(I - S)$ is closed, from the Hahn-Banach theorem there is an $x^* \in \mathbf{X}^*$ such that $<z, x^*> \neq 0$ and that $<(I-S)x, x^*> = 0$ for all $x \in \mathbf{X}$. Hence $<Sx, x^*> = <x, x^*>$ for all $x \in \mathbf{X}$. For $n = 1, 2, \cdots$, set

$$x^n := T^n x^0 = S^n x^0 + z + Sz + S^2 z + \cdots + S^{n-1} z.$$

Since $<S^k x, x^*> = <x, x^*>$ for $k = 1, 2, \cdots, x \in \mathbf{X}$, it follows that

$$<x^n, x^*> = <x^0, x^*> + n <z, x^*>, \quad n = 1, 2, \cdots.$$

This is a contradiction since the right side is bounded for $n \geq 1$ and $<z, x^*> \neq 0$. Therefore, T has a fixed point.

Denote by $\mathcal{F}_T$ the set of the fixed points of the affine map T on $\mathbf{X}$. Then $\mathcal{F}_T = y + N(I-S)$ for some $y \in \mathcal{F}_T$ is an affine space. Define $\dim \mathcal{F}_T := \dim N(I-S)$. Denote by $\Phi_+(\mathbf{X})$ the family of semi-Fredholm operators on $\mathbf{X}$, which is defined in Section 4.1 of Appendices. If $I - S \in \Phi_+(\mathbf{X})$, then Theorem 2.24 can be easily refined as follows :

Theorem 2.25 *Assume that there is an $x^0 \in \mathbf{X}$ such that $\{x^0, Tx^0, T^2x^0, \cdots\}$ is bounded in $\mathbf{X}$. If $I - S \in \Phi_+(\mathbf{X})$, then $\mathcal{F}_T \neq \emptyset$ and $\dim \mathcal{F}_T$ is finite.*

Remark 2.21 Let S be a bounded linear operator on $\mathbf{X}$.

i) If S is an α-contraction operator on $\mathbf{X}$, then $r_e(S) < 1$, where $r_e(S)$ stands for the radius of the esssential spectrum of S, see (4.1).

ii) If $r_e(S) < 1$, then one is a nomal point of S.

iii) If one is a nomal point of S, then $I - S \in \Phi_+(\mathbf{X})$.

iv) If $I - S \in \Phi_+(\mathbf{X})$, then $\mathcal{R}(I - S)$ is closed.

2.4.2. Decomposition of Solution Operators

In this section we will assume that A is the generator of a C_0-semigroup of linear operators $T(t)$ on the space E such that $\|T(t)\| \leq M_w e^{wt}, t \geq 0$, and $L(t, \cdot)$ is the family of bounded linear operators from $\mathcal{C} := C([-r, 0], E)$ to E. Then, according to Section 1.2, for every ϕ in $\mathcal{C}$ and $\sigma \in \mathbf{R}$ there exists a unique mild solution $u(t, \sigma, \phi) = u(t, \sigma, \phi, 0)$ to the following equation

$$u'(t) = Au(t) + L(t, u_t), \; \forall t \geq \sigma. \tag{2.147}$$

Define the solution operator $U(t, \sigma)$ on $\mathcal{C}$ for $t \geq \sigma$ by $U(t, \sigma)\phi = u_t(\sigma, \phi) := u(t + \cdot, \sigma, \phi)$. It may be seen that this is a family of bounded linear operators, and $\|U(t, \sigma)\|$ is bounded for t in every compact interval of $[\sigma, \infty)$. Put $v(t, \sigma, \phi) = T(t-\sigma)\phi(0)$ for $t > \sigma$, and $v(t, \sigma, \phi) = \phi(t-\sigma)$ for $\sigma - r \leq t \leq \sigma$. Define $z(t, \sigma, \phi) = u(t, \sigma, \phi) - v(t, \sigma, \phi)$ for $t \geq \sigma - r$; that is,

$$z(t, \sigma, \phi) = \begin{cases} \int_\sigma^t T(t-s)L(s, u_s(\sigma, \phi))ds & t \geq \sigma \\ 0 & \sigma - r \leq t \leq \sigma. \end{cases} \tag{2.148}$$

Then $z(t, \sigma, \phi)$ is continuous for $t \geq \sigma - r$. For $t \geq \sigma$ define an operator $K(t, \sigma)$ by $K(t, \sigma)\phi = z_t(\sigma, \phi)$ for $\phi \in \mathcal{C}$. Then it is a bounded linear operator on $\mathcal{C}$, and $U(t, \sigma)$ is decomposed as

$$U(t, \sigma) = \widehat{T}(t - \sigma) + K(t, \sigma), \tag{2.149}$$

where $\widehat{T}(t - \sigma)\phi = v_t(\sigma, \phi)$ for $\phi \in \mathcal{C}$; that is,

$$(\widehat{T}(t)\phi)(\theta) = \begin{cases} T(t+\theta)\phi(0) & \text{for } t+\theta \geq 0 \\ \phi(t+\theta) & \text{for } t+\theta \leq 0, \end{cases} \quad (2.150)$$

for $\theta \in [-r, 0]$.

We can easily prove the following :

Theorem 2.26 $\widehat{T}(t)$ *is a C_0-semigroup on $\mathcal{C}$.*

Proposition 2.6 *i)*

$$\|\widehat{T}(t)\| \leq M_w e^{\max\{0,w\}t} \text{ for } t \geq 0. \quad (2.151)$$

ii) If $\|T(t)\| \leq M_w e^{-wt}, w > 0$, *for* $t \geq 0$, *then*

$$\|\widehat{T}(t)\| \leq \begin{cases} M_w e^{-w(t-r)} & \text{for } t > r \\ M_w & \text{for } t \leq r. \end{cases} \quad (2.152)$$

Set

$$K(t) = K(t, 0). \quad (2.153)$$

We will give an estimate of $\|K(t)\|$ by using Theorem 1.12.

Proposition 2.7 *Let M_w, w be the constants in Inequality (1.19). Let K be the operator defined by Relation (2.153). Then*

$$\|K(t)\| \leq M_w e^{\max\{0,w\}t} \left(\exp\left(\int_0^t M_w \|L(r)\| dr \right) - 1 \right).$$

Proof. Set $z = \max\{0, w\}$. Let $t \geq 0$. Then from Theorem 1.12 we have

$$\begin{aligned} |(K(t)\phi)(\theta)| &\leq \int_0^t M_w e^{z(t-s)} \|L(s)\| |u_s(0,\phi)| ds \\ &\leq \int_0^t M_w e^{z(t-s)} \|L(s)\| |\phi| M_w e^{ws} \exp\left(\int_0^s M_w \|L(r)\| dr \right) ds. \end{aligned}$$

Since $w \leq z$, we have that

$$\begin{aligned} |(K(t)\phi)(\theta)| &\leq |\phi| M_w e^{zt} \int_0^t M_w \|L(s)\| \exp\left(\int_0^s M_w \|L(r)\| dr \right) ds \\ &\leq |\phi| M_w e^{zt} \left(\exp\left(\int_0^t M_w \|L(r)\| dr \right) - 1 \right), \end{aligned}$$

and hence

$$\|K(t)\phi\| \leq |\phi| M_w e^{zt} \left(\exp\left(\int_0^t M_w \|L(r)\| dr \right) - 1 \right).$$

Lemma 2.27 *Let a and w be positive constants, and let $f, u : [0,d] \to \mathbf{R}$ be nonnegative continuous functions. Suppose that $f(t)$ is a nondecreasing function in t and that $u(t)$ satisfies the inequality*

$$u(t) \leq a \sup_{\max\{0,t-r\}\leq \tau \leq t} \int_0^\tau e^{-w(\tau-s)} u(s) ds + f(t). \qquad (2.154)$$

If $w > a$, then

$$u(t) \leq w f(t)/(w-a).$$

Proof. Set $v(t) := \sup\{u(s) : 0 \leq s \leq t\}$. Then from the inequality (2.154) we have

$$\begin{aligned}
u(t) &\leq av(t) \sup_{\max\{0,t-r\}\leq \tau \leq t} \int_0^\tau e^{-w(\tau-s)} ds + f(t) \\
&\leq av(t) \sup_{\max\{0,t-r\}\leq \tau \leq t} (1-e^{-w\tau})/w + f(t) \\
&\leq av(t)(1-e^{-wt})/w + f(t) \\
&\leq av(t)/w + f(t),
\end{aligned}$$

and hence,

$$v(t) \leq av(t)/w + f(t).$$

This implies the desired inequality for $u(t)$.

Proposition 2.8 *Suppose that $\|T(t)\| \leq M_w e^{-wt}$, $M_w, w > 0$, for $t \geq 0$ and that $\|L\|_\infty := \sup_{t \geq 0} \|L(t)\| < \infty$. Then, for Equation(2.147), if $w > M_w \|L\|_\infty, t \geq 0$*

$$\|K(t)\| \leq M_w^2 \|L\|_\infty (1-e^{-wt})/(w - M_w \|L\|_\infty) := k(t)$$

Proof. Set $a = M_w \|L\|_\infty$. Using the decomposition (2.149) of the solution operator for Equation(2.147), we have

$$\begin{aligned}
|K(t)\phi| &\leq M_w \sup_{\max\{0,t-r\}\leq \tau \leq t} \int_0^\tau e^{-w(\tau-s)} |L(s, U(s,0)\phi)| ds \\
&\leq a \sup_{\max\{0,t-r\}\leq \tau \leq t} \{\int_0^\tau e^{-w(\tau-s)} |K(s)\phi| ds + \int_0^\tau e^{-w(\tau-s)} |\widehat{T}(s)\phi| ds\}.
\end{aligned}$$

Furthermore, it follows from the estimate (2.152) of $\widehat{T}(t)$ that

$$\int_0^\tau e^{-w(\tau-s)} |\widehat{T}(s)\phi| ds \leq \int_0^\tau e^{-w(\tau-s)} M_w |\phi| ds \leq M_w (1-e^{-w\tau}) |\phi|/w.$$

If we set $u(t) = |K(t)\phi|$, then

$$u(t) \leq a \sup_{\max\{0,t-r\}\leq \tau \leq t} \int_0^\tau e^{-w(\tau-s)} u(s) ds + a M_w (1-e^{-wt}) |\phi|/w.$$

Using Lemma 2.27, we have the estimates described in the proposition.

Proposition 2.9 *If $(T(t))_{t\geq 0}$ is a compact C_0-semigroup, or if $L(t,\cdot)$ is a compact operator for $t \geq \sigma$, then $K(t,\sigma)$ is a compact operator.*

Proof. Let $S := \{\phi \in C([-r,0], E) : |\phi| \leq 1\}$, and $\mathcal{U}$ be the family of functions $L(s, u_s(\sigma, \phi))$ of variable $s \in [\sigma, \infty)$ with the index $\phi \in S$. Since $\|U(t,\sigma)\|$ is locally bounded, $\mathcal{U}$ is uniformly bounded on every compact set of $[\sigma, \infty)$. Applying Theorem 4.8, we have

$$\alpha(K(t,\sigma)S) \leq \gamma_T \sup\{\alpha(\{z(t+\theta,\sigma,\phi) : \phi \in S\}) : -r \leq \theta \leq r\}.$$

We see that $\{z(t,\sigma,\phi) : \phi \in S\}$ is relatively compact for eact $t \geq \sigma - r$. From Corollary 4.2 it suffices to see that the set $H(s) := \{T(t-s)L(s, u_s(\sigma,\phi)) : \phi \in S\}$ is relatively compact for $\sigma \leq s < t$. But this follows immediately from the assumption of the theorem.

2.4.3. Periodic Solutions and Fixed Point Theorems

Let the following equation

$$u'(t) = Au(t) + L(t, u_t) + f(t) \tag{2.155}$$

and

$$u'(t) = Au(t) + L(t, u_t) \tag{2.156}$$

be periodic in t with a period ω ; that is,

$$L(t+\omega, \phi) = L(t, \phi), \quad f(t+\omega) = f(t)$$

for $(t, \phi) \in \mathbf{R} \times \mathcal{C}$.

We are going to state the well-known connection between the existence of periodic solutions and that of fixed points of the linear affine map associated with the equation under consideration. By the uniqueness of solutions, if $\sigma < \tau$, and if $\tau - r \leq t$,

$$u(t, \sigma, \phi, f) = u(t, \tau, u_\tau(\sigma, \phi, f), f).$$

On the other hand, by the periodicity as well as the uniqueness, for $t \geq \sigma - r$,

$$u(t+\omega, \sigma+\omega, \phi, f) = u(t, \sigma, \phi, f).$$

Proposition 2.10 *The following conditions are equivalent for any $\sigma \in \mathbf{R}$;*

i) There exists a $\psi \in \mathcal{C}$ such that $u(t+\omega, \sigma, \psi, f) = u(t, \sigma, \psi, f)$ for $t \geq \sigma$,

ii) There exists an ω-periodic function $p(t)$ such that $p(t) = u(t, \tau, p_\tau, f)$ whenever $-\infty < \tau \leq t < \infty$,

iii) There exists a $\phi \in \mathcal{C}$ such that $u_\omega(0, \phi, f) = u_0(0, \phi, f)$.

Proof. Suppose that condition i) holds. Let $p(t)$ be a periodic function such that $p(t) = u(t, \sigma, \psi, f)$ for $t \geq \sigma$ and that $p(t + \omega) = p(t)$ for all $t \in \mathbf{R}$. Suppose that $\tau \leq t$, and take an integer k such that $\tau + k\omega - r \geq \sigma$. Then

$$p_\tau = p_{\tau+k\omega} = u_{\tau+k\omega}(\sigma, \psi, f).$$

We have that, for $t > \tau$,

$$p(t) = p(t + k\omega) = u(t + k\omega, \sigma, \psi, f)$$
$$= u(t + k\omega, \tau + k\omega, u_{\tau+k\omega}(\sigma, \psi, f), f) = u(t, \tau, p_\tau, f).$$

Hence condition i) implies ii).

Suppose that condition ii) holds. Set $\phi := p_0$. Then $u(t, 0, \phi, f) = p(t)$ for $t \geq -r$ and $u_\omega(0, \phi, f) = p_\omega = p_0 = \phi = u_0(0, \phi, f)$. Hence condition ii) implies iii).

Suppose that condition iii) holds. Then, for $t \geq -r$,

$$u(t + \omega, 0, \phi, f) = u(t + \omega, \omega, u_\omega(0, \phi, f), f) = u(t, 0, \phi, f)$$

Take an integer k such that $\sigma + k\omega \geq 0$. Set $\psi = u_{\sigma+k\omega}(0, \phi, f)$. Then, for $t \geq \sigma - r$,

$$\begin{aligned} u(t, \sigma, \psi, f) &= u(t + k\omega, \sigma + k\omega, \psi, f) \\ &= u(t + k\omega, \sigma + k\omega, u_{\sigma+k\omega}(0, \phi, f), f) \\ &= u(t + k\omega, 0, \phi, f). \end{aligned}$$

Hence, for $t \geq \sigma - r$,

$$u(t + \omega, \sigma, \psi, f) = u(t + \omega + k\omega, 0, \phi, f) = u(t + k\omega, 0, \phi, f) = u(t, \sigma, \psi, f),$$

that is, condition i) holds.

Since the solution $u(t, \sigma, \phi, f)$ is decomposed as

$$u(t, \sigma, \phi, f) = u(t, \sigma, \phi, 0) + u(t, \sigma, 0, f),$$

it follows that

$$u_t(\sigma, \phi, f) = u_t(\sigma, \phi, 0) + u_t(\sigma, 0, f).$$

Define a family of operators $U(t, \sigma), t > \sigma$, on $\mathcal{C}$ by $U(t, \sigma)\phi = u_t(\sigma, \phi, 0), \phi \in \mathcal{C}$. Then, it forms an ω-periodic process of bounded linear operators on $\mathcal{C}$. Since

$$u_t(\sigma, \phi, f) = U(t, \sigma)\phi + u_t(\sigma, 0, f),$$

the condition $u_\omega(0, \phi, f) = \phi$ becomes $U(\omega, 0)\phi + u_\omega(0, 0, f) = \phi$. Set $S := U(\omega, 0), \psi := u_\omega(0, 0, f)$ and define a linear affine map T on $\mathcal{C}$ by

$$T\phi = S\phi + \psi$$

for $\phi \in \mathcal{C}$. From the definition, $T\phi = u_\omega(0, \phi, f)$ and hence $T^n\phi = u_{n\omega}(0, \phi, f)$ for $n = 1, 2, \cdots$. From Proposition 2.10, we have the following proposition.

Proposition 2.11 *Equation (2.155) has an ω-periodic solution if and only if the above operator T on $\mathcal{C}$ has a fixed point.*

From the decomposition (2.149) of solution operator $U(t,\sigma)$ of Equation (2.156), $U(\omega) := U(\omega, 0)$ is expressed as $U(\omega) = \widehat{T}(\omega) + K(\omega)$.

Now we consider criteria for the existence and uniqueness of periodic solutions to Equation (2.155). Denote by $\mathcal{S}_L(\omega)$ the set of all ω-periodic solutions of Equation (2.155).

Combining Proposition 2.11 with Theorem 2.25, Theorem 4.1 and Theorem 4.5, respectively, are expressed as follows.

Theorem 2.27 *Assume that $I - \widehat{T}(\omega) \in \Phi_+(\mathcal{C})$ and that $K(\omega)$ is a compact operator. If Equation (2.155) has a bounded solution, then $\mathcal{S}_L(\omega) \neq \emptyset$ and $\dim \mathcal{S}_L(\omega)$ is finite.*

Theorem 2.28 *Assume that $I - \widehat{T}(\omega) \in \Phi_+(\mathcal{C})$. Let $\dim N(I - \widehat{T}(\omega)) = n$ and let c be a positive constant such that*

$$\|[\phi]\| \leq c|(I - \widehat{T}(\omega))\phi| \quad \text{for } \phi \in \mathcal{C},$$

where $[\phi] \in \mathcal{C}/N(I - \widehat{T}(\omega))$. Assume further that

$$\|K(\omega)\| \leq 1/2c(1 + \sqrt{n}).$$

If Equation (2.155) has a bounded solution, then $\mathcal{S}_L(\omega) \neq \emptyset$ and $\dim \mathcal{S}_L(\omega) \leq \dim N(I - \widehat{T}(\omega)) = n$.

Criteria for the existence and uniqueness of ω-periodic solutions to Equation (2.156) can be studied by using the following two approaches. The first one is to apply the proposition below which is an immediate consequence of Theorem 2.28.

Proposition 2.12 *Assume that $I - \widehat{T}(\omega) \in \Phi_+(\mathcal{C})$ and $\dim N(I - \widehat{T}(\omega)) = 0$. Let c be a positive constant such that*

$$|\phi| \leq c|(I - \widehat{T}(\omega))\phi| \quad \text{for } \phi \in \mathcal{C}.$$

Assume further that

$$\|K(\omega)\| \leq 1/2c.$$

If Equation (2.155) has a bounded solution, then it has a unique ω-periodic solution.

The second one is based on a well known result in the operator theory: *Let $F, K : \mathbf{X} \to \mathbf{X}$ be bounded linear operators, and $1 \in \rho(F)$. Then, if $1 \in \rho((I - F)^{-1}K)$, one has $1 \in \rho(F + K)$.*

Proposition 2.13 *Assume that $1 \in \rho(\widehat{T}(\omega))$. If $1 \in \rho((I - \widehat{T}(\omega))^{-1}K(\omega))$, then Equation (2.155) has a unique ω-periodic solution.*

Proof. From the above result we see that $1 \in \rho(U(\omega))$. Hence the assertion is proved by using Theorem 2.22 and Proposition 2.11.

2.4.4. Existence of Periodic Solutions: Bounded Perturbations

In this subsection we will apply Theorem 2.28 to study the existence of periodic solutions of Eq.(2.155). To this end, we first consider the operator $\widehat{T}(\omega)$.

Proposition 2.14 *Let $(T(t))_{t\geq 0}$ be a C_0-semigroup on E. Then $\phi \in N(I - \widehat{T}(\omega))$ if and only if $\phi(0) \in N(I - T(\omega))$ and $\phi(\theta) = T(n\omega + \theta)\phi(0), -r \leq \theta \leq 0$, whenever $n\omega \geq r$; in particular $\phi \in N(I - \widehat{T}(\omega))$ is a restriction to $[-r, 0]$ of an ω-periodic continuous function. Furthermore,*

$$\dim N(I - \widehat{T}(\omega)) = \dim N(I - T(\omega)).$$

Proof. If $x \in N(I - T(\omega))$, then $x = T(\omega)x$ and $T(t)x = T(t)T(\omega)x = T(t+\omega)x$ for $t \geq 0$; and vice versa.

Suppose that $\phi \in N(I - \widehat{T}(\omega))$. Since $\phi = \widehat{T}(\omega)\phi$, it follows that $\phi(0) = T(\omega)\phi(0)$ or $\phi(0) \in N(I - T(\omega))$. Since $\widehat{T}(t)$ is a C_0-semigroup, we see that $\widehat{T}(\omega)^n = \widehat{T}(n\omega), n = 1, 2, \cdots$. If $n\omega > r$, then $n\omega + \theta \geq 0$ for $\theta \in [-r, 0]$; and hence

$$(\widehat{T}(\omega)^n \psi)(\theta) = T(n\omega + \theta)\psi(0), \quad \theta \in [-r, 0].$$

Thus $\phi = \widehat{T}(\omega)^n \phi$ for $n \geq 1$; in particular, if $n\omega > r$, $\phi(\theta) = T(n\omega + \theta)\phi(0)$ for $\theta \in [-r, 0]$.

Conversely, suppose that $x \in N(I - T(\omega))$ and set $(\phi_n)(\theta) = T(n\omega + \theta)x, \theta \in [-r, 0]$ for n such that $n\omega \geq r$. Since $T(t+\omega)x = T(t)x$ for $t \geq 0$, ϕ_n is independent of $n > r/\omega$. Denote by ϕ this independent function. Then $\phi(0) = T(n\omega)x = x$. Let $\theta \in [-r, 0]$. If $\omega + \theta \geq 0$, then

$$\widehat{T}(\omega)\phi(\theta) = T(\omega + \theta)\phi(0) = T(\omega + \theta)x = T(n\omega + \theta)x = \phi(\theta).$$

For $\omega + \theta < 0$, one has

$$\begin{aligned}\widehat{T}(\omega)\phi(\theta) &= \phi(\omega + \theta) = \phi_n(\omega + \theta) \\ &= T(n\omega + \omega + \theta)x = T(n\omega + \theta)x \\ &= \phi_n(\theta) = \phi(\theta).\end{aligned}$$

Hence, $\widehat{T}(\omega)\phi = \phi$. It is obvious that this map from $N(I - T(\omega))$ to $N(I - \widehat{T}(\omega))$ is injective. Its surjectiveness follows from the first half of the proof.

To see that the range $\mathcal{R}(I - \widehat{T}(\omega))$ is a closed subspace, we solve the equation $(I - \widehat{T}(\omega))\phi = \psi$. Let p be a positive integer such that

$$(p-1)\omega < r \leq p\omega. \tag{2.157}$$

Set $I_p = [-r, -(p-1)\omega]$ and $I_k = [-k\omega, -(k-1)\omega]$ for $k = 1, 2, \cdots, p-1$ provided $p \geq 2$.

CHAPTER 2. SPECTRAL CRITERIA

Proposition 2.15 *The functions $\phi, \psi \in \mathcal{C}$ satisfy the equation $(I - \widehat{T}(\omega))\phi = \psi$ if and only if*

i) $(I - T(\omega))\phi(0) = \psi(0)$,

ii) $\phi(\theta) = \sum_{j=0}^{k-1} \psi(\theta + j\omega) + T(\theta + k\omega)\phi(0)$, $\quad \theta \in I_k, \quad k = 1, 2, \cdots, p.$

Proof. Suppose that $(I - \widehat{T}(\omega))\phi = \psi$. Then

$$\psi(\theta) = \begin{cases} \phi(\theta) - T(\theta + \omega)\phi(0) & \theta \in I_1 \\ \phi(\theta) - \phi(\theta + \omega) & \theta \in I_k, \quad k \geq 2. \end{cases}$$

Putting $\theta = 0$ in the first equation, we obtain the condition i). Solving this equation with respect to $\phi(\theta)$ on I_k successively for $k = 1, 2, \cdots, p$ we have the representation of $\phi(\theta)$ in the condition ii) as mentioned above. The value $\phi(-k\omega)$ is well defined for $k \geq 1$ because of the condition $(I - T(\omega))\phi(0) = \psi(0)$.

Conversely, if $\phi, \psi \in \mathcal{C}$ have the properties i) and ii), then it follows immediately that $(I - \widehat{T}(\omega))\phi = \psi$. The proof is complete.

Let the null space $N(I - T(\omega))$ be of finite dimension. Then it follows from Theorem 4.6 that there exists a closed subspace M of E such that $E = N \oplus M$, where $N = N(I - T(\omega))$. Let S_M be the restriction of $I - T(\omega)$ to M. Then

$$S_M := (I - T(\omega))|_M : M \to R(I - T(\omega))$$

is a continuous, bijective and linear operator. Thus there is the inverse operator S_M^{-1} of S_M. If $I - T(\omega) \in \Phi_+(E)$, M can be taken so that S_M^{-1} is continuous and that

$$\|S_M^{-1}\| \leq c_0(1 + \sqrt{n}), \tag{2.158}$$

where $n = \dim N(I - T(\omega))$, and c_0 is the constant such that

$$|[x]| \leq c_0|(I - T(\omega))x| \tag{2.159}$$

for $x \in E$ (see Theorem 4.6).

Put $\mathcal{D} = \{\psi \in \mathcal{C} : \psi(0) \in \mathcal{R}(I - T(\omega))\}$ and let $\psi \in \mathcal{D}$. Since $\mathcal{R}(S_M) = \mathcal{R}(I - T(\omega))$, $S_M^{-1}\psi(0)$ is well defined and $(I - T(\omega))S_M^{-1}\psi(0) = \psi(0)$. We define a function $(V\psi)(\cdot) : [-r, 0] \to E$ pointwise by

$$[V\psi](\theta) = \sum_{j=0}^{k-1} \psi(\theta + j\omega) + T(\theta + k\omega)S_M^{-1}\psi(0), \quad \theta \in I_k, \tag{2.160}$$

for $k = 1, 2, \cdots, p$, and $[V\psi](0) = S_M^{-1}\psi(0)$. Notice that $\mathcal{D}(V) = \mathcal{D}$.

Lemma 2.28 *The operator V defined by (2.160) is a linear operator from $\mathcal{D}(V)$ to $\mathcal{C}$.*

Proof. It is sufficient to prove that $V\psi \in \mathcal{C}$ for $\psi \in \mathcal{D}(V)$. Clearly, $V\psi$ is continuous in each interval I_k. Thus we prove that $(V\psi)(-k\omega) = (V\psi)(-k\omega - 0)$ for $k = 1, 2, \cdots p - 1$. Notice that $S_M^{-1}\psi(0) = T(\omega)S_M^{-1}\psi(0) + \psi(0)$. From the definition (2.160) of V, we have,

$$\begin{aligned}(V\psi)(-k\omega) &= \sum_{j=0}^{k-1}\psi(-k\omega + j\omega) + T(0)S_M^{-1}\psi(0) \\ &= \sum_{j=0}^{k-1}\psi(-k\omega + j\omega) + \psi(0) + T(\omega)S_M^{-1}\psi(0) \\ &= \sum_{j=0}^{k}\psi(-k\omega + j\omega) + T(-k\omega + (k+1)\omega)S_M^{-1}\psi(0) \\ &= \lim_{\theta \to -k\omega - 0}\{\sum_{j=0}^{k}\psi(\theta + j\omega) + T(\theta + (k+1)\omega)S_M^{-1}\psi(0)\}\end{aligned}$$

as required.

Lemma 2.29

$$\mathcal{R}(I - \widehat{T}(\omega)) = \mathcal{D}(V).$$

Proof. Let $\psi \in \mathcal{R}(I - \widehat{T}(\omega))$. Then there is a $\phi \in \mathcal{C}$ such that $[I - \widehat{T}(\omega)]\phi = \psi$. From Proposition 2.15 we see that $(I - T(\omega))\phi(0) = \psi(0)$; and hence, $\psi(0) \in \mathcal{R}(I - T(\omega))$. This implies that $\psi \in \mathcal{D}(V)$.

Conversely, if $\psi \in \mathcal{D}(V)$, then $\psi(0) \in \mathcal{R}(I - T(\omega))$. Lemma 2.28 means that $V\psi \in \mathcal{C}$. Hence it follows from Proposition 2.15 that $[I - \widehat{T}(\omega)]V\psi = \psi$. Therefore the proof is completed.

Proposition 2.16 $I - T(\omega) \in \Phi_+(E)$ *if and only if* $I - \widehat{T}(\omega) \in \Phi_+(\mathcal{C})$.

Proof. If $\mathcal{R}(I - T(\omega))$ is closed in E, then $\mathcal{D}(V)$ is closed in $\mathcal{C}$. Indeed, if $\phi_n \in \mathcal{D}(V) \to \phi$ as $n \to \infty$, then $\phi_n(0) \to \phi(0)$ as $n \to \infty$. This implies that $\phi \in \mathcal{D}(V)$. From this fact and Lemma 2.29 it follows that $\mathcal{R}(I - \widehat{T}(\omega))$ is closed in $\mathcal{C}$.

Conversely, we assume that $\mathcal{R}(I - \widehat{T}(\omega))$ is closed in $\mathcal{C}$. Then it follows from Lemma 2.29 that $\mathcal{D}(V)$ is closed in $\mathcal{C}$. Let $x_n \in \mathcal{R}(I - T(\omega)) \to x$ as $n \to \infty$. Then there exist a sequence $\{\varphi_n\} \subset \mathcal{C}$ of constant functions and $\varphi \in \mathcal{C}$ such that $\varphi_n(0) = x_n, \varphi_n \in \mathcal{D}(V)$ and $\varphi_n \to \varphi$ as $n \to \infty$ in $\mathcal{C}$. Hence we have $\varphi(0) = x$ and $\varphi \in \mathcal{D}(V)$. This implies that $x \in \mathcal{R}(I - T(\omega))$. The remainder follows from Proposition 2.14 ; and hence, the proof is complete.

Proposition 2.17 $1 \in \rho(T(\omega))$ *if and only if* $1 \in \rho(\widehat{T}(\omega))$.

Proof. It is sufficient to prove that if $1 \in \rho(T(\omega))$, then $1 \in \rho(\widehat{T}(\omega))$. Since $\mathcal{R}(I - T(\omega)) = E$, we have $\mathcal{D}(V) = \mathcal{C}$. Hence it follows from Lemma 2.29 that $\mathcal{R}(I - \widehat{T}(\omega)) = \mathcal{D}(V) = \mathcal{C}$. Lemma 2.14 implies that $1 \in \rho(\widehat{T}(\omega))$.

Next, we give criteria for the existence of periodic solutions to Equation (2.155) by using Theorem 2.28.

Proposition 2.18 *If there exists a positive constant c such that*

$$|V\psi| \leq c|\psi| \quad \text{for all } \psi \in \mathcal{D}(V), \tag{2.161}$$

where V is defined by (2.160), then

$$|[\phi]| \leq c|(I - \widehat{T}(\omega))\phi| \quad \text{for all } \phi \in \mathcal{C} :$$

as a result, the range $\mathcal{R}(I - \widehat{T}(\omega))$ is closed.

Proof. Take the quotient space $\mathcal{C}/N(I - \widehat{T}(\omega))$. Suppose that $(I - \widehat{T}(\omega))\phi = \psi$. Then $\psi \in \mathcal{R}(I - \widehat{T}(\omega))$ and $V\psi \in \mathcal{C}$, because of Lemma 2.28. Hence, we have $\psi = (I - \widehat{T}(\omega))V\psi$ and $[\phi] = [V\psi]$. Using these facts and the condition (2.161), we see that

$$|[\phi]| \leq |V\psi| \leq c|\psi| = c|(I - \widehat{T}(\omega))\phi|.$$

We note that $\mathcal{R}(I - \widehat{T}(\omega))$ is closed if and only if there is a positive constant c such that $|[\phi]| \leq c|(I - \widehat{T}(\omega))\phi|$ for all $\phi \in \mathcal{C}$, cf. Lemma 4.1. This prove the proposition.

Theorem 2.29 *Suppose that $I - T(\omega) \in \Phi_+(E)$. Let $\dim N(I - T(\omega)) = n$, and c_0 be a positive constant such that*

$$|[x]| \leq c_0|(I - T(\omega))x|$$

for $x \in E$. Then $I - \widehat{T}(\omega) \in \Phi_+(\mathcal{C})$, $\dim N(I - \widehat{T}(\omega)) = n$ and

$$|[\phi]| \leq (p + m_\omega c_0(1 + \sqrt{n}))|(I - \widehat{T}(\omega))\phi| \quad \text{for } \phi \in \mathcal{C},$$

where p is the positive integer satisfying Inequality (2.157) and $m_\omega := \sup\{\|T(t)\| : 0 \leq t \leq \omega\}$.

Proof. Take the closed subspace M so that S_M^{-1} is continuous and the estimate (2.158) holds. Suppose that $\psi(0) \in \mathcal{R}(I - T(\omega))$ and that $V\psi \in \mathcal{C}$. Then, for $\theta \in I_k, k = 1, 2, \cdots, p$,

$$|V\psi(\theta)| \leq k|\psi| + m_\omega \|S_M^{-1}\| |\psi(0)| \leq (p + m_\omega \|S_M^{-1}\|)|\psi|.$$

This implies that

$$|V\psi| \leq (p + m_\omega \|S_M^{-1}\|)|\psi|. \tag{2.162}$$

From this, the estimate (2.158) and Proposition 2.18, we have the conclusion.

We are now in a position to prove a criterion for the existence of ω-periodic solution to Eq.(2.155)

Theorem 2.30 *Assume that $I - T(\omega) \in \Phi_+(E)$ satisfies the conditions in Theorem 2.29, and that Equation (2.155) has a bounded solution. If the inequality*

$$\|K(\omega)\| \leq 1/2(1 + \sqrt{n})(p + m_\omega c_0(1 + \sqrt{n})), \tag{2.163}$$

is satisfied, then $\mathcal{S}_L(\omega) \neq \emptyset$ and $\dim \mathcal{S}_L(\omega) \leq \dim N(I - T(\omega)) = n$.

Using Proposition 2.7, we have the following one.

Corollary 2.21 *The inequality (2.163) in the theorem can be replaced by the following inequality*

$$M_w e^{w\omega} \left(\exp \left(\int_0^\omega M_w \|L(r)\| dr \right) - 1 \right) \leq 1/2(1 + \sqrt{n})(p + m_\omega c_0(1 + \sqrt{n})).$$

2.4.5. Existence of Periodic Solutions : Compact Perturbations

We consider the existence of periodic solutions for Equation (2.155) by using Theorem 2.27. In particular, in order to check the condition $I - \widehat{T}(\omega) \in \Phi_+(\mathcal{C})$, we will compute the α-measure of the operator $\widehat{T}(t)$ as well as the radius $r_e(\widehat{T}(t))$ of the essential spectrum of $\widehat{T}(t)$.

For a subset $D \subset E$ we denote by $T(\cdot)D$ the family of functions $T(t)x$ defined for $t \in [0, \infty)$ with a parameter $x \in D$, and by $T(\cdot)D|[a,b]$ its restriction to $[a,b]$. We note that $\alpha(\Omega(0)) \leq \alpha(\Omega)$ for a bounded set $\Omega \subset \mathcal{C}$, where $\Omega(0) = \{\phi(0) : \phi \in \Omega\}$. Put $\delta_T := \max\{1, \gamma_T\}$, where $\gamma_T = \limsup_{t \to 0} \|T(t)\|$.

Lemma 2.30 *Let $D \subset E$ be bounded. Then the following assertions hold true:*

(1) If $(T(t))_{t \geq 0}$ is a C_0-semigroup on E,

$$\alpha(T(\cdot)D|[a,b]) \leq \sup_{a \leq s \leq b} \|T(s)\| \alpha(D), \quad b > a \geq 0.$$

(2) If $(T(t))_{t \geq 0}$ be a C_0-semigroup on E such that $T(t)x \in \mathcal{D}(A)$ for all $x \in E$ and $t > 0$. If $b > \varepsilon > 0$, then

  (i) $\alpha(T(\cdot)D|[\varepsilon, b]) \leq \sup_{\varepsilon \leq s \leq b} \alpha(T(s))\alpha(D)$.
  (ii) $\alpha(T(\cdot)D|[0, b]) \leq \max\{\sup_{0 \leq s \leq \varepsilon} \|T(s)\|, \sup_{\varepsilon \leq s \leq b} \alpha(T(s))\}\alpha(D)$.

(3) Let $(T(t))_{t \geq 0}$ be a compact C_0-semigroup on E. If $0 < \varepsilon < b$, then

  (i) $\alpha(T(\cdot)D|[\varepsilon, b]) = 0$.
  (ii) $\alpha(T(\cdot)D|[0, b]) \leq \delta_T \alpha(D)$.

Proof. The assertion (1) is derived directly from the definition of $\alpha(T(\cdot)D|[\varepsilon, b])$. Set $\mathcal{H} = T(\cdot)D|[\varepsilon, b]$. We claim that $\omega(t, \mathcal{H}) = 0$ for $t > 0$ (see Section 4.2 for the deinition) if $T(t)$ has the property in (2) or (3). It is clear in the case (3). Consider the case (2). Put $M := \sup\{|x| : x \in D\}$. Then M is finite, and $|T(s)x - T(t)x| \leq \|T(s) - T(t)\|M$ for $s, t \geq 0, x \in D$. Since $T(t)x \in D(A), t > 0, x \in E$, it follows

that $T(t)$ is continuous for $t > 0$ in the uniform operator topology, cf. [179, p.52]. This implies that $\omega(t, \mathcal{H}) = 0$ for $t > 0$.
Now, from Lemma 4.2 it follows that $\alpha(\mathcal{H}) = \sup_{\varepsilon \leq s \leq b} \alpha(\mathcal{H}(s))$. Since $\alpha(\mathcal{H}(s)) \leq \alpha(T(s))\alpha(D)$, we have the properties (2) (i) and (3) (i). Since

$$\alpha(T(\cdot)D|[0,b]) \leq \max\{\alpha(T(\cdot)D|[0,\varepsilon]), \alpha(T(\cdot)D|[\varepsilon,b])\},$$

we obtain the properties (2) (ii) and (3) (ii).

Lemma 2.31 *The following assertions are valid:*

(1) If $(T(t))_{t\geq 0}$ be a C_0-semigroup on E, then

$$\alpha(\widehat{T}(t)) \leq \begin{cases} \sup_{0 \leq s \leq t} \|T(s)\|, & r \geq t \geq 0 \\ \sup_{t-r \leq s \leq t} \|T(s)\|, & t > r. \end{cases}$$

(2) If $(T(t))_{t\geq 0}$ be a C_0-semigroup on E such that $T(t)x \in \mathcal{D}(A)$ for all $x \in E$ and $t > 0$, then

$$\alpha(\widehat{T}(t)) \leq \begin{cases} \max\{\delta_T, \sup_{0 < s \leq t} \alpha(T(s))\}, & r \geq t > 0 \\ \sup_{t-r \leq s \leq t} \alpha(T(s)), & t > r. \end{cases}$$

(3) Let $(T(t))_{t\geq 0}$ be a compact C_0-semigroup for $t > t_0$ on E. Then
  (i) If $t_0 \geq r$,

$$\alpha(\widehat{T}(t)) \leq \begin{cases} \sup_{0 \leq s \leq t} \|T(s)\|, & r > t \geq 0 \\ \sup_{t-r \leq s \leq t} \|T(s)\|\delta_T, & t_0 \geq t \geq r \\ \sup_{t_0 - r \leq s \leq t_0} \|T(s)\|\delta_T, & t_0 + r \geq t > t_0 \\ 0, & t > t_0 + r. \end{cases}$$

  (ii) If $r > t_0$,

$$\alpha(\widehat{T}(t)) \leq \begin{cases} \sup_{0 \leq s \leq t} \|T(s)\|, & t_0 \geq t \geq 0 \\ \sup_{0 \leq s \leq t_0} \|T(s)\|\delta_T, & r \geq t > t_0 \\ \sup_{t-r \leq s \leq t_0} \|T(s)\|\delta_T, & t_0 + r \geq t > r \\ 0, & t > t_0 + r. \end{cases}$$

Proof. For a bounded set $\Omega \subset \mathcal{C}$ and for $t \geq 0$, we have

$$\alpha(\widehat{T}(t)\Omega) \leq \begin{cases} \max\{\alpha(T(\cdot)\Omega(0)|[0,t]), \alpha(\Omega|[t-r,0])\}, & r \geq t > 0 \\ \alpha(T(\cdot)\Omega(0)|[t-r,t]), & t > r. \end{cases}$$

From this we can easily obtain the conclusion of the lemma. For example, we now show the relation

$$\alpha(\widehat{T}(t)) \leq \sup_{0 \leq s \leq t_0} \|T(s)\|\delta_T \quad \text{for } t \in (t_0, r]$$

in the assertion 3)(ii). Using Lemma 2.30 we have that for any $\varepsilon > 0$,

$$\begin{aligned}
\alpha(\widehat{T}(t)\Omega) & \\
&\leq \max\{\alpha(\Omega|[t-r,0]), \alpha(T(\cdot)\Omega(0)|[0,t_0+\varepsilon]), \alpha(T(\cdot)\Omega(0)|[t_0+\varepsilon,t])\} \\
&\leq \max\{\alpha(\Omega), \alpha(T(\cdot)\Omega(0)|[0,t_0+\varepsilon])\} \\
&\leq \max\{1, \sup_{0\leq s\leq t_0+\varepsilon}\|T(s)\|\}\alpha(\Omega) \\
&\leq \max\{\sup_{0\leq s\leq t_0}\|T(s)\|, \sup_{t_0\leq s\leq t_0+\varepsilon}\|T(s)\|\}\alpha(\Omega) \\
&\leq \max\{\sup_{0\leq s\leq t_0}\|T(s)\|, \|T(t_0)\|\sup_{0\leq s\leq \varepsilon}\|T(s)\|\}\alpha(\Omega) \\
&\leq \max\{\sup_{0\leq s\leq t_0}\|T(s)\|, \|T(t_0)\|\delta_T\}\alpha(\Omega) \\
&\leq \sup_{0\leq s\leq t_0}\|T(s)\|\delta_T\alpha(\Omega).
\end{aligned}$$

This implies the described inequality

Proposition 2.19 *i) If $(T(t))_{t\geq 0}$ is a compact C_0-semigroup on E, then $\widehat{T}(t)$ is a compact C_0-semigroup for $t > r$ on $\mathcal{C}$,*

ii) If $(T(t))_{t\geq 0}$ is a compact C_0-semigroup for $t > t_0$ on E, then $\widehat{T}(t)$ is a compact C_0-semigroup for $t > t_0 + r$ on $\mathcal{C}$.

To show the existence of fixed points of T we will estimate the radius of the essential spectrum of the solution operator of Equation (2.155). Suppose that a function $g : [0,\infty) \to [0,\infty)$ is loacally bounded, and submultiplicative (that is, $g(t+s) \leq g(t)g(s)$ for $t,s \geq 0$). Then, it is well known that

$$\lim_{t\to\infty} t^{-1}\log g(t) = \inf_{t>0} t^{-1}\log g(t),$$

which may be $-\infty$, but not be $+\infty$. This quantity is called the *type number* of the function $g(t)$. We denote respectively by

$$\tau, \tau^\nu, \hat{\tau}, \hat{\tau}^\nu,$$

the type numbers of the functions

$$\alpha(T(t)), \|T(t)\|, \alpha(\widehat{T}(t)), \|\widehat{T}(t)\|,$$

provided that $(T(t))_{t\geq 0}$ is a C_0-semigroup on E.

In view of the Nussbaum formula we notice that

$$r_e(\widehat{T}(t)) = e^{\hat{\tau}t}, \quad t > 0.$$

Thus, if $\hat{\tau} < 0$, then $I - \widehat{T}(\omega) \in \Phi_+(\mathcal{C})$ (see Remark 2.21). Now we will give several conditions that $\hat{\tau}$ is negative.

CHAPTER 2. SPECTRAL CRITERIA

Theorem 2.31 *i) Let $(T(t))_{t\geq 0}$ be a C_0-semigroup on E. If τ^ν is negative, then $\hat{\tau}^\nu < 0$; and hence, $\hat{\tau} < 0$.*

ii) Let $(T(t))_{t\geq 0}$ be a C_0-semigroup on E such that $T(t)x \in \mathcal{D}(A)$ for all $x \in E$ and $t > 0$. If $\tau < 0$, then $\hat{\tau} < 0$.

iii) If $(T(t))_{t\geq 0}$ be a compact C_0-semigroup on E or a compact C_0-semigroup for $t > t_0$ on E, then $\hat{\tau} < 0$.

Proof. It is sufficient to prove only the assertion i). Since τ^ν is negative, there is a μ such that $-\tau^\nu > \mu > 0$ and $\|T(t)\| \leq M_\mu e^{-\mu t}$ for all $t \geq 0$. Using the estimate (2.152), we see that

$$\begin{aligned}\hat{\tau}^\nu &= \lim_{t\to\infty} \frac{1}{t} \log \|\widehat{T}(t)\| \\ &\leq \lim_{t\to\infty} \frac{1}{t} \log M_\mu e^{-\mu(t-r)} \\ &= -\mu < 0,\end{aligned}$$

from which we have the assertion i). The remainder follows from Lemma 2.31, cf. [206, Theorem 4.8].

Proposition 2.20 *Let $(T(t))_{t\geq 0}$ be a compact C_0-semigroup, or $L(t,\cdot)$ be a compact operator for each $t \in \mathbf{R}$. Then*

$$r_e(U(t,\sigma)) = r_e(\widehat{T}(t-\sigma)) = \exp(\hat{\tau}(t-\sigma)), \quad t > \sigma.$$

Proof. From the assumption and Proposition 2.9, it follows that $K(t,\sigma)$ is compact. Hence we have $\alpha(U(t,\sigma)^n) = \alpha(\widehat{T}(t-\sigma)^n), n = 1, 2, \cdots$, which implies the formula in the proposition from the Nussbaum formula.

We are now in a position to give criteria for the existence of periodic solutions of Equation (2.155).

Theorem 2.32 *Assume that at least one of the following conditions is satisfied:*

i) $(T(t))_{t\geq 0}$ is a compact C_0-semigroup on E,

ii) $L(t,\cdot)$ is a compact operator for each $t \in \mathbf{R}$ and $\tau^\nu < 0$,

iii) $(T(t))_{t\geq 0}$ is a compact C_0-semigroup for $t > t_0$ on E and $L(t,\cdot)$ is a compact operator for each $t \in \mathbf{R}$.

iv) $(T(t))_{t\geq 0}$ is a C_0-semigroup on E such that $T(t)x \in \mathcal{D}(A)$ for all $x \in E$ and $t > 0$, $L(t,\cdot)$ is a compact operator for each $t \in \mathbf{R}$, and $\tau < 0$.

Then 1 is a normal point of $U(\omega, 0)$, and hence, the following results hold true:

i) In the case where $1 \in \rho(U(\omega,0))$, Equation (2.155) has a unique ω-periodic solution.

ii) *In the case where 1 is a normal eigenvalue of $U(\omega, 0)$, if Equation (2.155) has an bounded solution, then $S_L(\omega) \neq \emptyset$ and $\dim S_L(\omega)$ is finite.*

Proof. From Theorem 2.31, Proposition 2.20 and Theorem 2.27 follows easily the conclusion of the theorem.

Finally, based on Theorem 2.23 we will give another proof of the assertion i) of the above theorem.

Lemma 2.32 *Assume that $(T(t))_{t\geq 0}$ is a compact C_0-semigroup and that Equation (2.155) has a bounded solution $u(t) := u(t, \sigma, \phi, f)$. Then the set $O := \{u(t) : t \geq \sigma\}$ is relatively compact in E, $u(t)$ is uniformly continuous for $t \geq \sigma$, and the set $\mathcal{O} := \{u_t : t \geq \sigma\}$ is relatively compact in $\mathcal{C}$.*

Proof. For the sake of simplicity of notations, we assume $\sigma = 0$, and set $F(t, \phi) := L(t, \phi) + f(t)$. Let $h > 0$. Since $u(t, 0, \phi, f) = u(t, t - h, u_{t-h}(0, \phi, f), f)$ whenever $t \geq h$, it follows that, for $t \geq h$,

$$u(t) = T(h)u(t-h) + \int_{t-h}^{t} T(t-s)F(s, u_s)ds.$$

Since $T(h)$ is compact and $\{u(t-h) : t \geq h\}$ is bounded, the set $\{T(h)u(t-h) : t \geq h\}$ is relatively compact. Since $u(t)$ is bounded, there exists a $B \geq 0$ such that $|F(s, u_s)| \leq B$ for $s \geq 0$. Thus the norm of the above integral term is not greater than $h\gamma_h B$, where $\gamma_h := \sup\{\|T(t)\| : 0 \leq t \leq h\}$. Hence we have that $\alpha(O_h) \leq 2h\gamma_h B$, where $O_h := \{u(t) : t \geq h\}$. Since the diameter of the set $O \setminus O_h$ converges to zero as $h \to 0$, we have that $\alpha(O) = 0$; hence $\overline{O}$ is compact.

As computed above, we have that

$$|u(t) - u(t-h)| \leq |T(h)u(t-h) - u(t-h)| + h\gamma_h B.$$

Since $T(t)$ is uniformly continuos on compact sets, $|u(t-h) - T(h)u(t-h)| \to 0$ as $h \to 0$ uniformly for $t > 0$. Hence we have the second assertion in the lemma. In addition the third assertion follows from this and Lemma 4.2.

Theorem 2.33 *Assume that $(T(t))_{t\geq 0}$ is a compact C_0-semigroup and that the equation (2.155) has a bounded solution. Then it has an ω-periodic solution.*

Proof. Let $u(t, \sigma, \phi^0, f)$ be a bounded solution. Take an integer k such that $k\omega > \sigma$ and define $v(t) := u(t + k\omega, \sigma, \phi^0, f)$ for $t \geq -r$. Since for $t \geq -r$,

$$u(t + k\omega, \sigma, \phi^0, f) = u(t + k\omega, k\omega, u_{k\omega}(\sigma, \phi^0, f), f) = u(t, 0, u_{k\omega}(\sigma, \phi^0, f), f),$$

$v(t)$ is a bounded solution satisfying the equation for $t \geq 0$. Hence we can assume that $\sigma = 0$. Define $T\phi = P\phi + \psi$ for $\phi \in \mathcal{C}$ as before. Set $\mathcal{D} = \{T^n \phi^0 : n = 0, 1, 2, \cdots\}$. Then $T\mathcal{D} \subset \mathcal{D}$ and T is a continuous, linear affine map. Notice $T^n \phi^0 = u_{n\omega}(0, \phi^0, f)$ for $n = 1, 2, \cdots$. Since $\{u_t : t \geq 0\}$ is relatively compact as proved above, Theorem 2.23 implies the conclusion of the theorem.

2.4.6. Uniqueness of Periodic Solutions I

In this subsection we will apply Proposition 2.12 to study the existence of bounded solutions, the existence and uniqueness of ω-periodic solutions to Equation (2.155). First, we state a general result on the uniqueness of ω-periodic solutions to Equation (2.155), which follows from Proposition 2.12 and Theorem 2.30.

Theorem 2.34 *Assume that $I - T(\omega) \in \Phi_+(E)$ and $\dim N(I - T(\omega)) = 0$. Let c_0 be a constant such that*

$$|x| \leq c_0 |(I - T(\omega))x| \quad \text{for all } x \in E.$$

Assume further that the inequality

$$\|K(\omega)\| \leq 1/2(p + m_\omega c_0)$$

is satisfied. Then, if Equation (2.155) has a bounded solution, it has a unique ω-periodic solution.

For each $\phi \in \mathcal{C}$ take the space $BC(\phi)$, the set of bounded, continuous functions $x : [0, \infty) \to E$ such that $x(0) = \phi(0)$. This is a complete metric space with the metric $d(x, y) = \|x - y\|_\infty := \sup\{|x(t) - y(t)| : t \geq 0\}$. Define an operator F_ϕ on $BC(\phi)$ by

$$(F_\phi x)(t) = T(t)\phi(0) + \int_0^t T(t-s)\left(L(s, \tilde{x}_s) + f(s)\right) ds, \quad t \geq 0, \tag{2.164}$$

where $\tilde{x}(t) = \phi(t)$ for $-r \leq t \leq 0$ and $\tilde{x}(t) = x(t)$ for $t \geq 0$. Set

$$\|f\|_\infty = \sup\{|f(t)| : t \geq 0\}.$$

Proposition 2.21 *Suppose that $\|L\|_\infty := \sup_{t \geq 0} \|L(t)\|$ and $\|f\|_\infty$ are finite, and that $(T(t))_{t \geq 0}$ is a C_0-semigroup on E such that there exist $M_w, w > 0$ for which $\|T(t)\| \leq \tilde{M}_w e^{-wt}$ for $t \geq 0$. If $M_w \|L\|_\infty < w$, then every solution of Equation (2.155) is bounded.*

Proof. It suffices to show that F_ϕ is a contraction for each $\phi \in \mathcal{C}$. It is obvious that $(F_\phi x)(t)$ is continuous for $t \geq 0$. Since $|\tilde{x}_s| \leq \max\{\|x\|_\infty, |\phi|\}, x \in BC(\phi)$, and since

$$|F_\phi x(t)| \leq M_w e^{-wt}|\phi(0)| + \int_0^t M_w e^{-w(t-s)}\left[\|L\|_\infty \max\{\|x\|_\infty, |\phi|\} + \|f\|_\infty\right] ds,$$

we have $\|F_\phi x\|_\infty \leq M_w|\phi(0)| + M_w w^{-1}\left[\|L\|_\infty \max\{\|x\|_\infty, |\phi|\} + \|f\|_\infty\right]$; that is, $F_\phi x \in BC(\phi)$. In the similar manner, we have also that $\|F_\phi x - F_\phi y\|_\infty \leq w^{-1} M_w \|L\|_\infty \|x - y\|_\infty$. Therefore, if $M_w \|L\|_\infty < w$, then F_ϕ is a contraction on $BC(\phi)$, and has a unique fixed point, z, in $BC(\phi)$. Then $\tilde{z}$ is an bounded solution of Equation (2.155).

We recall the function $k(t)$ in Proposition 2.8 :

$$k(t) = M_w^2 \|L\|_\infty (1 - e^{-wt})/(w - M_w\|L\|_\infty) \text{ for } t \geq 0.$$

The following result is derived from Proposition 2.12.

Theorem 2.35 *Suppose that* $\|T(t)\| \leq M_w e^{-wt}$ *for* $t \geq 0$. *Let* $c > 0$ *be the constant such that* $|\phi| \leq c|(I - \widehat{T}(\omega))\phi|$ *for all* $\phi \in \mathcal{C}$. *If the condition*

$$2ck(\omega) < 1 \text{ and } w > M_w\|L\|_\infty,$$

is satisfied, then Equation (2.155) has a unique ω-periodic solution v, and

$$\|v\|_\infty \leq \frac{M_w}{w - M_w\|L\|_\infty}\|f\|_\infty. \tag{2.165}$$

Proof. From the assumption on $\|T(t)\|$ and Proposition 2.17 it follows that $1 \in \rho(\widehat{T}(\omega))$, which means that the constant c in the theorem exists. Since $w - M_w\|L\|_\infty$ is positive, Equation (2.155) has bounded solutions from Proposition 2.21. To show the existence of ω-periodic solutions of Equation (2.155), we will estimate $\|K(\omega)\|$. From Proposition 2.8 and the condition in the theorem, we have

$$\|K(\omega)\| \leq k(t) < \frac{1}{2c}.$$

On the other hand, since the type number of $(T(t))_{t \geq 0}$ is negative, by Proposition 2.14,

$$dim N(I - \widehat{T}(\omega)) = dim N(I - T(\omega)) = 0.$$

Therefore, by Proposition 2.12, Equation (2.155) has a unique ω-periodic solution, $v(t)$.

Set $\psi = v_0$. Then v is the fixed point of F_ψ defined by (2.164). Obviously, $|v_s| \leq \|v\|_\infty$ for $s \geq 0$. Then we have that $|L(s, v_s)| \leq \|L\|_\infty \|v\|_\infty$, and that $|v(t)| = |F_\psi v(t)| \leq M_w e^{-wt}|\psi(0)| + M_w w^{-1}(\|L\|_\infty \|v\|_\infty + \|f\|_\infty)$ for $t \geq 0$. Hence, for $t \geq 0$ and for $n = 1, 2, \cdots$,

$$|v(t)| = |v(t + n\omega)| \leq M_w e^{-w(t+n\omega)}|\psi(0)| + M_w w^{-1}(\|L\|_\infty \|v\|_\infty + \|f\|_\infty).$$

Taking the limit as $n \to \infty$, we have that

$$|v(t)| \leq M_w w^{-1}(\|L\|_\infty \|v\|_\infty + \|f\|_\infty)$$

for $t \geq 0$, which implies that $\|v\|_\infty \leq M_w w^{-1}(\|L\|_\infty \|v\|_\infty + \|f\|_\infty)$. Hence v satisfies the inequality in the theorem and the proof is complete.

Theorem 2.36 *Let* $\|T(t)\| \leq M_w e^{-wt}$ *for* $t \geq 0$ *and let p be a nonnegative integer such that $(p-1)\omega < r \leq p\omega$. Assume that $w\omega > \log M_w$. If*

$$\frac{2(p + M_w)k(\omega)}{1 - M_w e^{-w\omega}} < 1 \text{ and } w > M_w\|L\|_\infty, \tag{2.166}$$

then Equation (2.155) has a unique ω-periodic solution.

Proof. Now we will compute the value of the constant c in Theorem 2.35. Since $w\omega > \log M_w$, we have $\|T(\omega)\| \leq M_w e^{-w\omega} < 1$, and hence,

$$\|S_M^{-1}\| = \|(I - T(\omega))^{-1}\| \leq 1/(1 - \|T(\omega)\|) \leq 1/(1 - M_w e^{-w\omega}). \tag{2.167}$$

Combining the relation (2.162) and (2.167) we get

$$\begin{aligned} c &\leq p + m_\omega \|S_M^{-1}\| \\ &\leq p + M_w/(1 - M_w e^{-w\omega}) \\ &\leq (p + M_w)/(1 - M_w e^{-w\omega}). \end{aligned}$$

Therefore, from the assumption we see that all conditions of Theorem 2.35 are satisfied, and the proof is complete.

Remark 2.22 Notice that the above theorem can be shown by applying Theorem 2.34. We also notice that in stead of (2.166) we can impose a little weaker condition

$$\frac{2(p - pM_w e^{-w\omega} + M_w)k(\omega)}{1 - M_w e^{-w\omega}} < 1 \text{ and } w > M_w \|L\|_\infty.$$

Remark 2.23 Let $\|T(t)\| \leq e^{-wt}, w > 0$, in Theorem 2.36. Then (2.166) becomes

$$\frac{2(p+1)\|L\|_\infty}{w - \|L\|_\infty} < 1 \text{ and } w > \|L\|_\infty.$$

2.4.7. Uniqueness of Periodic Solutions II

This subsection is devoted to applications of Proposition 2.13 to the study of criteria for the uniqueness of periodic solutions to Equation (2.155).

Theorem 2.37 *Assume that $1 \in \rho(T(\omega))$. Then, Equation (2.155) has a unique ω-periodic solution provided that*

$$1 \in \rho((I - \widehat{T}(\omega))^{-1} K(\omega)). \tag{2.168}$$

Proof. From Proposition 2.17 we have that if $1 \in \rho(T(\omega))$, then $1 \in \rho(\widehat{T}(\omega))$. The theorem now follows from Proposition 2.13.

We now find a sufficient condition for the condition (2.168).

Lemma 2.33 *Assume that $\|\widehat{T}(\omega)\| < 1$. Then, the condition (2.168) holds provided that*

$$\|K(\omega)\| < 1 - \|\widehat{T}(\omega)\|. \tag{2.169}$$

Proof. From the condition (2.169) we have

$$\|((I - \widehat{T}(\omega))^{-1}K(\omega)\| \leq \|((I - \widehat{T}(\omega))^{-1}\|\|K(\omega)\|$$
$$\leq \frac{\|K(\omega)\|}{1 - \|\widehat{T}(\omega)\|} < 1.$$

This fact implies the condition (2.168).

Lemma 2.34 *Let* $\|T(t)\| \leq M_w e^{-wt}, w > 0$. *Assume that* $\omega > r$ *and* $w(\omega - r) > \log M_w$. *Then the inequality (2.169) in Lemma 2.33 can be replaced by the following inequality:*

$$\frac{\|K(\omega)\|}{1 - M_w e^{-w(\omega - r)}} < 1. \tag{2.170}$$

Proof. From assmptions and the estimate (2.152) of $\widehat{T}(t)$ we have $\|\widehat{T}(\omega)\| \leq M_w e^{-w(\omega - r)} < 1$. Hence, the inequality (2.169) follows from the inequality (2.170). This proves the lemma.

Combining Lemma 2.34 with Proposition 2.8, we can obtain the following result.

Theorem 2.38 *Assume that* $\|T(t)\| \leq M_w e^{-wt}$ *for* $t \geq 0$, $\omega > r$ *and* $w(\omega - r) > \log M_w$. *Futhermore, assume that*

$$\frac{k(\omega)}{1 - M_w e^{-w(\omega - r)}} < 1 \quad \text{and} \quad w > M_w \|L\|_\infty.$$

Then Equation (2.155) has a unique ω-periodic solution.

In general, Theorem 2.36 and Theorem 2.38 are independent of each other, which will be shown in Remark 2.24. Hence, summarizing those results, we have the following result.

Theorem 2.39 *Suppose that* $\|T(t)\| \leq M_w e^{-wt}$ *for* $t \geq 0$, $\omega > r$, $w > M_w \|L\|_\infty e^{wr}$ *and* $w(\omega - r) > \log M_w$. *Moreover, assume that*

$$\min\{\frac{2(p + M_w)}{1 - M_w e^{-w\omega}}, \frac{1}{1 - M_w e^{-w(\omega - r)}}\}k(\omega) < 1. \tag{2.171}$$

Then Equation (2.155) has a unique ω-periodic solution.

2.4.8. An Example

In this subsection, we will consider the existence of periodic solutions of a partial differential-integral equation as an example of applications of Theorem 2.36 and Theorem 2.38.

Denote by $E = C[-\infty, \infty]$, the space of all continuous real valued functions $u(x)$, defined on $(-\infty, \infty)$, such that $\lim_{x \to -\infty} u(x)$ and $\lim_{x \to +\infty} u(x)$ exist. Then, with norm $\|u\| = \sup_{-\infty < x < +\infty} |u(x)|$, E becomes a Banach space.

We consider the following initial value problem

$$\frac{\partial u(t,x)}{\partial t} = \frac{\partial^2 u(t,x)}{\partial x^2} - \alpha u(t,x) + b(t,x) \int_{t-r}^{t} e^{-c(t-s)} u(s,x) ds$$
$$+ f(t,x), \quad (2.172)$$

$$u(\theta, x) = \phi(\theta, x), -r \leq \theta \leq 0, \phi \in \mathcal{C}.$$

It is well known that the linear operators A and A_0, defined by

$$Au = \frac{d^2 u}{dx^2} - \alpha u \text{ for } u \in \mathcal{D}(A), \quad A_0 u = \frac{d^2 u}{dx^2} \text{ for } u \in \mathcal{D}(A_0)$$

and

$$\mathcal{D}(A) = \mathcal{D}(A_0) = \{u \in E : \frac{d}{dx} u, \frac{d^2}{dx^2} u \in E\},$$

are the infinitesimal generators of C_0-semigroups $T(t)$ and $T_0(t)$ on E, respectively, cf.[63, Chapter VIII]. Hence,

$$\|T_0(t)\| = 1, T(t) = e^{-\alpha t} T_0(t), \text{ and } \|T(t)\| = e^{-\alpha t}$$

for all $t \geq 0$.

Assume that

i) (C-1) $\alpha > 0$ and $c > 0$,

ii) (C-2) $b(t,x)$ and $f(t,x) : \mathbf{R} \times \mathbf{R} \to \mathbf{R}$ are continuous and ω-periodic functions in t such that $b(t, \cdot), f(t, \cdot) \in E, t \in \mathbf{R}$.

Put $\|b(t)\| = \sup_{-\infty < x < \infty} |b(t,x)|, \|b\|_\infty = \sup_{0 \leq t \leq \omega} \|b(t)\|$. Similarly, we define $\|f(t)\|$ and $\|f\|_\infty$ for $f(t,x)$. Set

$$B(t, \phi)(x) = b(t,x) \int_{-r}^{0} e^{c\theta} \phi(\theta, x) d\theta, \quad \phi \in \mathcal{C}.$$

Then we have

$$|B(t,\phi)(x)| \leq \|b(t)\| \int_{-r}^{0} e^{c\theta} |\phi| \, d\theta \leq \frac{\|b(t)\|}{c} |\phi|,$$

and hence, $\|B\|_\infty \leq \|b\|_\infty / c$. Therefore, from Theorem 2.36 and the estimate (2.165) we have the following result.

Theorem 2.40 *Asssume that the conditions (C-1) and (C-2) are satisfied. Let p be a positive integer such that $(p-1)\omega < r \leq p\omega$. If*

$$\frac{2(p+1)\|b\|_\infty}{c\alpha - \|b\|_\infty} < 1 \text{ and } c\alpha > \|b\|_\infty, \tag{2.173}$$

then Equation (2.172) has a unique ω-periodic solution v, and

$$\|v\|_\infty \leq \frac{c}{c\alpha - \|b\|_\infty}\|f\|_\infty. \tag{2.174}$$

The following theorem follows immediately from Theorem 2.38.

Theorem 2.41 *Asssume that $\omega > r$ and the conditions (C-1) and (C-2) are satisfied. If*

$$\frac{\|b\|_\infty(1 - e^{-\alpha\omega})}{(1 - e^{-\alpha(\omega-r)})(c\alpha - \|b\|_\infty)} < 1 \text{ and } c\alpha > \|b\|_\infty, \tag{2.175}$$

then Equation (2.172) has a unique ω-periodic solution v with the estimate (2.174).

Remark 2.24 For simplicity, let $\omega > r$, $c = 1$ and $\alpha = 1 > \|b\|_\infty$. Then we compare the condition (2.173) in Theorem 2.40 with the condition (2.175) in Theorem 2.41. Since $\omega > r$, we have $p = 1$. Hence, using the condition (2.171) in Theorem 2.39 we can obtain the following fact : if

$$\|b\|_\infty < \max\{\frac{1}{4+1}, \frac{1}{\frac{1-e^{-\omega}}{1-e^{-(\omega-r)}}+1}\},$$

then Equation (2.172) has a unique ω-periodic solution. However, observe that

$$\lim_{\omega\downarrow r}\frac{1-e^{-\omega}}{1-e^{-(\omega-r)}} = +\infty > 4;$$

and

$$\lim_{\omega\to+\infty}\frac{1-e^{-\omega}}{1-e^{-(\omega-r)}} = 1 < 4.$$

This shows that the condition (2.173) in Theorem 2.40 and the condition (2.175) in Theorem 2.41 are independent of each other.

2.4.9. Periodic Solutions in Equations with Infinite Delay

Consider the periodic equation with infinite delay on the phase space $\mathcal{B}$:

$$u'(t) = Au(t) + L(t, u_t) + f(t). \tag{2.176}$$

$L(t, \phi), f(t)$ has the same periodic property as in the previous section, but $\phi \in \mathcal{B}$. In the following, we say that a solution $u(t)$ is E-bounded if the E-norm $|u(t)|$ is bounded in $t \geq 0$. Theorem 2.33 is extended to Equation (2.176) as follows.

Theorem 2.42 *Let $\mathcal{B}$ be a fading memory space. Assume that $T(t)$ is compact for $t > 0$ and that the equation (2.176) has an E-bounded solution. Then it has an ω-periodic solution.*

We need the following axiom (D) to apply the fixed point theorem by Chow and Hale together with the semi-Fredholm property of $\widehat{T}(\omega)$ on the phase space $\mathcal{B}$.

(D) $|\phi^1 - \phi^2|_\mathcal{B} = 0$ for $\phi^1, \phi^2 \in \mathcal{B}$ if and only if $\phi^1(\theta) = \phi^2(\theta)$ for $\theta \in (-\infty, 0]$.

The operator $\widehat{T}(t), t \geq 0$, on $\mathcal{B}$ is defined in the similar manner to the case of finite delay, but the variable θ is taken in the interval $(-\infty, 0]$.

Lemma 2.35 *If $(T(t))_{t \geq 0}$ is a C_0-semigroup on E, $N(I - \widehat{T}(\omega))$ consists of ω-periodic continuous functions ϕ given by*

$$\phi(\theta) = T(\theta + n\omega)\phi(0), \theta \in [-n\omega, 0], n = 1, 2, \cdots,$$

where $\phi(0) \in N(I - T(\omega))$, and

$$\dim N(I - \widehat{T}(\omega)) = \dim N(I - T(\omega)).$$

Denote by $\mathcal{S}_L(\omega)$ the set of ω-periodic solutions for Equation (2.176). Define the operator $K(t)$ on the space $\mathcal{B}$ by Equation(2.153), in the same way as in the finite delay case.

Theorem 2.43 *Suppose that $\mathcal{B}$ is a fading memory space with Axioms (C), (D), $I - T(\omega) \in \Phi_+(E)$ and that there exists a positive constant c such that*

$$\|[\phi]\| \leq c|(I - \widehat{T}(\omega))\phi|_\mathcal{B} \quad (2.177)$$

for $\phi \in \mathcal{B}$. Let $n = \dim N(I - T(\omega))$. If $\|K(\omega)\| < 1/2c(1 + \sqrt{n})$, and if Eq.(2.176) has an E-bounded solution, then $\mathcal{S}_L(\omega) \neq \emptyset$ and $\dim \mathcal{S}_L(\omega) \leq n$.

For the general fading memory space, Condition (2.177) is assumed for $R(I - \widehat{T}(\omega))$ to be closed. If we assume more conditions on the space $\mathcal{B}$, the semi-Fredholm propery of $I - T(\omega)$ is inherited as follows.

Theorem 2.44 *If $\mathcal{B} = UC_g$ is a uniform fading memory space and if $I - T(\omega) \in \Phi_+(E)$, then $I - \widehat{T}(\omega) \in \Phi_+(\mathcal{B})$.*

Theorem 2.45 *Assume that $\mathcal{B} = UC_g$ is a uniform fading memory space and at least one of the following conditions is satisfied :*

i) $(T(t))_{t \geq 0}$ is a compact C_0-semigroup on E.

ii) $L(t, \cdot)$ is a compact operator for each $t \in \mathbf{R}$ and $I - T(\omega) \in \Phi_+(E)$.

If Equation(2.176) has an E-bounded solution, then it has an ω-periodic solution.

Theorem 2.46 *Suppose that UC_g is a uniform fading memory space, $I - T(\omega) \in \Phi_+(E)$ such that Condition (2.159) holds and that $\|T(t)\| \leq M_w e^{wt}$ for $t \geq 0$. Let $n := \dim N(I - T(\omega))$. Let M_0, ε_0 be the positive constants such that $\|S_0(t)\| \leq M_0 e^{-\varepsilon_0 t}$ for $t \geq 0$. Set $J = \sup_{\theta \leq 0} 1/g(\theta)$, and*

$$c = M_0/(1 - e^{-\varepsilon_0 \omega}) + \sup\{\|T(t)\| : 0 \leq t \leq \omega\} c_0 (1 + \sqrt{n}),$$

where c_0 is the constant in Condition (2.159).

Suppose that $\|L\|_\infty := \sup\{\|L(t)\| : 0 \leq t \leq \omega\}$ satisfies the condition

$$JM_w \|L\|_\infty \int_0^\omega e^{(a+w_+)(\omega-s)} \|\widehat{T}(s)\| ds < 1/2c(1 + \sqrt{n}),$$

where $w_+ = \max\{w, 0\}$, and that Equation (2.176) has an E-bounded solution. Then $S_L(\omega) \neq \emptyset$ and $\dim S_L(\omega) \leq n$.

As in the case of finite delay, these results are applicable to the existence and uniqueness of periodic solutions of Equation (2.176). See [206] for the details.

2.5. BOUNDEDNESS AND ALMOST PERIODICITY IN DISCRETE SYSTEMS

Let us consider the difference equation

$$x_{n+1} = Ax_n + f_n, n \in \mathbf{Z}, \tag{2.178}$$

where A is a bounded linear operator on a Banach space $\mathbf{X}$, $f_n, n \in \mathbf{Z}$ is a sequence in $\mathbf{X}$ and is almost periodic.

We will study spectral criteria for the existence of almost periodic solutions for Eq.(2.178) and then apply them to study similar problems for evolution equations of the form

$$\frac{dx}{dt} = A(t)x + f(t), \tag{2.179}$$

where $A(t)$ is a (in general unbounded) linear operator on $\mathbf{X}$ which is periodic and f is almost periodic. Similar applications will be made for functional differential equations with infinite delay

$$\frac{dx}{dt} = Ax + F(t)x_t + f(t), \tag{2.180}$$

where A is the infinitesimal generator of a C_0-semigroup of linear operators, $F(t)$ is a periodic family of continuous linear operators, x_t is segment of the solution $x(\cdot)$.

In the simplest case our problem is concerned with the existence of fixed points of affine operators which arises naturally when one considers the existence of periodic solutions of Eq.(2.179) using its monodromy operator (with f periodic). In this case, it is well known that there is a close connection between the boundedness and periodicity which is sometimes called Massera-type problem.

One of the main advantages of the approach proposed in this section is that it is simple and does not make use of sophisphicated techniques as in the previous sections. In principle, this method can be applied to various kinds of functional equations.

The main results of this section are stated in Theorems 2.47, 2.48, 2.49, 2.51. To prove them we will first consider the notion of spectrum of bounded sequences. As is shown in next subsection we have to introduce a little stronger notion of spectrum to extend our results in Sections 2.1., 2.3. to discrete systems (Theorems 2.47, 2.48). In Theorem 2.49 we will apply our results obtained in subsection 3 to study the Massera-type problem of periodic solutions to functional differential equations with infinite delay. On the other hand Theorem 2.51 will give a spectral criterion for the existence of almost periodic solutions to evolution equations.

2.5.1. Spectrum of Bounded Sequences and Decomposition

This subsection will be devoted entirely to the notion of spectrum of bounded sequences. Almost all results of this subsection can be proved in the same way as in Chapter 1. We will discuss only some particular points which allow us to prove analogs of the results in Sections 2.1. and 2.2..

We will denote by $l_\infty(\mathbf{X})$ the space of all two-sided sequences with sup-norm, i.e.,

$$l_\infty(\mathbf{X}) := \{(x_n)_{n\in\mathbf{Z}} : x_n \in \mathbf{X}, \sup_{n\in\mathbf{Z}} \|x_n\| < \infty\}.$$

First we will make precise the definition of the spectrum of a bounded sequence $g := \{g_n\}_{n\in\mathbf{Z}}$ in $\mathbf{X}$ used in this section. Recall that the set of all bounded sequences in $\mathbf{X}$ forms a Banach space $l_\infty(\mathbf{X})$ with norm $\|g\| := \sup_n \|g_n\|_\mathbf{X}$. We will denote by $S(k)$ the k-translation in $l_\infty(\mathbf{X})$, i.e., $(Sg)_n = g_{n+k}, \forall g, n$.

Definition 2.16 *The subset of all λ of the unit circle $\Gamma := \{z \in \mathbf{C} : |z| = 1\}$ at which*

$$\hat{g}(\lambda) := \begin{cases} \sum_{n=0}^\infty \lambda^{-n-1} S(n)g, & \text{for } |\lambda| > 1, \\ -\sum_1^\infty \lambda^{n-1} S(-n)g, & \text{for } |\lambda| < 1 \end{cases}$$

has no holomorphic extension, is said to be the spectrum of the sequence $g := \{g_n\}_{n\in\mathbf{Z}}$ and will be denoted by $\sigma(g)$ [3].

We list below some properties of spectrum of $\{g_n\}$.

Proposition 2.22 *Let $g := \{g_n\}$ be a two-sided bounded sequence in $\mathbf{X}$. Then the following assertions hold:*

i) $\sigma(g)$ is closed,

---
[3] In Definition 2.16 our notion of spectrum is somewhat stronger than that in [217].

ii) If g^n is a sequence in $l_\infty(\mathbf{X})$ converging to g such that $\sigma(g^n) \subset \Lambda$ for all $n \in \mathbf{N}$, where Λ is a closed subset of the unit circle, then $\sigma(g) \subset \Lambda$,

iii) If $g \in l_\infty(\mathbf{X})$ and A is a bounded linear operator on the Banach space $\mathbf{X}$, then $\sigma(Ag) \subset \sigma(g)$, where the sequence $(Ag)_n := Ag_n, \forall n \in \mathbf{Z}$.

iv) Let the space $\mathbf{X}$ not contain any subspace which is isomorphic to c_0 and $x \in l_\infty(\mathbf{X})$ be a sequence such that $\sigma(x)$ is countable. Then x is almost periodic.

Proof. i) From the definition it is obvious that the set of regular points (at which $\hat{g}(z)$ has analytic continuation) is open. Hence, $\sigma(g)$ is closed.
ii) The proof can be taken from that of [185, Theorem 0.8, pp.21-22]. In fact, from the assumption, for every positive ε

$$\lim_{n \to \infty} \sum_{k=-\infty}^{\infty} e^{-\varepsilon|k|} \|g_k^n - g_k\| = 0. \tag{2.181}$$

This yields that $\widehat{g_n}(\lambda) \to \hat{g}(\lambda)$ as $n \to \infty$ uniformly on every compact subsets of $\mathbf{C}\backslash S^1$ (S^1 denotes the unit circle). From the uniform convergence of g_n to g, without loss of generality we can assume that $\sup_n \|g_n(k) - g(k)\| \leq \|g\|$. Hence, we can assume that $\|g_n\| \leq 2\|g\|, \forall n$. Thus

$$|\widehat{g_n}(\lambda)| \leq \{ 2\|g\|/(|\lambda|-1), \forall |\lambda| > 1, 2\|g\|/(1-|\lambda|), \forall |\lambda| < 1 = \frac{2\|g\|}{|1-|\lambda||}. \tag{2.182}$$

Now let $\rho_0 \in (\mathbf{C}\backslash\Lambda)$. Since Λ is closed, obviously that $dist(\rho_0, \Lambda) > 0$. Thus we can choose $r > 0$ such that $0 < r < min(dist(\rho_0, \Lambda), 1/4)$. By assumption, $\rho_0 \notin \sigma(g_n), \forall n$. This means that $\widehat{g_n}(\lambda)$ is analytic in $B_r(\rho_0)$ for all n. Using exactly the argument of the proof of [185, Proposition 0.8, p. 21] for the sequence $\widehat{F_n}(\xi) := \widehat{g_n}(e^\xi)$, we can show that

$$|\widehat{g_n}(\lambda)| \leq M, \forall n \in \mathbf{N}, \lambda \in \mathcal{U},$$

where $\mathcal{U}$ is a neiborhood of ρ_0. Thus, by Montel's Theorem (see e.g. [48, p. 149]) the family $\widehat{g_n}(\lambda)|_\mathcal{U}$ is normal, i.e. every subsequence of it contains a subsubsequence which converges in $C(\mathcal{U}, l_\infty(\mathbf{X}))$. Obviously, since this sequence converges to $\hat{g}(\lambda)$ pointwise, the limit function here should be $\hat{g}(\lambda)$. Since $H(\mathcal{U})$ is closed in $C(\mathcal{U}, l_\infty(\mathbf{X}))$ and since $\widehat{g_n}(\lambda)$ is analytic in $\mathcal{U}$ this shows that $\hat{g}(\lambda)$ is analytic in $\mathcal{U}$. Hence $\rho_0 \notin \sigma(g)$, so $\sigma(g) \subset \Lambda$.
iii) The assertion is obvious.
iv) The proof can be done in the same way as in [137, Chap. 6] or [8, Section 3].

In view of Proposition 2.22 if Λ is a closed subset of the unit circle, then the set of all bounded sequences in $g \in l_\infty(\mathbf{X})$ such that $\sigma(g) \subset \Lambda$ forms a closed subspace of $l_\infty(\mathbf{X})$ which we will denote by $\Lambda(\mathbf{Z}, \mathbf{X})$.

The following lemma will be used in the next subsections. We will denote by $\mathcal{M}_g$ the closure of the subspace of $l_\infty(\mathbf{X})$ spanned by all elements $S(n)g, n \in \mathbf{Z}$.

Lemma 2.36 *We can define the spectrum of a given sequence* $x = \{x_n\}$ *as* $\sigma(S|_{\mathcal{M}_x})$.

Proof. First note that $\mathcal{M}_x$ is invariant under all translations $S(k)$. It is easy to see that $\|S|_{\mathcal{M}_x}\| = \|S^{-1}|_{\mathcal{M}_x}\| = 1$. Hence $\sigma(S|_{\mathcal{M}_x}) \subset \Gamma$. We will use the following indentity

$$(I - A)^{-1} = \sum_{n=0}^{\infty} A^n, \qquad (2.183)$$

for any bounded linear operator A such that $\|A\| < 1$. By assumptions and by definition for $|z| > 1$ we have

$$\begin{aligned}
\hat{x}(z) &= \sum_{n=0}^{\infty} z^{-n-1} S(n) x \\
&= z^{-1} \sum_{n=0}^{\infty} z^{-n} S(n) x \\
&= z^{-1}(I - z^{-1} S)^{-1} x \\
&= (z - S)^{-1} x. \qquad (2.184)
\end{aligned}$$

Similarly, for $|z| < 1$, by using in addition the indentity $I - (I - zS^{-1})^{-1} = (I - z^{-1}S)^{-1}$ we can show that

$$\hat{x}(z) = (z - S)^{-1} x.$$

Thus it is obvious that if $z_0 \in \rho(S|_{\mathcal{M}_x})$, then z_0 is a regular point of x. Conversely, suppose that z_0 is a regular point of $\hat{x}$. We will prove that the mapping $z_0 - S|_{\mathcal{M}_x} : \mathcal{M}_x \to \mathcal{M}_x$ is one-to-one and onto to establish $z_0 \in \rho(S|_{\mathcal{M}_x})$. From the above calculation it follows that for any $y \in \mathcal{M}_x$,

$$\hat{y}(z) = (z - S|_{\mathcal{M}_x})^{-1} y \text{ or } (z - S|_{\mathcal{M}_x}) \hat{y}(z) = y$$

whenever $|z| \neq 1$. Hence, for every $y \in \mathcal{M}_x$ we get $(z - S|_{\mathcal{M}_x}) \hat{y}(z) = y$ on $\mathcal{U}$ (because of the analyticity of the function $\hat{y}(z)$ on $\mathcal{U}$). In particular, the mapping $z_0 - S|_{\mathcal{M}_x} : \mathcal{M}_x \to \mathcal{M}_x$ is "onto". Furthermore, we show that this mapping is one-to-one. Indeed, if $(z_0 - S|_{\mathcal{M}_x}) a = 0$ for an $a = \{a(n)\} \in \mathcal{M}_x$, then $a(n+1) = z_0 a(n) \; \forall n \in \mathbf{Z}$, and hence, $a(n) = z_0^n a(0) \; \forall n \in \mathbf{Z}$. Then

$$\hat{a}(z) = \frac{a}{z - z_0} \qquad (\forall |z| \neq 1).$$

Since $\hat{a}$ is analytic in $\mathcal{U}$ because of $a \in \mathcal{M}_x$, we have $a = 0$, as required. This shows that $z_0 \in \rho(S|_{\mathcal{M}_x})$, completing the proof of the lemma.

Corollary 2.22 *Let* $x = \{x_n\}$ *be an element of* $l_\infty(\mathbf{X})$ *such that* $x_n = x_{n+1} = c \neq 0, \forall n \in \mathbf{Z}$ *if and only if* $\sigma(x) = \{1\}$. *Similarly,* $x \in l_\infty(\mathbf{X})$ *such that* $x_n = -x_{n+1} \neq 0, \forall n \in \mathbf{Z}$ *if and only if* $\sigma(x) = \{-1\}$.

Proof. If $x_n = c \neq 0, \forall n \in \mathbf{Z}$, then it is easy to compute $\sigma(x) = \{1\}$. Conversely, let $\sigma(x) = \{1\}$. Then by Lemma 2.36, $\sigma(S_1) = \{1\}$, where S_1 is the restriction of S to $\mathcal{M}_x$. In view of Gelfand's Theorem $S_1 = I_{\mathcal{M}_x}$ which completes the proof. For the second assertion note that in this case $x_n = x_{n+2}$. Using the previous argument for $S(2)$ we get the assertion.

Corollary 2.23 *Let $\Lambda(\mathbf{Z}, \mathbf{X})$ denote the subspace of $l_\infty(\mathbf{X})$ consisting of all sequences x such that $\sigma(x) \subset \Lambda$ for given closed subset Λ of the unit circle. Then the translation S leaves $\Lambda(\mathbf{Z}, \mathbf{X})$ invariant and its restriction to $\Lambda(\mathbf{Z}, \mathbf{X})$ which is denoted by S_Λ has the property that*

$$\sigma(S_\Lambda) = \Lambda. \qquad (2.185)$$

Proof. For $x_0 \neq 0$ put $x_n = \lambda^n x_0, \lambda \in \Lambda$. It is easy to see that $\sigma(x) = \{\lambda\}$ and $\lambda \in \sigma(S_\Lambda)$. Now we prove the converse, i.e., $\mathbf{C}\backslash\Lambda \subset \mathbf{C}\backslash\sigma(S_\Lambda) = \rho(S_\Lambda)$. To this end, suppose that $\lambda_0 \in \mathbf{C}\backslash\Lambda$. By definition of $\Lambda(\mathbf{Z}, \mathbf{X})$, we have that for every $y \in \Lambda(\mathbf{Z}, \mathbf{X})$, $\lambda_0 \notin \sigma(y)$. To show that $\lambda_0 \in \rho(S_\Lambda)$ we will prove that for every $y \in \Lambda(\mathbf{Z}, \mathbf{X})$ there exists a unique solution $x \in \Lambda(\mathbf{Z}, \mathbf{X})$ such that

$$\lambda_0 x - Sx = y . \qquad (2.186)$$

In fact, first the existence of such a solution x is obvious in view of Lemma 2.36. Now we show the uniqueness of such a solution x. Equivalently, we show that equation $\lambda_0 w - Sw = 0$ has only the trivial solution in $\Lambda(\mathbf{Z}, \mathbf{X})$. In fact, since $w, 0$ belong to $\mathcal{M}_y$ in view of Lemma 2.36 $w = 0$ is the unique solution to the above equation. Hence, $\lambda_0 \in \rho(S_\Lambda)$.

The following result will relate the Bohr spectrum of an almost periodic function $f(t)$ and the spectrum of the sequence $f(n), n \in \mathbf{Z}$. Before stating this result we recall the Approximation Theorem saying that for every $\mathbf{X}$-valued almost periodic function $f(\cdot)$ there exists a sequence of trigonometric polynomials

$$P_n(t) = P_n := \sum_{j=1}^{N(n)} a_j e^{i\lambda_j t}, t \in \mathbf{R}$$

which converges uniformly on the real line to the function f (for more information see [137]). Obviously, every almost periodic function is bounded and uniformly continuous. Moreover, for every $\lambda \in \mathbf{R}$ the following limit

$$a(\lambda) := \lim_{T\to\infty} \frac{1}{2T} \int_{-T}^{T} e^{-\lambda \xi} f(\xi) d\xi$$

exists and there are at most countably many reals λ such that $a(\lambda) \neq 0$. We define Bohr spectrum of f as the set $\sigma_b(f) := \{\lambda : a(\lambda) \neq 0\}$ and use the following notation $\sigma(f) := \overline{e^{i\sigma_b(f)}}$. Note that [137] the exponents of the approximate trigonometric polynomials of an almost periodic function can be chosen from its Bohr spectrum.

Proposition 2.23 *Let $f(\cdot)$ be an almost periodic function in Bohr's sense on the real line with Bohr spectrum $\sigma_b(f)$. Then the spectrum of the sequence $x := \{f(n), n \in \mathbf{Z}\}$ satisfies*

$$\sigma(x) \subset \sigma(f) := \overline{e^{i\sigma_b(f)}}. \tag{2.187}$$

Proof. Let $P_n := \sum_{j=1}^{N(n)} a_j e^{i\lambda_j t}$ be a sequence of trigonometric polynomials which approximates the almost periodic function $f(\cdot)$ with $\lambda_j \in \sigma_b(f)$. Then as in the proof of Corollary 2.23 we have $\sigma(Q_n) = \{e^{i\lambda_j}, j = 1, \cdots, N(n)\}$, where $Q_n(k) := P_n(k), \forall k \in \mathbf{Z}$. Hence $\sigma(Q_n) \subset \overline{e^{i\sigma_b(f)}} := \sigma(f)$ and in view of Proposition 2.22 $\sigma(x) \subset \sigma(f)$.

2.5.2. Almost Periodic Solutions of Discrete Systems

We consider in this subsection the following equation

$$x_{n+1} = Ax_n + f_{n+1}, n \in \mathbf{Z}, x \in \mathbf{X} \tag{2.188}$$

where $A \in L(\mathbf{X})$, $f = \{f_n\} \in l_\infty(\mathbf{X})$. Re-writing (2.188) in another form we have

$$x = AS^{-1}x + f, \tag{2.189}$$

where for the sake of simplicity we denote also by A the operator of multiplication by A $\{x_n\} \mapsto \{Ax_n\}$. We are in a position to prove the following theorem in which we will denote $\sigma_\Gamma(A) = \sigma(A) \cap \Gamma$.

Theorem 2.47 *Let $\sigma(f) \cap \sigma_\Gamma(A) = \varnothing$. Then Eq.(2.188) has a unique solution $x \in l_\infty(\mathbf{X})$ such that $\sigma(x) \subset \sigma(f)$.*

Proof. Let us consider the subspace $M(f)$ of $l_\infty(\mathbf{X})$ consisting of all sequences w such that $\sigma(w) \subset \sigma(f)$. Obviously, $M(f)$ is closed and invariant under S and A. We will denote by S_f the restriction of the translation S to $M(f)$. If $\sigma(f) \cap \sigma_\Gamma(A) = \varnothing$, then $1 \notin \sigma(A) \cdot \sigma(S_f^{-1})$. Using the commutativeness of A and S_f and $\sigma(AS_f^{-1}) \subset \sigma(A) \cdot \sigma(S_f^{-1})$ we have $1 \notin \sigma(AS_f^{-1})$ (see [193, Theorem 11.23, p. 280]). This shows that there is a unique $x \in M(f)$ which is a solution to Eq.(2.188).

Remark 2.25 In the particular case when $\sigma_\Gamma(A) = \varnothing$ Eq.(2.188) has an exponential dichotomy (see e.g. [17]) and the existence of such unique solutions x can be easily proved. Here our Theorem 2.47 deals with the critical case allowing the equations under question to have no exponential dichotomy. The condition in Theorem 2.47 will be called *nonresonant* to distinguish it from that in our next theorem in which we even allow the intersection of $\sigma_\Gamma(A) \cap \Gamma$ to be nonempty.

We now consider the resonant case in which $\sigma(f) \cap \sigma_\Gamma(A)$ may not be empty. We will prove the discrete version of the spectral decomposition in Section 3 which will yield the existence of bounded solutions with specific spectral properties.

Lemma 2.37 *Let x be any bounded solution to Eq.(2.188). Then the following estimates hold true:*

i)
$$\sigma(x) \subset \sigma_\Gamma(A) \cup \sigma(f) \, . \tag{2.190}$$

ii)
$$\sigma(x) \supset \sigma(f) \, . \tag{2.191}$$

Proof. (i) A simple computation shows that
$$\hat{x}(z) = z^{-1}Ax_{-1} + z^{-1}A\hat{x}(z) + \hat{f}(z), \tag{2.192}$$

for all $|z| \neq 1$. Hence, if $z_0 \in \mathbf{C} \setminus (\sigma_\Gamma(A) \cup \sigma(f))$, then z_0 is a regular of x, i.e. (2.190) is proved.
(ii) Conversely, from (2.192) if z_0 is a regular point of $\hat{x}$, it is obvious that z_0 is also a regular point of f. Hence (2.191) is proved.

Theorem 2.48 *Let Eq.(2.188) have a bounded solution x. Moreover, let the following condition be satisfied:*
$$\sigma_\Gamma(A) \setminus \sigma(f) \text{ is closed.} \tag{2.193}$$

Then there is a bounded solution x_f to Eq.(2.188) such that $\sigma(x_f) = \sigma(f)$.

Proof. Let us denote by $\Lambda := \sigma_\Gamma(A) \cup \sigma(f)$. By Lemma 2.37, since the sets $\sigma_\Gamma(A) \setminus \sigma(f)$ and $\sigma(f)$ are disjoint compact, using the Riesz spectral projection we see that there is a projection Q in $\mathcal{M}_x$ such that
$$\sigma(QSQ) \subset \sigma(f) \, , \ \sigma((I-Q)S(I-Q)) \subset \sigma_\Gamma(A) \setminus \sigma(f). \tag{2.194}$$

Moreover, Q is invariant under S. Hence, $x = Qx + (I-Q)x$. We now show that $\sigma(Qx) \subset \sigma(f)$. In fact, we have $\mathcal{M}_{Qx} \subset ImQ$ because of the invariance of ImQ under S. On the other hand, $ImQ = Q\mathcal{M}_x \subset \mathcal{M}_{Qx}$. Thus, $\mathcal{M}_{Qx} = ImQ$ and $\sigma(Qx) = \sigma(QSQ) \subset \sigma(f)$. Actually, we have proved that the space $\Lambda(\mathbf{Z}, \mathbf{X})$ is split into the direct sum of two spectral subspaces consisting of all sequences with spectra contained in $\sigma(f)$ and $\sigma_\Gamma(A) \setminus \sigma(f)$, respectively. Note that this splitting is invariant under translations and multiplication by a bounded operator. Hence, denoting the projections by P_1, P_2, respectively, we have
$$\begin{aligned} P_1 x &= P_1(AS(-1)x + f) \\ &= AS(-1)P_1 x + P_1 f \\ &= AS(-1)P_1 x + f. \end{aligned} \tag{2.195}$$

Hence $x_f := P_1 x$ is the spectral component we are looking for.

Remark 2.26 Note that Theorem 2.48 is a discrete version of [169, Theorem 3.3].

CHAPTER 2. SPECTRAL CRITERIA

We are now in a position to find spectral criteria for the existence of periodic and almost periodic solutions to Eq.(2.179) and Eq.(2.180) which will be then applicable to differential equations. We recall that a bounded sequence x is said to be almost periodic if it belongs to the following subspace of $l_\infty(\mathbf{X})$

$$APS(\mathbf{X}) := \overline{span\{\lambda \cdot z, \lambda \in \Gamma, z \in \mathbf{X}\}} \tag{2.196}$$

Corollary 2.24 *Let the assumptions of Theorem 2.48 be satisfied. Moreover, let the spectrum $\sigma(f)$ be countable and the space $\mathbf{X}$ not contain c_0. Then, there is an almost periodic solution to Eq.(2.188).*

Proof. The corollary is an immediate consequence of Theorem 2.48 and Proposition 2.22.

Corollary 2.25 *Let the assumptions of Theorem 2.47 be satisfied. Moreover, let the sequence f be almost periodic. Then, there is an almost periodic solution x_f to Eq.(2.188) which is unique if one requires that $\sigma(x_f) \subset \sigma(f)$.*

Proof. The proof can be done in the same way as in that of Theorem 2.47 in the framework of $APS(\mathbf{X})$.

2.5.3. Applications to Evolution Equations

Although the results of the previous subsection should have independent interests, we now discuss several applications of our results to study the existence of (almost) periodic solutions to differential equations.

Boundedness and periodic solutions to abstract functional differential equations

First we will consider the Massera-type problem with respect to mild solutions of the abstract functional differential equation

$$\frac{dx}{dt} = Ax + F(t)x_t + f(t). \tag{2.197}$$

More precisely, we consider Eq.(2.197) in the following setting: If $x : (-\infty, a) \to \mathbf{X}$, then a function $x_t : (-\infty, 0] \to \mathbf{X}, t \in (-\infty, a)$, is defined by $x_t(\theta) = x(t+\theta), \theta \in (-\infty, 0]$. Let $\mathcal{B}$ be a Banach space, consisting of functions $\psi : (-\infty, 0] \to \mathbf{X}$, which satisfies some axioms listed below:

(B-1) If a function $x : (-\infty, \sigma + a) \to \mathbf{X}$ is continuous on $[\sigma, \sigma + a)$ and $x_\sigma \in \mathcal{B}$, then

  (a) $x_t \in \mathcal{B}$ for all $t \in [\sigma, \sigma + a)$ and x_t is continuous in $t \in [\sigma, \sigma + a)$;

(b) $H^{-1}|x(t)| \leq |x_t|_\mathcal{B} \leq K(t-\sigma)\sup\{|x(s)| : \sigma \leq s \leq t\} + M(t-\sigma)|x_\sigma|_\mathcal{B}$ for all $t \in [\sigma, \sigma+a)$, where $H > 0$ is constant, $K : [0,\infty) \to [0,\infty)$ is continuous, $M : [0,\infty) \to [0,\infty)$ is measurable, locally bounded and they are independent of x.

(B-2) The space $\mathcal{B}$ is complete.

We always assume that Eq.(2.197) satisfies the following hypothesis H:

(H-1) $A : \mathcal{D}(A) \subset \mathbf{X} \to \mathbf{X}$ is the infinitesimal generator of a C_0-semigroup $(T(t))_{t\geq 0}$ on $\mathbf{X}$;

(H-2) $B : R \times \mathcal{B} \to \mathbf{X}$ is continuous and $B(t,\cdot) : \mathcal{B} \to \mathbf{X}$ is linear;

(H-3) $F : R \to \mathbf{X}$ is continuous.

We recall that *mild solutions* to Eq.(2.197) are defined to be solutions to the integral equation

$$x(t) = T(t-s)x(s) + \int_s^t T(t-\xi)[F(\xi)x_\xi + f(\xi)]d\xi, \forall t \geq s. \qquad (2.198)$$

Under the hypothesis H and the above listed axioms on the phase space $\mathcal{B}$ the following Cauchy problem has always a unique solution

$$\begin{cases} x(t) = T(t-s)x(s) + \int_s^t T(t-\xi)[F(\xi)x_\xi + f(\xi)]d\xi, \forall t \geq s \\ x_s = \phi \in \mathcal{B}. \end{cases} \qquad (2.199)$$

We refer the reader to Section 4 for more information on this kind of equations as well as the problem we are concerned with. Let $F(t), f(t)$ be both periodic with period, say 2π. We call *monodromy operator* of the corresponding homogeneous equation of Eq.(2.198) the operator $V(2\pi, 0)$ which maps every $\phi \in \mathcal{B}$ to $x_{2\pi} \in \mathcal{B}$, where $x(\cdot)$ is the solution of the Cauchy problem (2.199).

Theorem 2.49 *Suppose that $F(t), f(t)$ are both periodic with the same periods, say, 2π. Furthermore, we assume that 1 is an isolated point of the part of the spectrum on the unit circle of the monodromy operator corresponding to the homogeneous equation of Eq.(2.198). Then, Eq.(2.197) has a mild 2π-periodic solution if and only if it has a mild solution bounded on the whole line.*

Proof. It is easy to see that Eq.(2.198) gives rise to the following equation

$$u_t = V(t, t-2\pi)u_{t-2\pi} + g_t, \forall t \in \mathbf{R}, \qquad (2.200)$$

where $V(t,s)$ is the evolution operator generated by the corresponding homogeneous equation of Eq.(2.198) in the phase space $\mathcal{B}$ and g_t is 2π-periodic function in $\mathcal{B}$. Hence, we get the boundedness of the sequence $\{u_{2\pi n}, n \in \mathbf{Z}\}$. Moreover, $u(t)$ is 2π-periodic if and only if $\{u_{2\pi n}, n \in \mathbf{Z}\}$ is a constant sequence. Thus we are in a position to apply Theorem 2.48. Note that 2π-periodic solutions of the continuous equations correspond to constant solutions of discrete equations.

Remark 2.27 If $\mathcal{B} = BUC(\mathbf{R}_-, \mathbf{X})$ we can see that from the boundedness of u in the future, i.e., $\sup_{t>0} \|u(t)\| < \infty$ we get the boundedness on the whole line. Hence, in the assumptions of the above theorem we can weaken the boundedness condition to require only the boundedness in the future. We refer the reader to Section 4 for more information on the Massera-type problem for functional differential equations with infinite delay with conditions guaranteeing that there are only finitely many points of the spectrum of the monodromy operator on the unit circle (hence Theorem 2.49's conditions are satisfied).

Now we will show that the above approach can be easily extended to the Massera-type problem for anti-periodic solutions. Recall that an **X**-valued function on the real line $f(t)$ is said to be τ-anti-periodic if and only if $f(t) = -f(t+\tau), \forall t \in \mathbf{R}$.

Theorem 2.50 *Suppose that $F(t)$ is periodic with period, say, 2π and $f(t)$ is 2π-anti-periodic. Furthermore, we assume that -1 is an isolated point of the part of the spectrum on the unit circle of the monodromy operator corresponding to the homogeneous equation of Eq.(2.198). Then, Eq.(2.197) has a mild 2π-anti-periodic solution if and only if it has a mild solution bounded on the whole line.*

Proof. The theorem can be proved in the same manner as the previous one, so the details are omitted.

Almost periodic solutions to periodic evolution equations

We now consider in this subsection conditions for the existence of almost periodic mild solutions to Eq.(2.179). Once Eq.(2.179) is well-posed this problem is actually reduced to find conditions for the existence of almost periodic solutions to the following more general equations

$$x(t) = U(t,s)x(s) + \int_s^t U(t,\xi)f(\xi)d\xi, \forall t \geq s, \qquad (2.201)$$

where $(U(t,s))_{t\geq s}$ is a 1-evolution process (see definition in Section 1).

We will relate our above results to continuous case of evolution equations via the well-known procedure of extension of almost periodic sequences (see [67, Section 9.5]) which can be sumarized in the following propositon.

Proposition 2.24 *Let $(U(t,s))_{t\geq s}$ be a 1-periodic strongly continuous evolutionary process and f be almost periodic. Assume that u is a solution on the real line of Eq.(2.201). Then if the sequence $\{u(n), n \in \mathbf{Z}\}$ is almost periodic, the solution u is almost periodic as well.*

Proof. First, note that the function

$$w(t) := su(n) + (1-s)u(n+1), \text{ if } t = sn + (1-s)(n+1),$$

as a function defined on the real line, is almost periodic. Also, the function taking t into $g(t) := (w(t), f(t))$ is almost periodic (see [137, p.6]). As is known, the sequence

$\{g(n)\} = \{(w(n), f(n)\}$ is almost periodic. Hence, for every positive ε the following set is relatively dense (see [67, p. 163-164])

$$T := \mathbf{Z} \cap T(g, \varepsilon), \qquad (2.202)$$

where $T(g, \varepsilon) := \{\tau \in \mathbf{R} : \sup_{t \in \mathbf{R}} \|g(t + \tau) - g(t)\| < \varepsilon\}$, i.e., the set of ε periods of g. Hence, for every $m \in T$ we have

$$\|f(t + m) - f(t)\| < \varepsilon, \forall t \in \mathbf{R}, \qquad (2.203)$$
$$\|u(n + m) - u(n)\| < \varepsilon, \forall n \in \mathbf{Z}. \qquad (2.204)$$

Since u is a solution to Eq.(2.201), we have

$$\|u(n + m + s) - u(n + s)\| \leq \|U(s, 0)(u(n + m) - u(n))\| +$$
$$+ \left\| \int_0^s U(s, \xi)(f(n + \xi + m) - f(n + \xi))d\xi \right\|$$
$$\leq Ne^\omega \|u(n + m) - u(n)\| + Ne^\omega \sup_{t \in \mathbf{R}} \|f(m + t) - f(t)\|.$$

In view of (2.203) and (2.204) m is a εN-period of the function u. Finally, since εT is relatively dense, we see that u is an almost periodic solution of Eq.(2.201).

Consider the function

$$g(t) = \int_t^{t+1} U(t + 1, \xi) f(\xi) d\xi, t \in \mathbf{R}. \qquad (2.205)$$

In view of Lemma 2.1 g is almost periodic. Moreover, by Lemma 2.3 $\sigma(g) \subset \sigma(f)$.

Theorem 2.51 *Let the inhomogeneous evolution equation (2.201) have a bounded solution v and f be almost periodic. Moreover, with g defined as above let the the following conditions be satisfied:*

i)
$$\sigma_\Gamma(P) \backslash \sigma(g) \text{ is closed.} \qquad (2.206)$$

ii) *$\sigma(g)$ is countable and $\mathbf{X}$ does not contain c_0.*

Then there is an almost periodic solution to Eq.(2.201). Moreover, if

$$\sigma_\Gamma(P) \cap \sigma(f) = \varnothing. \qquad (2.207)$$

Then the existence of almost periodic solution to Eq.(2.201) is guaranteed without the assumption on the existence of a bounded solution v.

Proof. From the 1-periodicity of the process $(U(t, s)_{t \geq s})$ we can apply Theorem 2.48 to the discrete equation

$$x_{n+1} = Px_n + g(n) \tag{2.208}$$

where P is the monodromy operator of the corresponding homogeneous equation. In fact, as a result we can find an almost periodic solution to the above equation, say, $w_n, n \in \mathbf{Z}$. It is easy to see that this implies the existence of a bounded uniformly continuous solution on the whole line to the inhomogeneous continuous evolution equation (2.201), say, $v(\cdot)$. In fact

$$v(t) = U(t,n)w_n + \int_n^t U(t,\xi)f(\xi)d\xi, \forall t \in [n, n+1), n \in \mathbf{Z}. \tag{2.209}$$

Obviously, $v(n+1) = w_{n+1}, \forall n \in \mathbf{Z}$. The uniform continuity follows easily from the boundedness of the sequence w and the periodicity of the evolutionary process $(U(t,s))_{t \geq s}$. We now apply Proposition 2.24 to see that $v(t)$ should be almost periodic. If (2.207) holds, then we can apply Theorem 2.47 as in this case $\sigma_\Gamma(P) \cap \sigma(g) = \emptyset$. Hence, we can repeat the above argument to show the existence of an almost periodic solution to Eq.(2.201).

Remark 2.28 In Corollary 2.11 the condition for the existence of almost periodic solutions is that $\sigma_\Gamma(P) \backslash \sigma(f)$ is closed. Comparing this with that of Theorem 4.2 we see a little difference. A question as whether $\sigma(g) = \sigma(f)$ holds is still open to us, although we believe that this should be true. Another aspect of our Theorem 2.51 is we do not need the condition on the uniform continuity of a solution v.

2.6. BOUNDEDNESS AND ALMOST PERIODIC SOLUTIONS OF SEMILINEAR EQUATIONS

The method we have used in the previous sections can be extended to nonlinear equations. The first question we are faced with is how to associate to a given nonlinear evolution equation evolution semigroups in suitable function spaces. The next one is to find common fixed points of these semigroups in chosen function spaces. It turns out that using evolution semigroups we can not only give a simple proofs of previous results in the finite dimensional case, but also extend them easily to the infinite dimensional case.

2.6.1. Evolution Semigroups and Semilinear Evolution Equations

In the first three subsections we are mainly concerned with the existence of almost periodic semilinear evolution equations of the form

$$\frac{dx}{dt} = Ax + f(t, x, x_t) \tag{2.210}$$

where A is the infinitesimal generator of a C_0-semigroup $(T(t))_{t\geq 0}$ and f is an everywhere defined continuous operator from $\mathbf{R} \times \mathbf{X} \times C$ to $\mathbf{X}$. Note that our method used in this subsection applies to nonautonomous equations with almost periodic coefficients, not restricted to periodic or autonomous equations as in the previous sections.

Throughout this section we will denote by $C = BUC((-\infty, 0], \mathbf{X})$ the space of all uniformly continuous and bounded functions from $(-\infty, 0]$ to $\mathbf{X}$, by $\mathbf{X}$ a given Banach space and by x_t the map $x(t+\theta) = x_t(\theta)$, $\theta \in (-\infty, 0]$, where $x(\cdot)$ is defined on $(-\infty, a]$ for some $a > 0$.

Almost periodic solutions of differential equations without delay

In this subsection we will deal with almost periodic mild solutions to the semilinear evolution equation

$$\frac{dx}{dt} = Ax + f(t,x), \quad x \in \mathbf{X} \tag{2.211}$$

where $\mathbf{X}$ is a Banach space, A is the infinitesimal generator of a C_0-semigroup of linear operators $(S(t))_{t\geq 0}$ of type ω, i.e.

$$\|S(t)x - S(t)y\| \leq e^{\omega t}\|x - y\|, \; \forall\, t \geq 0, \; x,y \in \mathbf{X},$$

and B is an everywhere defined continuous operator from $\mathbf{R} \times \mathbf{X}$ to $\mathbf{X}$. Hereafter, recall that by a mild solution $x(t), t \in [s, \tau]$ of equation (2.211) we mean a continuous solution of the integral equation

$$x(t) = S(t-s)x + \int_s^t S(t-\xi)B(\xi, x(\xi))d\xi, \; \forall s \leq t \leq \tau. \tag{2.212}$$

Definition 2.17 (condition H4). Equation (2.211) is said to satisfy *condition H4* if

i) A is the infinitesimal generator of a linear semigroup $(S(t))_{t\geq 0}$ of type ω in $\mathbf{X}$,

ii) B is an everywhere defined continuous operator from $\mathbf{R} \times \mathbf{X}$ to $\mathbf{X}$,

iii) For every fixed $t \in \mathbf{R}$, the operator $(-B(t,\cdot) + \gamma I)$ is accretive in $\mathbf{X}$.

The following condition will be used frequently:

Definition 2.18 (condition H5). Equation (2.211) is said to satisfy *condition H5* if for every $u \in AP(\mathbf{X})$ the function $B(\cdot, u(\cdot))$ belongs to $AP(\mathbf{X})$ and the operator B_* taking u into $B(\cdot, u(\cdot))$ is continuous.

The main point of our study is to associate with equation (2.211) an evolution semigroup which plays a role similar to that of the monodromy operator for equations with periodic cofficients. Hereafter we will denote by $U(t,s)$, $t \geq s$, the evolution operator corresponding to equation (2.211) which satisfies the assumptions of Theorem 1.11, i.e. $U(t,s)x$ is the unique solution of equation (2.212).

CHAPTER 2. SPECTRAL CRITERIA

Proposition 2.25 *Let the conditions H4 and H5 be satisfied. Then with equation (2.211) one can associate an evolution semigroup $(T^h)_{h\geq 0}$ acting on $AP(\mathbf{X})$, defined as*

$$[T^h v](t) = U(t, t-h)v(t-h), \forall h \geq 0, \ t \in \mathbf{R}, v \in AP(\mathbf{X}).$$

Moreover, this semigroup has the following properties:

i) $T^h, h \geq 0$ *is strongly continuous, and*

$$T^h u = S^h u + \int_0^h S^{h-\xi} B_*(T^\xi u) d\xi, \ \forall h \geq 0, u \in AP(\mathbf{X}),$$

where $(S^h u)(t) = S(h)u(t-h), \forall h \geq 0, t \in \mathbf{R}, u \in AP(\mathbf{X})$.

ii)
$$\|T^h u - T^h v\| \leq e^{(\omega+\gamma)h}\|u - v\|, \ \forall h \geq 0, u, v \in AP(\mathbf{X}).$$

Proof. We first look at the solutions to the equation

$$w(t) = S^{t-a}z + \int_a^t S^{t-\xi} B_*(w(\xi))d\xi \ \ \forall z \in AP(\mathbf{X}), t \geq a \in \mathbf{R}. \tag{2.213}$$

It may be noted that $(S^h)_{h\geq 0}$ is a strongly continuous semigroup of linear operators in $AP(\mathbf{X})$ of type ω. Furthermore, for $\lambda > 0, \lambda\gamma < 1$ and $u, v \in AP(\mathbf{X})$, from the accretiveness of the operators $-B(t, \cdot) + \gamma I$ we get

$$\begin{aligned}
(1 - \lambda\gamma)\|x - y\| &= (1 - \lambda\gamma)\sup_t \|u(t) - v(t)\| \\
&= \sup_t (1 - \lambda\gamma)\|u(t) - v(t)\| \\
&\leq \sup_t \|u(t) - v(t) - \lambda[B(t, u(t)) - B(t, v(t))]\| \\
&= \|u - v - \lambda(B_* u - B_* v)\|.
\end{aligned} \tag{2.214}$$

This shows that $(-B_* + \gamma I)$ is accretive. In virtue of Theorem 1.11 there exists a semigroup $(T^h)_{h\geq 0}$ such that

$$T^h u = S^h u + \int_0^h S^{h-\xi} B_* T^\xi u \, d\xi,$$

$$\|T^h u - T^h v\| \leq e^{(\omega+\gamma)h}\|u - v\|, \ \forall h \geq 0, u, v \in AP(\mathbf{X}).$$

From this we get

$$[T^h u](t) = [S^h u](t) + \int_0^h [S^{h-\xi} B_*(T^\xi u)](t)d\xi, \forall t \in \mathbf{R}.$$

Thus

$$[T^h u](t) = S(h)u(t-h) + \int_0^h S(h-\xi)[B_*(T^\xi u)](t-h+\xi)d\xi$$

$$= S(h)u(t-h) + \int_0^h S(h-\xi)B(t+\xi-h, [T^u](t+\xi-h))d\xi$$

$$= S(h)u(t-h) + \int_{t-h}^t S(t-\eta)B(\eta, [T^{\eta-(t-h)}u](\eta))d\eta.$$

If we denote $[T^{t-s}u](t)$ by $x(t)$, we get

$$x(t) = S(t-s)z + \int_s^t S(t-\xi)B(\xi, x(\xi))d\xi, \forall t \geq s, \qquad (2.215)$$

where $z = u(s)$. Consequently, from the uniqueness of mild solutions of Equation (2.211) we get $[T^{t-s}u](t) = x(t) = U(t,s)u(s)$ and $[T^h u](t) = U(t, t-h)u(t-h)$ for all $t \geq s, u \in AP(Q)$. This completes the proof of the proposition.

The main idea underlying our approach is the following assertion.

Corollary 2.26 *Let all assumptions of Proposition 2.25 be satisfied. Then a mild solution $x(t)$ of Equation (2.210), defined on the whole real line $\mathbf{R}$, is almost periodic if and only if it is a common fixed point of the evolution semigroup $(T^h)_{h\geq 0}$ defined in Proposition 2.25.*

Proof. Suppose that $x(t)$, defined on the real line $\mathbf{R}$, is an almost periodic mild solution of equation (2.211). Then from the uniqueness of mild solutions we get

$$x(t) = U(t, t-h)x(t-h) = [T^h x](t), \ \forall t \in \mathbf{R}.$$

This shows that x is a fixed point of T^h for every $h > 0$. Conversely, suppose that $y(\cdot)$ is any common fixed point of $T^h, h \geq 0$. Then

$$y(t) = [T^{t-s}y](t) = U(t,s)y(s), \ \forall t \geq s.$$

This shows that $y(\cdot)$ is a mild solution of equation (2.211).

We now apply Corollary 2.26 to find sufficient conditions for the existence of almost periodic mild solutions of equation (2.211).

Corollary 2.27 *Let all conditions of Proposition 1 be satisfied. Furthermore, let $\omega + \mu$ be negative and $-B_* - \mu I$ be accretive. Then there exists a unique almost periodic mild solution of Equation (2.211).*

Proof. It is obvious that there exists a unique common fixed point of the semigroup $(T^h)_{h\geq 0}$. The assertion now follows from Corollary 2.26.

Remark 2.29 i) A particular case in which we can check the accretiveness of $-B_* - \mu I$ is $\omega + \gamma < 0$. In fact, this follows easily from the above estimates for $\|u - v\|$ (see the estimate (2.214)).

ii) It is interesting to "compute" the infinitesimal generator of the evolution semigroup $(T^h)_{h \geq 0}$ determined by Proposition 2.25. To this purpose, let us recall the operator L which relates a mild solution u of the equation $\dot{x} = Ax + f(t)$ to the forcing term f by the rule $Lu = f$ (see Section 2.1 and 2.2 for more discussion on this operator). From the proof of Proposition 2.25 it follows in particular that *the infinitesimal generator $\mathcal{G}$ of the evolution semigroup $(T^h)_{h \geq 0}$ is $-L + B_*$*.

iii) We refer the reader to Sections 1 and 2 for information on semilinear equations with small nonlinear terms. In this case we can consider the semilinear equations under question as nonlinearly perturbed ones (see Corollary 2.4 and Theorem 2.11). It can be seen that u is a mild solution of Eq.(2.211) if and only if $(-L + B_*)u = 0$, so it is the fixed point of the semigroup $(T^h)_{h \geq 0}$.

iv) Let B_* act on the function space $\Lambda(\mathbf{X}) \cap AP(\mathbf{X})$. Then by the same argument as in the proof of Proposition 2.25 we can prove that the evolution semigroup $(T^h)_{h \geq 0}$ leaves $\Lambda(\mathbf{X}) \cap AP(\mathbf{X})$ invariant. This will be helpful if we want to discuss the spectrum of the unique almost periodic solution in Corollary 2.27.

Almost periodic solutions of differential equations with delays

In this subsection we apply the results of the previous subsection to study the existence of almost periodic mild solutions of the equation

$$\frac{dx}{dt} = Ax + f(t, x, x_t) \quad (2.210)$$

where A is defined as in the previous subsection, and f is an everywhere defined continuous mapping from $\mathbf{R} \times \mathbf{X} \times C$ to $\mathbf{X}$. Hereafter we call a continuous function $x(t)$ defined on the real line $\mathbf{R}$ a mild solution of equation (2.210) if

$$x(t) = S(t-s)x(s) + \int_s^t S(t-\xi)f(\xi, x(\xi), x_\xi)d\xi, \quad \forall t \geq s.$$

We should emphasize that our study is concerned only with the existence of almost periodic mild solutions of equation (2.210), and not with mild solutions in general. Consequently, the conditions guaranteeing the existence and uniqueness of mild solutions of equation (2.210) as general as in [205] are not supposed to be *a priori* conditions[4].

Definition 2.19 (condition H6). Equation (2.210) is said to satisfy *condition H6* if the following is true:

---
[4]We refer the reader to [205] for more information on the existence and uniqueness of mild solutions of equations of the form (2.210).

i) For every $g \in AP(\mathbf{X})$ the mapping $F(t,x) = f(t,x,g_t)$ satisfies conditions $H4$ and $H5$ with the same constant γ,

ii) There exists a constant μ with $\omega - \mu < 0$ such that $-(\mu I + F_*)$ is accretive for every $g \in AP(\mathbf{X})$,

iii) $[x-y, f(t,x,\phi) - f(t,y,\phi')] \leq \gamma \|x-y\| + \delta \|\phi - \phi'\|, \forall t \in \mathbf{R}, x, y \in \mathbf{X}, \phi, \phi' \in C$.

Theorem 2.52 *Let condition H6 hold. Then for δ sufficiently small (see the estimate (2.219) below), equation (2.210) has an almost periodic mild solution.*

Proof. First we fix a function $g \in AP(\mathbf{X})$. In view of Proposition 2.25 we observe that the equation
$$\frac{dx}{dt} = Ax + F(t,x)$$
has a unique almost periodic mild solution, where $F(t,x) = f(t,x,g_t)$. We denote this solution by Tg. Thus, we have defined an operator T acting on $AP(\mathbf{X})$. We now prove that T is a strict contraction mapping. In fact, let us denote by $U(t,s)$ and $V(t,s)$ the Cauchy operators

$$U(t,s)x = S(t-s)x + \int_s^t S(t-\xi)f(\xi, U(\xi,s)x, g_\xi)d\xi, \quad (2.216)$$

$$V(t,s)x = S(t-s)x + \int_s^t S(t-\xi)f(\xi, V(\xi,s)x, h_\xi)d\xi, \quad (2.217)$$

for given $g, h \in AP(\mathbf{X}), x \in \mathbf{X}, t \geq s$.

Putting $u(t) = U(t,s)x, v(t) = V(t,s)x$ for given s, x, from the assumptions we have
$$[u(t) - v(t), f(t, u(t), g_t) - f(t, v(t), h_t)] \leq m(t, \|u(t) - v(t)\|),$$
where $m(t, \|u(t) - v(t)\|) = \gamma \|u(t) - v(t)\| + \delta \|h - g\|$. Using this we get

$$\|u(t) - v(t)\| \leq \|u(t-\eta) - v(t-\eta)\| + \eta m(t, \|u(t) - v(t)\|) +$$
$$+ \int_{t-\eta}^t \|S(t-\xi)f(\xi, u(\xi), h_\xi) - f(t, u(t), h_t)\|d\xi +$$
$$+ \int_{t-\eta}^t \|S(t-\xi)f(\xi, v(\xi), g_\xi) - f(t, v(t), g_t)\|d\xi.$$

Now let us fix arbitrary real numbers $a \leq b$. Since the functions $S(t-\xi)f(\xi, u(\xi), h_\xi)$ and $S(t-\xi)f(\xi, v(\xi), g_\xi)$ are uniformly continuous on the set $a \leq \xi \leq t \leq b$, for every $\varepsilon > 0$ there exists an $\eta_0 = \eta_0(\varepsilon)$ such that

$$\|S(t-\xi)f(\xi, u(\xi), h_\xi) - f(t, u(t), h_t)\| < \varepsilon,$$
$$\|S(t-\xi)f(\xi, v(\xi), g_\xi) - f(t, v(t), g_t)\| < \varepsilon,$$

for all $\|t - \xi\| < \eta_0$ and $t \leq \xi \in [a,b]$. Hence, denoting $\|u(t) - v(t)\|$ by $\alpha(t)$, for $\eta < \eta_0$ we have

$$\alpha(t) - e^{\omega\eta}\alpha(t-\eta) \leq \eta m(t,\alpha(t)) + 2\eta\varepsilon. \tag{2.218}$$

Applying this estimate repeatedly, we get

$$\alpha(t) - e^{\omega(t-s)} \leq \sum_{i=1}^{n} e^{\omega(t-t_i)} m(t_i, \alpha(t_i))\Delta_i + 2\varepsilon \sum_{i=1}^{n} e^{\omega(t-t_i)}\Delta_i,$$

where $t_0 = s < t_1 < t_2 < ... < t_n = t$ and $|t_i - t_{i-1}| = \Delta_i$. Thus, since ε is arbitrary, and since the function m is continuous, we get

$$\alpha(t) - e^{\omega(t-s)}\alpha(s) \leq \int_s^t e^{\omega(t-\xi)} m(\xi, \alpha(\xi)) d\xi$$
$$= \int_s^t e^{\omega(t-\xi)}(\gamma\alpha(\xi) + \delta\|h-g\|)d\xi.$$

Applying Gronwall's inequality we get

$$\alpha(t) \leq e^{(\gamma+\omega)(t-s)}\alpha(s) + e^{\gamma(t-s)+\omega t}\left(\frac{e^{-\omega s} - e^{-\omega t}}{\omega}\right)\delta\|h-g\|.$$

Because of the identity $\alpha(s) = \|u(s) - v(s)\| = \|U(s,s)x - V(s,s)x\| = 0$, from the above estimate we obtain

$$\sup_{t-1 \leq \xi \leq t} \|U(\xi, t-1)x - V(\xi, t-1)x\| \leq \frac{e^{\gamma+\omega} - e^{\gamma}}{\omega}\delta\|h-g\|.$$

Now let us denote by $T_h^t, T_g^t, t \geq 0$ the respective evolution semigroups corresponding to equations (2.216) and (2.217). Since Th and Tg are defined as the unique fixed points u_0, v_0 of T_h^1, T_g^1, respectively, we have

$$\|Th - Tg\| = \|u_0 - v_0\| = \|T_h^1 u_0 - T_g^1 v_0\| \leq$$
$$\leq \|T_h^1 u_0 - T_g^1 u_0\| + \|T_g^1 u_0 - T_g^1 - v_0\|$$
$$\leq \frac{e^{\gamma+\omega} - e^{\gamma}}{\omega}\delta\|h-g\| + e^{\omega-\mu}\|u_0 - v_0\|$$
$$= N\delta\|h-g\| + e^{\omega-\mu}\|Th - Tg\|,$$

where $N = (e^{\gamma+\omega} - e^{\gamma})/\omega$. Finally, we have

$$\|Th - Tg\| \leq \frac{e^{\gamma}(e^{\omega} - 1)}{\omega(1 - e^{\omega-\mu})}.$$

Thus, if the estimate

$$\delta < \frac{\omega(1 - e^{\omega-\mu})}{e^{\gamma}(e^{\omega} - 1)} \tag{2.219}$$

holds true, then T is a strict contraction mapping in $AP(\mathbf{X})$. By virtue of the Contraction Mapping Principle T has a unique fixed point. It is easy to see that this fixed point is an almost periodic mild solution of equation (2.210). This completes the proof of the theorem.

Remark 2.30 i) In case $\omega = 0, \gamma = -\mu$ we get the estimate

$$\delta < e^\mu - 1 = \mu + \mu^2/2 + ...$$

which guarantees the existence of the fixed point of T.

ii) If $\omega + \gamma < 0$, then we can choose $\mu = -\gamma$, and therefore we get the accretiveness condition on $-(F_* + \mu I)$. However, in general, the condition $\omega + \gamma < 0$ is a very strong restriction on the coefficients of equation (2.210), if f depends explicitly on t.

Examples

In applications one frequently encounters functions f from $\mathbf{R} \times \mathbf{X} \times C \to \mathbf{X}$ of the form

$$f(t, x, g_t) = F(t, x) + G(t, g_t), \ \forall t \in \mathbf{R}, \ x \in \mathbf{X}, \ g_t \in C,$$

where F satisfies condition ii) of Definition 2.19 and $G(t, y)$ is Lipschitz continuous with respect to $y \in C$, i.e.

$$\|G(t, y) - G(t, z)\| \leq \delta \|y - z\|, \ \forall t \in \mathbf{R}, \ y, z \in C$$

for some positive constant δ. With f in this form numerous examples of partial functional differential equations fitting into our abstract framework can be found (see e.g. [22], [129], [142], [179] and in [226]).

In order to describe a concrete example we consider a bounded domain Ω in $\mathbf{R}^n$ with smooth boundary $\partial\Omega$ and suppose that

$$A(x, D)u = \sum_{|\alpha| \leq 2m} a_\alpha(x) D^\alpha u$$

is a strongly elliptic differential operator in Ω. Then, defining the operator

$$Au = A(x, D)u, \ \forall u \in D(A) = W^{2m,2}(\Omega) \cap W_0^{m,2}(\Omega)$$

we know from [179, Theorem 3.6] that the operator $-A$ is the infinitesimal generator of an analytic semigroup of contractions on $L^2(\Omega)$. Now let $f, g : \mathbf{R} \times \Omega \times \mathbf{R} \to \mathbf{R}$ be Lipschitz continuous and define the operators $F(t, w)(x) = f(t, x, w(x))$ and $G(t, w)(x) = g(t, x, w(x))$ where $t \in \mathbf{R}$, $x \in \Omega$ and $w \in L^2(\Omega)$. Then, for any positive constant r, the boundary value problem

$$\frac{\partial u(t, x)}{\partial t} = A(x, D)u(t, x) + f(t, x, u(t, x)) + g(t, x, u(t-r, x)) \text{ in } \Omega,$$
$$u(t, x) = 0 \text{ on } \partial\Omega$$

fits into the abstract setting of equation (2.210).

2.6.2. Bounded and Periodic Solutions to Abstract Functional Differential Equations with Finite Delay

Let us consider the following functional evolution equation with finite delay

$$u'(t) + Au(t) = f(t, u_t), \ t > 0, \ u(s) = \phi(s), \ s \in [-r, 0], \quad (2.220)$$

in a Banach space $(\mathbf{X}, \|\cdot\|)$, with A the generator of a strongly continuous semigroup of linear operators and f a continuous T-periodic function which is Lipschitz continuous in $\phi \in C$. Here $r > 0$ is a constant. We recall that in this subsection the notation $C([-r, 0], X)$ stands for the space of continuous functions from $[-r, 0]$ to X with the sup-norm, $\|\phi\|_C = \max_{s \in [-r, 0]} \|\phi(s)\|$, and define $u_t \in C([-r, 0], X)$ by $u_t(s) = u(t + s)$, $s \in [-r, 0]$, for a function u.

In this subsection, we will study the relationship between the bounded solutions and the periodic solutions of Eq.(2.220). For this purpose, we will use the solution map P defined along the solution by the formula

$$P\phi = u_T(\cdot, \phi), \ \phi \in C([-r, 0], X),$$
$$(i.e., (P\phi)(s) = u_T(s, \phi) = u(T + s, \phi), \ s \in [-r, 0],),$$

where $u(\cdot, \phi)$ is the solution of Eq.(2.220) with the initial function ϕ.

It is possible to carry the idea and techniques used in finite dimensional spaces to general Banach spaces by using fundamental results of solution operators one of which is the compactness of the above defined operator P (see Theorem 1.13).

Recall that every continuous solution to the following equation

$$\begin{cases} u(t) = T(t-a)\phi(0) + \int_a^t T(t-s)f(s, u_s)ds, 0 \le t, \\ u_0 = \phi \end{cases}, \quad (2.221)$$

is called *mild solution* to Eq.(2.220) on $[0, +\infty)$. Under the above assumption, the Cauchy problem 2.221 has a unique solution on $[0, +\infty)$. We recall below some standard definitions.

Definition 2.20 Mild solutions of Eq.(2.220) are *bounded* if for each $B_1 > 0$, there is a $B_2 > 0$, such that $\|\phi\|_C \le B_1$ and $t \ge 0$ imply $\|u(t, \phi)\| < B_2$ $\forall$ solution u of Eq.(2.221).

Definition 2.21 Mild solutions of Eq.(2.220) are *ultimate bounded* if there is a bound $B > 0$, such that for each $B_3 > 0$, there is a $K > 0$, such that $\|\phi\|_C \le B_3$ and $t \ge K$ imply $\|u(t, \phi)\| < B$ $\forall$ solution u of Eq.(2.221).

Lemma 2.38 *(Horn's Fixed Point Theorem) Let $E_0 \subset E_1 \subset E_2$ be convex subsets of Banach space Z, with E_0 and E_2 compact subsets and E_1 open relative to E_2. Let $P : E_2 \to Z$ be a continuous map such that for some positive integer m,*

$$P^j(E_1) \subset E_2, \ 1 \le j \le m - 1,$$
$$P^j(E_1) \subset E_0, \ m \le j \le 2m - 1.$$

Then P has a fixed point in E_0.

Proof. The proof can be found in [110].

Now we study the periodic solutions for the finite delay evolution equation Eq.(2.220). As is well known, a fixed point of P corresponds to a periodic solution. Hence, we will find conditions suitable to apply Horn Fixed Point Theorem so that P has fixed points.

Theorem 2.53 *Let the above assumptions be satisfied. If the mild solutions of Eq.(2.220) are bounded and ultimate bounded, then Eq.(2.220) has a T-periodic mild solution.*

Proof. Let the map P be defined as above. We see that

$$P^m(\phi) = u_{mT}(\phi), \quad \phi \in C([-r,0]). \tag{2.222}$$

Next, let $B > 0$ be the bound in the definition of ultimate boundedness. Using boundedness, there is $B_2 > B$ such that $\{\|\phi\|_C \leq B, t \geq 0\}$ implies $\|u(t,\phi)\| < B_2$. And also, there is $B_4 > 2B_2$ such that $\{\|\phi\|_C \leq 2B_2, t \geq 0\}$ implies $\|u(t,\phi)\| < B_4$. Next, using ultimate boundedness, there is a positive integer m such that $\{\|\phi\|_C \leq 2B_2, t \geq (m-2)T\}$ implies $\|u(t,\phi)\| < B$. These imply

$$\begin{aligned}
\|P^{i-1}\phi\|_C &= \|u((i-1)T+\cdot,\phi)\|_C < B_4 \\
&\quad \text{for } i = 1,2,3,\ldots \text{ and } \|\phi\|_C \leq 2B_2,
\end{aligned} \tag{2.223}$$

$$\begin{aligned}
\|P^{i-1}\phi\|_C &= \|u((i-1)T+\cdot,\phi)\|_C < B \\
&\quad \text{for } i \geq m \text{ and } \|\phi\|_C \leq 2B_2.
\end{aligned} \tag{2.224}$$

Now let

$$\begin{aligned}
H &\equiv \{\phi \in C([-r,0],X) : \|\phi\|_C < B_4\}, & E_2 &\equiv \overline{(cov.(P(H)))}, \\
K &\equiv \{\phi \in C([-r,0],X) : \|\phi\|_C < 2B_2\}, & E_1 &\equiv K \cap E_2, \\
G &\equiv \{\phi \in C([-r,0],X) : \|\phi\|_C < B\}, & E_0 &\equiv \overline{(cov.(P(G)))},
\end{aligned} \tag{2.225}$$

where $cov.(F)$ is the convex hull of the set F defined by $cov.(F) = \{\sum_{i=1}^n \lambda_i f_i : n \geq 1, f_i \in F, \lambda_i \geq 0, \sum_{i=1}^n \lambda_i = 1\}$. Thus E_0, E_1 and E_2 are convex subsets and E_1 is open relative to E_2.

Note that $G \subset H$, thus $E_0 \subset E_2$. Next, we show that $E_0 \subset K$. For this, note that from the definition of B_2, $\|Pu\| < B_2$ for $u \in G$. Thus for $e \in cov.(P(G))$, one has $e = \sum \lambda_i e_i$ with $e_i \in P(G)$, and $\|e\| \leq \sum \lambda_i \|e_i\| \leq \sum \lambda_i B_2 = B_2$. This shows $E_0 \subset K$. Therefore, $E_0 \subset K \cap E_2 = E_1 \subset E_2$. Next, as H and G are bounded subsets and P is a compact operator, $P(G)$ and $P(H)$ are precompact subsets. Now, it is known that $cov.(P(G))$ and $cov.(P(H))$ are also precompact subsets. Therefore E_0 and E_2 are compact subsets. From (2.223) and (2.224), one has

$$\begin{aligned}
P^i(E_1) &\subset P^i(K) = PP^{i-1}(K) \subset P(H) \subset E_2, \quad i = 1,2,3,\ldots, \tag{2.226} \\
P^i(E_1) &\subset P^i(K) = PP^{i-1}(K) \subset P(G) \subset E_0, \quad i \geq m. \tag{2.227}
\end{aligned}$$

Since the continuity of P follows from Theorem 1.2 we are now in a position to apply the Horn's Fixed Point Theorem to get a T-periodic mild solution of Eq.(2.220).

2.7. ALMOST PERIODIC SOLUTIONS OF NONLINEAR EVOLUTION EQUATIONS

2.7.1. Nonlinear Evolution Semigroups in $AP(\Delta)$

In this section we will consider the following fully nonlinear evolution equation

$$0 \in \frac{du}{dt} + A(t)u, \qquad (2.228)$$

where $A(t)$ denotes a (possibly nonlinear multivalued) operator from $D(A(t)) \subset \mathbf{X}$ to $\mathbf{X}$, where $(\mathbf{X}, \|\cdot\|)$ is a given Banach space and $D(A(t))$ stands for the domain of a given operator $A(t)$ and $R(A(t))$ for the range of $A(t)$. It will be assumed in this subsection that $A(t)$ is almost periodic in the sense of Bohr (see condition C2 below).

Our method is based on fundamental results of Crandall-Pazy [52] on the existence of evolution operator $U(t,s)$ to Eq.(2.228) and the semigroup of nonlinear operators associated with it defined as

$$[T^h v](t) = U(t, t-h)v(t-h), t \in \mathbf{R}, \qquad (2.229)$$

where $v \in AP(\Delta)$, here $AP(\Delta)$ denotes the space of $\mathbf{X}$-valued almost periodic functions v with $v(t) \in \Delta \subset \mathbf{X}$ (for the precise definition of Δ see condition A2 below). Unfortunately, in this section we cannot say anything about the strong continuity of this semigroup as for the previous sections. However, this suggests to state similar results.

The main result of this section is Theorem 2.55 in which a sufficient condition is found for all generalized solutions to Eq.(2.228) to approach a unique almost periodic one. As we consider the general case of $A(t)$, e.g. $D(A(t))$ may depent on t, the almost periodic dependence of $A(t)$ on t will be stated in terms of the Yosida approximant of $A(t)$, but not on $A(t)$. Proposition 2.27 will be useful to compare this with the usual form of t-dependence and especially for Lipschitz perturbation. We give two examples on the applications of our result to functional and partial differential equations. Our condition on Eq.(2.228) will be imposed on the coefficient-operator $A(t)$ for Eq.(2.228) to have almost periodic evolution operator $U(t,s)$ [5] which has some advantages when we consider the perturbed equation (see Proposition 2.27 and Corollary 2.29 and Remarks which follows) and the relation between the frequency of the almost periodic solution and that of the coefficient-operator $A(t)$.

In what follows we introduce the notion of *admissibility* of the coefficient-operator $B(t)$ for a special case which will be frequently used later on.

---
[5]see e.g. [21], [53], [115], [130] for related results

Definition 2.22 A single-valued operator $B(t)$ is said to be *admissible (with respect to Δ)* if it satisfies the following conditions:

i) $B(t) : D(B(t)) \to \mathbf{X}$, and for some subset $\Delta \subset \cap_t D(B(t))$, $B(t)x$ is continuous with respect to $(t,x) \in \mathbf{R} \times \Delta$,

ii) $\sup_t \|B(t)x_0\| < +\infty$ for some $x_0 \in \Delta$,

iii) $\|B(t)x - B(t)y\| \leq L\|x-y\|, \forall x,y \in \Delta$.

Throughout the section we will need the following

Lemma 2.39 *Let $B(t)$ be admissible (with respect to Δ). Moreover, let $B(t)x$ be almost periodic in t for every fixed $x \in \Delta$. Then for every almost periodic function $v \in AP(\Delta)$, the function $B(t)v(t)$ is almost periodic in t.*

Proof. Let us consider a sequence of translates $B(t+\tau_n)x, x \in K$, where K is a compact subset of Δ. We now show that there is a subsequence $\{\tau_{n_k}\}$ such that $B(t+\tau_n)x$ converges uniformly in $x \in K$. To this end, by using the diagonal method we can choose $\{\tau_{n_k}\}$ so that the sequence $\{B(t+\tau_{n_k}x\}$ converges at each x from a countable subset K_1 that is dense in K. From the inequality

$$\begin{aligned}
\|B(t+\tau_{n_k})x - B(t+\tau_{n_j})x\| &\leq \|B(t+\tau_{n_k})x - B(t+\tau_{n_k})x_1\| + \\
&\leq \|B(t+\tau_{n_k})x_1 - B(t+\tau_{n_j})x_1\| + \\
&\quad + \|B(t+\tau_{n_j})x_1 - B(t+\tau_{n_j})x\| \\
&\leq 2L\|x-x_1\| + \\
&\quad + \|B(t+\tau_{n_k})x_1 - B(t+\tau_{n_j})x_1\|
\end{aligned}$$

it follows that for any positive ε there exists a natural number $N(x_1,\varepsilon)$ such that

$$\|B(t+\tau_{n_k})x - B(t+\tau_{n_j})x\| < \varepsilon \quad (n_k, n_j \geq N) \tag{2.230}$$

for all $x \in U(x_1,\varepsilon) := \{x : \|x-x_1\| < \varepsilon/4L\}$. The open balls $U(x_1,\varepsilon)$ cover the set K. Hence, one can pick up finitely many such balls. This implies immediately the uniform convergence.

Now suppose that $v(\cdot)$ is an almost periodic function with values in Δ. Then we have to prove that for any sequence $\{\tau_k\}$ there is a subsequence $\{\tau_{n_k}\}$ such that the function $B(t+\tau_{n_k})v(t+\tau_{n_k})$ is uniformly convergent in t. To this end, first define $K := \overline{\{y \in \Delta : y = v(s) \text{ for some s }\}}$. Obviously, K is compact. Consequently, by choosing the subsequence $\{\tau_{n_k}\}$ such that $v(t+\tau_{n_k})$ is convergent uniformly in t and $B(t+\tau_{n_k})y$ is convergent uniformly for $t \in \mathbf{R}, y \in K$ we get the convergence of $B(t+\tau_{n_k})v(t+\tau_{n_k})$.

Next we recall the basic notions and results from the Crandall-Pazy theory on the existence and uniqueness of generalized solutions to Eq.(2.228). Below we shall always identify a multivalued operator A with its graph in $\mathbf{X} \times \mathbf{X}$.

Definition 2.23 A subset A of $\mathbf{X} \times \mathbf{X}$ is said to be in *class $\mathcal{A}(\omega)$* if for each $\lambda > 0$ such that $\lambda \omega < 1$ and each pair $[x_i, y_i] \in A, i = 1, 2$, we have

$$\|(x_1 + \lambda y_1) - (x_2 + \lambda y_2)\| \geq (1 - \lambda \omega)\|x_1 - x_2\| \tag{2.231}$$

A is said to be *accretive* if $A \in \mathcal{A}(0)$.

Let us denote $J_\lambda = (I + \lambda A)^{-1}$. Below are the main properties of accretive sets.

Lemma 2.40 *Let ω be real and $A \in \mathcal{A}(\omega)$. If $\lambda \geq 0$ and $\lambda \omega < 1$, then the following statements hold:*

i) J_λ *is a (single-valued) function*

ii) $\|J_\lambda x - J_\lambda y\| \leq (1 - \lambda \omega)^{-1}\|x - y\| \quad \forall x, y \in D(J_\lambda)$.

In this section as a standing hypothesis we always assume that the following *condition A* are satisfied:

(A1) $A(t) \in \mathcal{A}(\omega)$ for all $t \in \mathbf{R}$

(A2) $\overline{D(A(t))} = \Delta$ is independent of t,

(A3) $R(I + \lambda A(t)) \subset \Delta$ for all $t \in \mathbf{R}$ and $0 < \lambda < \lambda_0$, where $\lambda_0 > 0$ is given,

Below we shall use the following notations: $J_\lambda(t)$ stands for the resolvent $(I + \lambda A(t))^{-1}$ and $A_\lambda(t)$ for the Yosida approximant $(I - J_\lambda(t))/\lambda$. As usual, since $D(A(t))$ may depend on t and then the function $A(t)x$ may not be defined for all t despite that $x \in D(A(t_1))$ for some t_1, the dependence on t of $A(t)$ is stated in terms of that of Yosida approximant $A_\lambda(t)$ and the resolvent $J_\lambda(t)$ as follows:

(C1) There is a bounded uniformly continuous function $f : \mathbf{R} \to \mathbf{X}$ and a monotone increasing function $L : [0, +\infty) \to [0, +\infty)$ such that

$$\|A_\lambda(t)x - A_\lambda(\tau)x\| \leq \|f(t) - f(\tau)\|L(\|x\|)$$

for all $t \in \mathbf{R}, 0 < \lambda < \lambda_0, x \in \Delta$,

(C2) For every fixed $x \in \Delta$ the function $A_\lambda(t)x$ is almost periodic in t.

Under the assumptions A1, A2, A3, C1 the following theorem holds true:

Theorem 2.54 *Let the conditions A1, A2, A3, C1 hold. Then*

$$U(t,s)x = \lim_{n \to +\infty} \prod_{i=1}^{n} J_{(t-s)/n}\left(s + i\frac{t-s}{n}\right)x \tag{2.232}$$

exists for $x \in \Delta$ and $s < t$. Moreover, $U(t,s)$ has the following properties:

i)
$$\|U(t,s)x - U(t,s)y\| \leq e^{\omega(t-s)}\|x-y\| \quad \forall s < t, \ x,y \in \Delta$$

ii) $U(s,s) = I, \ \forall s,$

iii) $U(t,s)U(s,r) = U(t,r)$ for all $r < s < t$,

iv) $U(t,s)x$ is continuous with respect to $s < t, x \in \Delta$.

v) For every fixed positive h the following limit
$$\lim_{n \to +\infty} \prod_{i=1}^{n} J_{h/n}(t + ih/n)x$$
is uniformly with respect to $t \in \mathbf{R}$.

Proof. The theorem is an immediate consequence of [52, Theorem 2.1].

We say that $u(t)$ is a *generalized solution* to (2.228) on $[s,T]$ if $u(t) = U(t,s)x$ for some x and $t \in [s,T]$.

We now show the existence the nonlinear semigroup (2.229) on $AP(\Delta)$.

Proposition 2.26 *Let conditions A1, A2, A3, C1, C2 hold. Then for every $h \geq 0$ and every almost periodic function $v(t)$ with values in Δ the function $U(t, t-h)v(t-h)$ is also almost periodic in t.*

Proof. By virtue of Lemma 2.39, it suffices to show that for every $x \in \Delta$ the function $w(t) = U(t+h,t)x$ is almost periodic. In fact, from conditions A2, C2 and Theorem 2.54, observe that for $0 < \lambda < \lambda_0$, $A_\lambda(t)$ is admissible. Now applying Lemma 2.39 we see that for every n sufficiently large (e.g $h/n < \lambda_0$) the function
$$\prod_{i=1}^{n} J_{h/n}\left(t + \frac{ih}{n}\right)x$$
is an almost periodic function. Furthermore, by Theorem 2.54, the convergence in the limit
$$U(t+h,t)x = \lim_{n \to +\infty} \prod_{i=1}^{n} J_{h/n}\left(t + \frac{ih}{n}\right)x \tag{2.233}$$
is uniformly in t. Thus, $U(t+h,t)x$ is almost periodic in t. The proof is complete.

Corollary 2.28 *Let the conditions of Proposition 2.26 be satisfied. Then a generalized solution of Equation (2.228) defined on the whole line is almost periodic if and only if it is a common fixed point of the semigroup $T^h, h \geq 0$ defined by (2.229).*

Proof. The corollary is an immediate consequence of Proposition 2.26.

2.7.2. Almost Periodic Solutions of Dissipative Equations

The main result of this section is stated in the following:

Theorem 2.55 *Let the conditions of Proposition 2.26 be satisfied. Moreover, let $\omega(t)$ be a function such that*

i) $\limsup_{|t|\to+\infty} \omega(t) < 0$, $\sup_{t\in\mathbf{R}} \omega(t) < +\infty$

ii) $A(t) \in \mathcal{A}(\omega(t))$ *for all* t.

Then Equation (2.228) has a unique generalized almost periodic solution $u(t)$ which is defined as a limit

$$u(t) = \lim_{m\to+\infty} U(t, t - mh)x$$

$$= \lim_{m\to+\infty} \lim_{n\to+\infty} \prod_{i=1}^{n} J_{mh/n}\left(t - mh + \frac{imh}{n}\right)x \quad (2.234)$$

for some sufficiently large h. Moreover, there exist positive constants N, μ such that every generalized solution $x(t)$ defined on $[s, +\infty)$ satisfies

$$\|u(t) - x(t)\| \le Ne^{-\mu(t-s)}\|u(s) - x(s)\| \quad \forall t \ge s. \quad (2.235)$$

Proof. By virtue of Proposition 2.26, the semigroup $(T^h)_{h\ge 0}$ is well defined in $AP(\Delta)$. From the assumptions it follows that there exist positive numbers μ, T such that $\omega(t) < -\mu < 0$, for $t \notin [-T, T]$. By Proposition 2.26, the semigroup $(T^h)_{h\ge 0}$ associated with the evolution operator $(U(t,s))_{t\ge s}$ defined by (2.229) acts on the function space $AP(\Delta)$. We now show that this semigroup has a common fixed point. To this end, by applying Theorem 2.54 to Eq.(2.228) in different intervals $(-\infty, -T], [-T, T], [T, +\infty)$ we have

$$\|U(t,s)x - U(t,s)y\| \le Ne^{\mu(t-s)}\|x - y\|, \quad \forall t \ge s, x, y \in \Delta, \quad (2.236)$$

where $N = e^{2(\mu+\eta)}, \eta = \sup_t \omega(t)$. Suppose h is any positive number such that $Ne^{-\mu h} < 1$. It is seen that T^h is a strict contraction. Thus, by the Contraction Mapping Theorem (see e.g. [18]) it has a unique fixed point. Furthermore, this fixed point is also a unique common fixed point for the semigroup $(T^h)_{h\ge 0}$. Now applying Corollary 2.28 we get the first assertion of the theorem. The estimate (2.235) follows from (2.236). The theorem is proved.

The following assertion on perturbation will be useful to compare our results with the previous ones.

Proposition 2.27 *Let condition C2 be satisfied for $A(t)$ with $\Delta = \mathbf{X}$. Moreover, let $B(t)$ be an admissible nonlinear operator with $D(B(t)) = \mathbf{X}$ such that for every fixed x the function $B(t)x$ is almost periodic. Then the operator $(A(t) + B(t))$ satisfies condition C2.*

Proof. Let us define the operators $B, J_\lambda^A : AP(\mathbf{X}) \to AP(\mathbf{X})$ by the following formulas
$$[Bu](t) = B(t)u(t), \quad \forall t$$
$$[J_\lambda^A u](t) = J_\lambda^A(t)u(t).$$

Consider the operator T_λ acting on the space $Lip(AP(\mathbf{X}))$ of Lipschitz continuous operators acting on $AP(\mathbf{X})$ defined as
$$T_\lambda \psi = J_\lambda^A [-\lambda B \psi - \lambda B J_\lambda^A].$$

From Lemma 2.40, and the assumptions it follows that
$$Lip(T_\lambda) \leq \frac{1}{(1-\lambda\omega)} \lambda Lip(B).$$

Thus, for $\lambda > 0$ sufficiently small $Lip(T_\lambda) < 1$. According to the Contraction Mapping Theorem there is a unique operator ϕ_λ such that $\phi_\lambda = T_\lambda \phi_\lambda$. Consequently,
$$\begin{aligned}[I + \lambda(A+B)][J_\lambda^A + \phi_\lambda] &= I + (I + \lambda A)\phi_\lambda + \lambda B J_\lambda^A + \lambda B \phi_\lambda \\ &= I + (-\lambda B \phi_\lambda - \lambda B J_\lambda^A + \lambda B J_\lambda^A + \lambda B \phi_\lambda) = 1\end{aligned}$$

On ther other hand, since $sup_t Lip(B(t)) < +\infty$ we can show easily that $[A(t) + B(t)] + (\omega - sup_t Lip(B(t)))I$ is accretive. So, for $\lambda > 0$ sufficiently small, we can well define the operators $[(I + \lambda(A(t) + B(t)))]^{-1}$ and J_λ^{A+B}. It is easily seen that
$$\begin{aligned}[J_\lambda^A + \phi_\lambda]u(t) &= J_\lambda^{A+B} u(t) \\ &= [(I + \lambda(A(t) + B(t)))]^{-1} u(t)\end{aligned}$$

for every $u \in AP(\mathbf{X})$. This shows that for every $u \in AP(\mathbf{X})$ the function $[(I + \lambda(A(t) + B(t)))]^{-1} u(t)$ is almost periodic.

Now we consider the perturbed equation
$$0 \in \frac{du}{dt} + A(t)u + B(t)u, \tag{2.237}$$

where $B(t)$ will be assumed to be admissible.

Corollary 2.29 *Let the conditions of Theorem 2.55 hold with $\Delta = \mathbf{X}$. Moreover, let the admissible operator $B(t)$ satisfy the following conditions*

i) $D(B(t)) = \mathbf{X}$, for every fixed $x \in \mathbf{X}$, $B(t)x$ is almost periodic in t,

ii) The perturbed equation (2.237) satisfies condition C1 (i.e. $(A(t) + B(t))$ satisfies C1).

CHAPTER 2. SPECTRAL CRITERIA

Then there exists a positive constant δ such that if $\sup_t \|B(t)\| < \delta$ the perturbed equation (2.236) has a unique almost periodic generalized solution $u(t)$. Moreover, any other generalized solution $v(t)$ defined on $[s, +\infty)$ of the perturbed equation approaches $u(t)$ at exponentially rate, i.e

$$\|u(t) - v(t)\| \leq N e^{-\eta(t-s)} \|u(s) - v(s)\|, \quad \forall\, t \geq s.$$

Proof. In fact, by [22, Theorem 3.2, p.158] one sees that there is a γ such that the operator $\gamma I + A(t) + B(t)$ is m-accretive. Now Proposition 2.27 and Theorem 2.55 apply.

Remark 2.31 i) As a particular case we can take an operator $A(t) = A(t + \tau)\ \forall t$, where τ is a given positive constant, and $\overline{D(A)} = \mathbf{X}$. Then condition C2 is satisfied for $A(t)$. According to Proposition 2.27 condition C2 is satisfied also for the perturbed equation (2.237) with $B(t)x = g(t)\ \forall t, x$, where $g(t)$ is an almost periodic function.

ii) In the above case, using the identity $J_\lambda^{A+B}(t)x = J_\lambda^A(x - \lambda B(t))$ one can show that condition ii) in Corollary 2.29 is satisfied without any additional assumption.

iii) In case the operator $A(t)$ in Eq.(2.237) is independent of t one can show that condition ii) is also automatically satisfied if $B(t)$ satisfies a condition similar to C1 but in the usual terms, i.e. there are functions h, L with similar properties as in the definition of condition C1 such that

$$\|B(t)x - B(\tau)x\| \leq |h(t) - h(\tau)| L(\|x\|), \quad \forall t, \tau, x.$$

In fact, by the proof of Proposition 2.27, $\phi_\lambda u(t) = Q(t)u(t)$ for some Lipschitz continous $Q(t)$. We now show that $Q(t)$ is also admissible. To this end, put $u(t) = x$ for a given $x \in \mathbf{X}, \forall t \in \mathbf{R}$. Then

$$\begin{aligned}
\|Q(t)0\| &= \left\| J_\lambda^A(0) + \frac{\lambda}{1-\lambda\omega} \|B(t)Q(t)0 + B(t)J_\lambda^A(0)\| \right. \\
&\leq \|J_\lambda^A(0)\| + \frac{\lambda}{1-\lambda\omega}\|B(t)J_\lambda^A(0)\| \\
&\quad + \frac{\lambda}{1-\lambda\omega}(\|B(t)0\| + K\|Q(t)0\|).
\end{aligned}$$

From this and the admissibility of $B(t)$ it can be easily seen that

$$\sup_t \|Q(t)0\| < \infty$$

Now we show that $A + B(t)$ satisfies condition $C1$. In fact

$$\|Q(t)x - Q(\tau)x\| \leq \frac{\lambda}{1-\lambda\omega}\|B(t)Q(t)x + B(t)J_\lambda^A(x) -$$

$$-B(\tau)Q(\tau)x + B(\tau)J_\lambda^A(x)\|$$
$$\leq \frac{\lambda}{1-\lambda\omega}\{|h(t) - h(\tau)|L(\|J_\lambda^A(x)\|) +$$
$$+ \|B(t)Q(t)x - B(t)Q(\tau)x\| +$$
$$+ \|B(t)Q(\tau)x - B(\tau)Q(\tau)x\|\}$$
$$\leq \frac{\lambda}{1-\lambda\omega}|h(t) - h(\tau)|L(\|J_\lambda^A(x)\|) +$$
$$+ \frac{\lambda}{1-\lambda\omega}\{K\|Q(t)x - Q(\tau)x\| +$$
$$+ |h(t) - h(\tau)|L(\|Q(\tau)x\|)\}.$$

From this and the admissibility of $B(t), Q(t)$ and the assumption that $B(t)$ satisfies condition C1

$$\|Q(t)x - Q(\tau)x\| \leq \frac{\lambda}{1-\lambda(\omega+K)}(|h(t) - h(\tau)|(L_1(\|x\|)),$$

where L_1 is a monotone increasing functions. This shows that $(A + B(t))$ satisfies condition $C1$.

2.7.3. An Example

Let $\mathbf{X} = L^2[0,1] = L^2$ be the set of Lebesgue integrabale real valued functions on $[0,1]$ with the usual norm $\|u\| = <u,u>^{1/2}$

$$<u,v> = \int_0^1 u(\xi), v(\xi)d\xi, u, v \in L^2.$$

Let $\sigma : \mathbf{R} \to \mathbf{R}$ be continuously differentiable, $\sigma(0) = 0$ and suppose that there are constants m and M such that $0 < m \leq \sigma'(u) \leq M < \infty$ for all $u \in \mathbf{R}$. Let

$$r(\cdot,\cdot,\cdot) : \mathbf{R} \times [0,1] \times \mathbf{R} \to \mathbf{R}$$

be in L^2, and for every fixed $(t,u) \in \mathbf{R} \times L^2$, $r(t, u(\cdot), \cdot)$ be L^2-almost periodic in t. Moreover, suppose that

i) $\sup_{t \in \mathbf{R}} \|r(t, 0, \cdot)\| < \infty$;

ii) $|r(t, x, \cdot) - r(t, y, \cdot)| \leq K|x - y|, \forall x, y \in \mathbf{R}$;

iii) $\|r(t, u(\cdot), \cdot) - r(\tau, u(\cdot), \cdot)\| \leq |h(t) - h(\tau)|L(\|u\|), \forall t, \tau, u \in L^2$,

where $h(\cdot)$ is a bounded, uniformly continuous function, L is a monotone increasing function $[0,\infty) :\to [0,\infty)$. Now applying Remark 2.31 (iii) following Corollary 2.29 we have

Theorem 2.56 *Let all assumptions made above be met. Then for the Lipschitz coefficient K sufficiently small each solution of the following equation*

$$\begin{cases} u_t(t,\xi) = (\sigma(u_\xi))_\xi + r(t,u,\xi), t \in \mathbf{R}, 0 < \xi < 1, \\ u(t,0) = u(t,1) = 0 \ \forall t \in \mathbf{R} \end{cases} \quad (2.238)$$

such that $u(t,\cdot) \in L^2$ for $t \geq 0$ approaches a unique generalized solution of (2.238) as $t \to \infty$.

Proof. Take $Au := (\sigma(u_\xi))_\xi$ with

$$D(A) = \{u \in L^2 : u(0) = u(1) = 0, \text{ with } u(\xi) \text{ and } u'(\xi) \\ \text{absolutely continuous on } [0,1], u''(\xi) \in L^2\}$$

and $[B(t)u](\xi) := r(t,u,\xi)$, $\forall \xi, t, u \in L^2$. As shown in [197, Section 4] A is m-accretive, we are now in a position to apply Remark 2.31 (iii) to finish the proof.

2.8. NOTES

The results of Section 1 are taken mainly from [167] and [160]. The topics of Section 1 have been discussed earlier in [186], [185], [218], [222] in the case of autonomous equations with methods quite different from what is presented here. For the periodic equations with Floquet representation the problem has been treated in [217]. Since even simplest infinite dimensional equations, such as functional and parabolic differential equations, have no Floquet representation, periodic equations should be treated directly. Evolution semigroups method has been used first in [167]. Theorem 2.2 is taken from [160]. An earlier version of it has been stated in [167]. Note that in [31] this result has been stated in more general terms, though its proof, based on the method of [167], seems to be incomplete. Other results of this section are the nonautonomous analogues of the corresponding results in [186], [222].

The approach of Section 2 has been first used in [160]. The results for abstract ordinary differential equations have been known earlier in [218], [222] in the case the operator A generates a C_0-semigroup. In [186], [185] Volterra equations have been considered. Similar conditions for the admissibility of function spaces $\Lambda(\mathbf{X})$ have been stated. Theorem 2.6 has an earliear version in [137, Lemma 3, p.93] for the case where the first order equation generates a C_0-semigroup and $\mathcal{F} = AP(\mathbf{X})$. More complete proofs can be found in [8], [26], [192]. In [26] the theorem was proved for arbitrary B-class $\mathcal{F}$ and A as the generator of a C_0-semigroup. The case of second order equations with $\mathcal{F} = AP(\mathbf{X})$ was considered in [8, Theorem 4.5, p.375]. In the case $\mathcal{F} = AP(\mathbf{X})$ the theorem was formulated in general terms in [221, the proof of Theorem 4.4] (for which, seemingly, additional conditions should be assumed).

The decomposition theorem 2.14 in Section 3 is taken from [169] and [70]. An earlier version of it for periodic solutions has been stated in [168]. The main results of this section extend a classical result in ordinary differential equations on periodic solution [4, Theorem 20.3, p. 278]. Earlier extensions of this result for solutions bounded on the positive line can be found in [147], [40], [45], [206]. To our view, the version of this section for equations defined on the positive line should be true. Since this area is currently very dynamic, the references should be updated in the near future.

The main results of Section 4 are taken from [206]. The main method is the applications of the fixed point theorems due to Chow and Hale. If the equation is described by the compact semigroup, the fixed point theorem is related to the well known theorem by Schawder. In other case, we use the theory of semi-Fredholm operator for the closedness of the operator for the fixed point.

Section 5 is taken from [170]. The method of using extensions of almost periodic sequences to study almost periodic solutions of differential equations can be read in [67, Section 9.5].

Sections 6 and 7 are taken from [18], [19] and [138]. Earlier results for finite dimensional or bounded case can be found in [149], [200], [123], [124], [147], [40].

CHAPTER 3

STABILITY METHODS FOR SEMILINEAR EVOLUTION EQUATIONS AND NONLINEAR EVOLUTION EQUATIONS

3.1. SKEW PRODUCT FLOWS OF PROCESSES AND QUASI - PROCESSES AND STABILITY OF INTEGRALS

Let $\mathcal{X}$ be a metric space with metric d and let $w : \mathbf{R}^+ \times \mathbf{R}^+ \times \mathcal{X} \mapsto \mathcal{X}, \mathbf{R}^+ := [0, \infty)$, be a function satisfying the following properties for all $t, \tau, s \in \mathbf{R}^+$ and $x \in \mathcal{X}$:

(p1) $w(0, s, x) = x$;

(p2) $w(t + \tau, s, x) = w(t, \tau + s, w(\tau, s, x))$.

Let $\mathcal{Y}$ be a nonempty closed set in $\mathcal{X}$. We call the mapping w a $\mathcal{Y}$-*quasi-process on $\mathcal{X}$* or simply a *quasi-process on $\mathcal{X}$*, if w satisfies the condition,

(p3)   the restricted mapping $w : \mathbf{R}^+ \times \mathbf{R}^+ \times \mathcal{Y} \mapsto \mathcal{X}$ is continuous,

together with the conditions (p1) and (p2). In case of $\mathcal{Y} = \mathcal{X}$, the concept of quasi-processes is identical with that of *processes* investigated in [53], [54] [77], [78] and [120]. We emphasize that the concept of processes does not fit in with the study of ρ-stabilities in functional differential equations in contrast with the concept of quasi-processes, as will be seen in Section 3.3. The ρ-stability is a useful tool in the study of the existence of almost periodic solutions for almost periodic systems.

Denote by W the set of all quasi-processes on $\mathcal{X}$. For $\tau \in \mathbf{R}^+$ and $w \in W$, we define the translation $\sigma(\tau)w$ of w by

$$(\sigma(\tau)w)(t, s, x) = w(t, \tau + s, x), \qquad (t, s, x) \in \mathbf{R}^+ \times \mathbf{R}^+ \times \mathcal{X},$$

and set $\gamma_\sigma^+(w) = \bigcup_{t \geq 0} \sigma(t)w$. Clearly $\gamma_\sigma^+(w) \subset W$. We denote by $H_\sigma(w)$ all functions $\chi : \mathbf{R}^+ \times \mathbf{R}^+ \times \mathcal{X} \mapsto \mathcal{X}$ such that for some sequence $\{\tau_n\} \subset \mathbf{R}^+$, $\{\sigma(\tau_n)w\}$ converges to χ pointwise on $\mathbf{R}^+ \times \mathbf{R}^+ \times \mathcal{X}$, that is,

$$\lim_{n \to \infty} (\sigma(\tau_n)w)(t, s, x) = \chi(t, s, x) \; \forall (t, s, x) \in \mathbf{R}^+ \times \mathbf{R}^+ \times \mathcal{X}.$$

The set $H_\sigma(w)$ is considered as a topological space with the pointwise convergence, and it is called the hull of w.

Consider a $\mathcal{Y}$-quasi-process w on $\mathcal{X}$ satisfying

(p4) $H_\sigma(w) \subset W$.

Clearly, $H_\sigma(w)$ is invariant with respect to the tanslation $\sigma(\tau), \tau \in \mathbf{R}^+$. For any $t \in \mathbf{R}^+$, we consider a function $\pi(t) : \mathcal{X} \times H_\sigma(w) \mapsto \mathcal{X} \times H_\sigma(w)$ defined by

$$\pi(t)(x,\chi) = (\chi(t,0,x), \sigma(t)\chi)$$

for $(x,\chi) \in \mathcal{X} \times H_\sigma(w)$. $\pi(t)$ is called *the skew product flow* of the quasi-process w, if the following property holds true:

(p5) $\pi(t)(y,\chi)$ is continuous in $(t,y,\chi) \in \mathbf{R}^+ \times \mathcal{Y} \times H_\sigma(w)$.

The skew product flow $\pi(t)$ is said to be $\mathcal{Y}$-*strongly asymptotically smooth* if, for any nonempty closed bounded set $B \subset \mathcal{Y} \times H_\sigma(w)$, there exists a compact set $J \subset \mathcal{Y} \times H_\sigma(w)$ with the property that $\{\pi(t_n)(y_n,\chi_n)\}$ has a subsequence which approaches to J whenever sequences $\{t_n\} \subset \mathbf{R}^+$ and $\{(y_n,\chi_n)\} \subset B$ satisfy $\lim_{n\to\infty} t_n = \infty$ and $\pi(t)(y_n,\chi_n) \in B$ for all $t \in [0,t_n]$. In case of $\mathcal{Y} = \mathcal{X}$, the $\mathcal{Y}$-strong asymptotic smoothness of $\pi(t)$ implies the asymptotic smoothness of $\pi(t)$ introduced in [77], [78]. Clearly, if $\pi(t)$ is completely continuous, then it is $\mathcal{X}$-strongly asymptotic smooth. In Section 3.3, we shall see that the $\mathcal{Y}$-strong asymptotic smoothness of $\pi(t)$ is ensured when w is a quasi-process generated by some functional differential equations.

Now we suppose that $H_\sigma(w)$ is sequentially compact and we discuss relationships between some stability properties for the quasi-process w and those for the "*limiting*" quasi-processes $\chi \in \Omega_\sigma(w)$; here $\Omega_\sigma(w)$ denotes the ω-limit set of w with respect to the translation semigroup $\sigma(t)$. A continuous function $\mu : \mathbf{R}^+ \mapsto \mathcal{Y}$ is called *an integral* of the quasi-process w on $\mathbf{R}^+$, if $w(t,s,\mu(s)) = \mu(t+s)$ for all $t,s \in \mathbf{R}^+$ (cf. [78, p. 80]). In the following, we suppose that there exists an integral $\mu : \mathbf{R}^+ \mapsto \mathcal{Y}$ of the quasi-process w on $\mathbf{R}^+$ such that the set $O^+(\mu) = \{\mu(t) : t \in \mathbf{R}^+\}$ is contained in $\mathcal{Y}$ and is relatively compact in $\mathcal{X}$. From (p5) we see that $\pi(\delta)(\mu(s),\sigma(s)w) = (w(\delta,s,\mu(s)),\sigma(s+\delta)w) = (\mu(s+\delta),\sigma(s+\delta)w)$ tends to $(\mu(s),\sigma(s)w)$ as $\delta \to 0^+$, uniformly for $s \in \mathbf{R}^+$; consequently, the integral μ must be uniformly continuous on $\mathbf{R}^+$. From Ascoli-Arzéla's theorem and the sequential compactness of $H_\sigma(w)$, it follows that for any sequence $\{\tau'_n\} \subset \mathbf{R}^+$, there exist a subsequence $\{\tau_n\}$ of $\{\tau'_n\}$, a $\chi \in H_\sigma(w)$ and a function $\nu : \mathbf{R}^+ \mapsto \mathcal{Y}$ such that $\lim_{n\to\infty} \sigma(\tau_n)w = \chi$ and $\lim_{n\to\infty} \mu(t+\tau_n) = \nu(t)$ uniformly on any compact interval in $\mathbf{R}^+$. We shall use the following expression. Let Y and Z be two pseudo-metric spaces, and $f, f_n, n = 1,2,\ldots$, be functions mapping Y and Z. If $f^n(\xi) \to f(\xi)$ in Z as $n \to \infty$, uniformly on any compact set in Y, then we write as

$$f^n(\xi) \to f(\xi) \quad \text{compactly on } Y,$$

for simplicity. Therefore, in the above case, we write as

$$(\mu^{\tau_n}, \sigma(\tau_n)w) \to (\nu,\chi) \quad \text{compactly,}$$

for simplicity. Denote by $H_\sigma(\mu,w)$ the set of all (ν,χ) such that $(\mu^{\tau_n},\sigma(\tau_n)w) \to (\nu,\chi)$ compactly for some sequence $\{\tau_n\} \subset \mathbf{R}^+$. In particular, we denote by $\Omega_\sigma(\mu,w)$ the set of all $(\nu,\chi) \in H_\sigma(\mu,w)$ for which one can choose a sequence $\{\tau_n\} \subset \mathbf{R}^+$ so that $\lim_{n\to\infty} \tau_n = \infty$ and $(\mu^{\tau_n}, \sigma(\tau_n)w) \to (\nu, \chi)$ compactly. We easily see that ν is an integral of the quasi-process χ on $\mathbf{R}^+$ whenever $(\nu, \chi) \in H_\sigma(\mu, w)$.

For any $x_0 \in \mathcal{X}$ and $\varepsilon > 0$, we set $V_\varepsilon(x_0) = \{x \in \mathcal{X} : d(x, x_0) < \varepsilon\}$. We will give the definition of stabilities for the integral μ of the quasi-process w.

Definition 3.1 The integral $\mu : \mathbf{R}^+ \mapsto \mathcal{Y}$ of the quasi-process w is said to be:

i) $\mathcal{Y}$-uniformly stable ($\mathcal{Y}$-US) (resp. $\mathcal{Y}$-uniformly stable in $\Omega_\sigma(w)$) if for any $\varepsilon > 0$, there exists a $\delta := \delta(\varepsilon) > 0$ such that $w(t, s, \mathcal{Y} \cap V_\delta(\mu(s))) \subset V_\varepsilon(\mu(t+s))$ for $(t, s) \in \mathbf{R}^+ \times \mathbf{R}^+$ (resp. $\chi(t, s, \mathcal{Y} \cap V_\delta(\nu(s))) \subset V_\varepsilon(\nu(t+s))$ for $(\nu, \chi) \in \Omega_\sigma(\mu, w)$ and $(t, s) \in \mathbf{R}^+ \times \mathbf{R}^+$);

ii) $\mathcal{Y}$-uniformly asymptotically stable ($\mathcal{Y}$-UAS) (resp. $\mathcal{Y}$-uniformly asymptotically stable in $\Omega_\sigma(w)$), if it is $\mathcal{Y}$-US (resp. $\mathcal{Y}$-US in $\Omega_\sigma(w)$) and there exists a $\delta_0 > 0$ with the property that for any $\varepsilon > 0$, there is a $t_0 > 0$ such that $w(t, s, \mathcal{Y} \cap V_{\delta_0}(\mu(s))) \subset V_\varepsilon(\mu(t+s))$ for $t \geq t_0$, $s \in \mathbf{R}^+$ (resp. $\chi(t, s, \mathcal{Y} \cap V_{\delta_0}(\nu(s))) \subset V_\varepsilon(\nu(t+s))$ for $(\nu, \chi) \in \Omega_\sigma(\mu, w)$ and $t \geq t_0$, $s \in \mathbf{R}^+$);

iii) $\mathcal{Y}$-attractive (resp. $\mathcal{Y}$-attractive in $\Omega_\sigma(w)$) if there is a $\delta_0 > 0$ such that for $y \in \mathcal{Y} \cap V_{\delta_0}(\mu(0))$ (resp. $y \in \mathcal{Y} \cap V_{\delta_0}(\nu(0))$ and $(\nu, \chi) \in \Omega_\sigma(\mu, w)$), $d(w(t, 0, y), \mu(t)) \to 0$ (resp. $d(\chi(t, 0, y), \nu(t)) \to 0$) as $t \to \infty$;

iv) $\mathcal{Y}$-weakly uniformly asymptotically stable ($\mathcal{Y}$-WUAS) in $\Omega_\sigma(w)$ if it is $\mathcal{Y}$-US in $\Omega_\sigma(w)$ and $\mathcal{Y}$-attractive in $\Omega_\sigma(w)$.

We assume the following property on $\mathcal{Y}$, w and μ:

(p6) There is a $\delta_1 > 0$ such that for any $s \in \mathbf{R}^+$ and $t_0 > 0$, $w(t_0, s, y) \in \mathcal{Y}$ whenever $y \in \mathcal{Y}$ and $w(t, s, y) \in V_{\delta_1}(\mu(t+s))$ for all $t \in (0, t_0]$.

If $\mathcal{Y} = \mathcal{X}$, then (p6) is clearly satisfied. In Section 3.3, we will give a nontivial example for which (p6) is satisfied.

Theorem 3.1 *Let $\mathcal{Y}$ be a closed set in a metric space $\mathcal{X}$ and suppose that w is a $\mathcal{Y}$-quasi-process on $\mathcal{X}$ for which $H_\sigma(w)$ is sequentially compact and that the skew product flow $\pi(t) : \mathcal{X} \times H_\sigma(w) \mapsto \mathcal{X} \times H_\sigma(w)$ of the quasi-process w is $\mathcal{Y}$-strongly asymptotically smooth. Also, suppose that $\mu : \mathbf{R}^+ \mapsto \mathcal{Y}$ is an integral of w on $\mathbf{R}^+$ such that $O^+(\mu)$ is a relative compact subset of $\mathcal{Y}$ and that (p4) and (p6) are satisfied. Then the following statements are equivalent: The integral μ is*

i) *$\mathcal{Y}$-UAS.*

ii) *$\mathcal{Y}$-US and $\mathcal{Y}$-attractive in $\Omega_\sigma(w)$.*

iii) *$\mathcal{Y}$-UAS in $\Omega_\sigma(w)$.*

iv) $\mathcal{Y}$-WUAS in $\Omega_\sigma(w)$.

Proof. We shall show that (ii) implies (i). We may choose δ_0 so that for $x \in V_{\delta_0}(\nu(0)) \cap \mathcal{Y}$ and $(\nu, v) \in \Omega_\sigma(\mu, w), d(v(t, 0, x), \nu(t)) \to 0$ as $t \to \infty$. Let $\eta_0 = \delta(\delta_0/2)/2$. Since $\mu(t)$ is $\mathcal{Y}$-US, in order to prove the $\mathcal{Y}$-UAS, it is sufficient to show the existence of t_0 for a given ε such that $w(t, s, \mathcal{Y} \cap V_{\eta_0}(\mu(s))) \subset V_{\delta(\varepsilon)}(\mu(t+s)), s \in \mathbf{R}^+$, for some $t \in [0, t_0]$. Suppose it is not the case. Then there exists sequences $\sigma_k, \sigma_k \geq 0$, and $x_k \in \mathcal{Y}$ such that $x_k \in \overline{V_{\eta_0}(\mu(\sigma_k))}$ but $w(t, \sigma_k, x_k) \notin V_{\delta(\varepsilon)}(\mu(t + \sigma_k))$ for $t \in [0, 2k]$. We can see $w(t, \sigma_k, x_k) \in V_{\delta_0/2}(\mu(t + \sigma_k))$ for all $t \geq 0$. Now we may assume $(\mu^{t_k}, \sigma(t_k)w) \to (\nu, v)$ compactly as $k \to \infty$, where $t_k = \sigma_k + k$ for some $(\nu, v) \in \Omega_\sigma(\mu, w)$. Since $\pi(t)$ is $\mathcal{Y}$-strongly asymptotically smooth, taking a subsequence if necessary, we can assume that $w(k, \sigma_k, x_k) \to x$ for some $x \in \mathcal{Y}$ as $k \to \infty$. Note that $x \in \mathcal{Y} \cap V_{\delta_0/2}(\nu(0)) \subset V_{\delta_0}(\nu(0))$ but $v(t, 0, x) \notin \overline{V_{\delta(\varepsilon)}(\nu(t))}$ for all $t \in \mathbf{R}^+$, which contradicts the $\mathcal{Y}$-attractivity in $\Omega_\sigma(w)$ of $\mu(t)$.

It is clear that (iii) implies (iv).

We shall show that (i) implies (iii). Let $\tau_k \to \infty$ as $k \to \infty$ and $(\mu^{\tau_k}, \sigma(\tau_k)w) \to (\nu, v) \in \Omega_\sigma(\mu, w)$, compactly. Let any $\sigma \in \mathbf{R}^+$ be fixed. If k is sufficiently large, we get

$$\nu(\sigma) \in V_{\delta(\varepsilon/2)/2}(\mu(\tau_k + \sigma)).$$

Let $y \in \mathcal{Y} \cap V_{\delta(\varepsilon/2)/2}(\nu(\sigma))$. Then $(\sigma(\tau_k)w)(t, \sigma, y) = w(t, \tau_k + \sigma, y) \in V_{\varepsilon/2}(\mu(t + \tau_k + \sigma))$ for $t \geq \sigma$, because $y \in V_{\delta(\varepsilon/2)}(\mu(\tau_k + \sigma))$. Since $(\sigma(\tau_k)w)(t, \sigma, y) \to v(t, \sigma, y)$ and $\mu(t + \tau_k + \sigma) \to \nu(t + \sigma), v(t, \sigma, y) \in \overline{V_{\varepsilon/2}(\nu(t + \sigma))}$ for all $t \geq \sigma$, which implies $\nu(t)$ is $\mathcal{Y}$-US.

Now we shall show $\nu(t)$ is $\mathcal{Y}$-UAS. Let $y \in \mathcal{Y} \cap V_{\delta_0/2}(\nu(\sigma))$ and $\nu(\sigma) \in V_{\delta_0/2}(\mu(\tau_k + \sigma))$. Since $\mu(t)$ is $\mathcal{Y}$-UAS, we have $w(t, \tau_k + \sigma, y) \in V_{\varepsilon/2}(\mu(t + \tau_k + \sigma))$ for $t \geq t_0(\varepsilon/2)$ because of $y \in V_{\delta_0}(\mu(\tau_k + \sigma))$. Hence $v(t, \sigma, y) \in \overline{V_{\varepsilon/2}(\nu(t + \sigma))}$ for $t \geq t_0(\varepsilon/2)$.

We shall show that (iv) implies (i). In order to establish this assertion, it suffices to show that (iv) yields the $\mathcal{Y}$-US of the integral μ. To do this by a contradiction, we assume that the integral μ is $\mathcal{Y}$-WUAS in $\Omega_\sigma(w)$, but not $\mathcal{Y}$-US. Then there exist an $\varepsilon > 0$, $\varepsilon < \min\{\delta_0, \delta_1\}$, and a sequence $\{(t_n, s_n, y_n)\}$ in $\mathbf{R}^+ \times \mathbf{R}^+ \times \mathcal{Y}$ such that $d(y_n, \mu(s_n)) \to 0$ as $n \to \infty$, $d(w(t_n, s_n, y_n), \mu(t_n + s_n)) = \varepsilon$ and $d(w(t, s_n, y_n), \mu(t + s_n)) < \varepsilon$ for $0 \leq t < t_n$, where δ_1 is the one ensured in (p6) and δ_0 is the one given for the $\mathcal{Y}$-attractivity in $\Omega_\sigma(w)$ of the integral μ. Take a positive constant $\gamma, \gamma < \varepsilon$, so that $\chi(t, s, \mathcal{Y} \cap V_\gamma(\nu(s))) \subset V_\varepsilon(\nu(t + s))$ for $(t, s) \in \mathbf{R}^+ \times \mathbf{R}^+$ and $(\nu, \chi) \in \Omega_\sigma(\mu, w)$, which is possible by the $\mathcal{Y}$-US in $\Omega_\sigma(w)$ of the integral μ. Since $d(y_n, \mu(s_n)) \to 0$ as $n \to \infty$, there exists a sequence $\{\tau_n\}$, $0 < \tau_n < t_n$, such that $d(w(\tau_n, s_n, y_n), \mu(\tau_n + s_n)) = \gamma/2$ and $d(w(t, s_n, y_n), \mu(t + s_n)) \geq \gamma/2$ for all $t \in [\tau_n, t_n]$.

We assert that $\tau_n \to \infty$ as $n \to \infty$. Suppose that the assertion is false. Then, without loss of generality, we may assume that $\tau_n \to \tau_0$ and $(\mu^{s_n}, \sigma(s_n)w) \to (\tilde{\mu}, \tilde{w})$ as $n \to \infty$, for some $\tau_0 < \infty$ and $(\tilde{\mu}, \tilde{w}) \in H_\sigma(\mu, w)$. From (p5) it follows that $\pi(\tau_n)(y_n, \sigma(s_n)w)$ tends to $\pi(\tau_0)(\tilde{\mu}(0), \tilde{w})$ in $\mathcal{Y} \times H_\sigma(w)$ as $n \to \infty$, which implies that $(\sigma(s_n)w)(\tau_n, 0, y_n) = w(\tau_n, s_n, y_n)$ tends to $\tilde{w}(\tau_0, 0, \tilde{\mu}(0)) = \tilde{\mu}(\tau_0)$ as

$n \to \infty$. On the other hand, since $d(w(\tau_n, s_n, y_n), \mu(\tau_n + s_n)) = \gamma/2$, we must get $d(\tilde{w}(\tau_0, 0, \tilde{\mu}(0)), \tilde{\mu}(\tau_0)) = \gamma/2$, a contradiction.

Now we may assume that $(\mu^{\tau_n + s_n}, \sigma(\tau_n + s_n)w) \to (\nu, \chi)$ as $n \to \infty$, for some $(\nu, \chi) \in \Omega_\sigma(\mu, w)$. Notice that $\pi(t)(y_n, \sigma(s_n)w) = (w(t, s_n, y_n), \sigma(t + s_n)w) \in V_{\delta_1}(\mu(t + s_n)) \times H(w)$ for $t \in [0, \tau_n]$. By virtue of (p6), we get $w(t, s_n, y_n) \in \mathcal{Y}$ for all $t \in [0, \tau_n]$. Since $\pi(t)$ is $\mathcal{Y}$-strongly asymptotically smooth, taking a subsequence if necessary, we can assume that $w(\tau_n, s_n, y_n) \to \tilde{y}$ for some $\tilde{y} \in \mathcal{Y}$ as $n \to \infty$. Note that $\tilde{y} \in V_\gamma(\nu(0))$. We first consider the case where the sequence $\{t_n - \tau_n\}$ has a convergent subsequence. Without loss of generality, we can assume that $t_n - \tau_n \to \tilde{t}$ as $n \to \infty$, for some $\tilde{t} < \infty$. Then (p5) implies that $\pi(t_n - \tau_n)(w(\tau_n, s_n, y_n), \sigma(\tau_n + s_n)w) = (w(t_n, s_n, y_n), \sigma(t_n + s_n)w)$ tends to $\pi(\tilde{t})(\tilde{y}, \chi)$ in $\mathcal{Y} \times H_\sigma(w)$ as $n \to \infty$. Letting $n \to \infty$ in the relation $d((\sigma(\tau_n + s_n)w)(t_n - \tau_n, 0, w(\tau_n, s_n, y_n)), \mu(t_n + s_n)) = d(w(t_n - \tau_n, s_n + \tau_n, w(\tau_n, s_n, y_n)), \mu(t_n + s_n)) = d(w(t_n, s_n, y_n), \mu(t_n + s_n)) = \varepsilon$, we get $d(\chi(\tilde{t}, 0, \tilde{y}), \tilde{\nu}(\tilde{t})) = \varepsilon$. This is a contradiction, because of $\chi(\tilde{t}, 0, \tilde{y}) \in \chi(\tilde{t}, 0, \mathcal{Y} \cap V_\gamma(\nu(0))) \subset V_\varepsilon(\nu(\tilde{t}))$. Thus we must have $\lim_{n \to \infty}(t_n - \tau_n) = \infty$. Now, letting $n \to \infty$ in the relation $d((\sigma(s_n + \tau_n)w)(t, 0, w(\tau_n, s_n, y_n)), \mu(t + \tau_n + s_n)) = d(w(t + \tau_n, s_n, y_n), \mu(t + \tau_n + s_n)) \leq \varepsilon$ for $t \in [0, t_n - \tau_n]$, we get $d(\chi(t, 0, \tilde{y}), \nu(t)) \leq \varepsilon < \delta_0$ for all $t \geq 0$. Then, from the $\mathcal{Y}$-attractivity in $\Omega_\sigma(w)$ of the integral μ, it follows that $d(\chi(t, 0, \tilde{y}), \nu(t)) \to 0$ as $t \to \infty$. On the other hand, since $d(w(t + \tau_n, s_n, y_n), \mu(t + \tau_n + s_n)) \geq \gamma/2$ for $t \in [0, t_n - \tau_n]$, we must get $d(\chi(t, 0, \tilde{y}), \nu(t)) \geq \gamma/2$ for all $t \geq 0$; hence $d(\chi(t, 0, \tilde{y}), \nu(t)) \not\to 0$ as $t \to \infty$, a contradiction.

When $\mathcal{X} = \mathcal{Y}$, the stabilities for the integral $\mu(t)$ of the process w is defined. For example, the integral $\mu(t)$ of the process w is uniformly stable in $\Omega_\sigma(w)$ (US in $\Omega_\sigma(w)$), if for any $\varepsilon > 0$, there exists a $\delta := \delta(\varepsilon) > 0$ such that $v(t, s, V_\delta(\nu(s))) \subset V_\varepsilon(\nu(t + s))$ for $(\nu, v) \in \Omega_\sigma(\mu, w)$ and $(t, s) \in \mathbf{R}^+ \times \mathbf{R}^+$. The other stabilities for the integral $\mu(t)$ of the process w are given in a similar way.

The next theorem follows from Theorem 3.1, immediately.

Theorem 3.2 *Suppose that w is a process on $\mathcal{X}$ for which $H_\sigma(w)$ is sequentially compact and that the skew product flow $\pi(t) : \mathcal{X} \times H_\sigma(w) \mapsto \mathcal{X} \times H_\sigma(w)$ of the process w is $\mathcal{X}$-strongly asymptotically smooth. Also, suppose that $\mu : \mathbf{R}^+ \mapsto \mathcal{X}$ is an integral of w on $\mathbf{R}^+$ such that $O^+(\mu)$ is a relative compact subset of $\mathcal{X}$. Then the following statements are equivalent:*

i) *The integral μ is UAS.*

ii) *The integral μ is US and attractive in $\Omega_\sigma(w)$.*

iii) *The integral μ is UAS in $\Omega_\sigma(w)$.*

iv) *The integral μ is WUAS in $\Omega_\sigma(w)$.*

3.2. EXISTENCE THEOREMS OF ALMOST PERIODIC INTEGRALS

3.2.1. Asymptotic Almost Periodicity and Almost Periodicity

A process $w : \mathbf{R}^+ \times \mathbf{R} \times \mathcal{X} \mapsto \mathcal{X}, \mathbf{R} = (-\infty, \infty)$, which satisfies (p1)–(p5) for $\mathcal{X} = \mathcal{Y}$ is said to be *almost periodic* if $w(t, s, x)$ is almost periodic in s uniformly with respect to t, x in bounded sets. Let w be an almost periodic process on $\mathcal{X}$. Bochner's theorem (Theorem 1.18) implies that $\Omega_\sigma(w) = H_\sigma(w)$ is a minimal set. Also, for any $v \in H_\sigma(w)$, there exists a sequence $\{\tau_n\} \subset \mathbf{R}^+$ such that $w(t, s + \tau_n, x) \to v(t, s, x)$ as $n \to \infty$, uniformly in $s \in \mathbf{R}$ and (t, x) in bounded sets of $\mathbf{R}^+ \times \mathcal{X}$. Consider a metric ρ^* on $H_\sigma(w)$ defined by

$$\rho^*(u, v) = \sum_{n=0}^{\infty} \frac{1}{2^n} \frac{\rho_n^*(u, v)}{1 + \rho_n^*(u, v)}$$

with $\rho_n^*(u, v) = \sup\{d(u(t, s, x), v(t, s, x)) : 0 \leq t \leq n, s \in \mathbf{R}, d(x, x_0) \leq n\}$, where x_0 is a fixed element in $\mathcal{X}$. Then $v \in H_\sigma(w)$ means that $\rho^*(\sigma(\tau_n)w, v) \to 0$ as $n \to \infty$, for some sequence $\{\tau_n\} \subset \mathbf{R}^+$. Let $\mu(t)$ be an integral of the process w.

Definition 3.2 *An integral $\mu(t)$ on $\mathbf{R}^+$ is said to be asymptotically almost periodic if it is a sum of a continuous almost periodic function $\phi(t)$ and a continuous function $\psi(t)$ defined on $\mathbf{R}^+$ which tends to zero as $t \to \infty$, that is*

$$\mu(t) = \phi(t) + \psi(t).$$

Let $\mu(t)$ be an integral on $\mathbf{R}^+$ such that the set $O^+(\mu)$ is relatively compact in $\mathcal{X}$. It is known (cf., e.g. [231, pp. 20–30]), that when $\mathcal{X} = \mathbf{R}^n$, $\mu(t)$ is asymptotically almost periodic if and only if it satisfies the following property:

(L) For any sequence $\{t'_n\}$ such that $t'_n \to \infty$ as $n \to \infty$ there exists a subsequence $\{t_n\}$ of $\{t'_n\}$ for which $\mu(t + t_n)$ converges uniformly on $\mathbf{R}^+$.

Indeed, using Bochner's criterion (Theorem 1.18) for almost periodic functions, we see that argument employed in [231, pp. 20–30] works even where $\mathcal{X}$ is any separable Banach space, under the condition $O^+(\mu)$ is relatively compact in $\mathcal{X}$. Therefore, the above equivalence holds true where $\mathcal{X}$ is a separable Banach space, too. Now, for an integral μ on $\mathbf{R}^+$ of the almost periodic process w we consider the following property:

(A) For any $\varepsilon > 0$, there exists a $\delta(\varepsilon) > 0$ such that $\nu(t) \in V_\varepsilon(\mu(t + \tau))$, for all $t \geq 0$, whenever $(\nu, v) \in \Omega_\sigma(\mu, w), \nu(0) \in V_{\delta(\varepsilon)}(\mu(\tau))$, and $\rho^*(\sigma(\tau)w, v) < \delta(\varepsilon)$ for some $\tau \geq 0$.

Theorem 3.3 *Assume that $\mu(t)$ is an integral on $\mathbf{R}^+$ of the almost periodic process w such that the set $O^+(\mu) = \{\mu(t) : t \in \mathbf{R}^+\}$ is relatively compact in $\mathcal{X}$. Then $\mu(t)$ is asymptotically almost periodic if and only if it has Property (A).*

Proof. Assume that $\mu(t)$ has Property (A), and let $\{t'_n\}$ be any sequence such that $t'_n \to \infty$ as $n \to \infty$. Then there exist a subsequence $\{t_n\}$ of $\{t'_n\}$ and a $(\nu, v) \in \Omega_\sigma(\mu, w)$ such that $(\mu^{t_n}, \sigma(t_n)w) \to (\nu, v)$ compactly on $\mathbf{R}$. For any $\varepsilon > 0$, there exists an $n_0(\varepsilon) > 0$ such that if $n \geq n_0(\varepsilon)$, then $\nu(0) \in V_{\delta(\varepsilon)}(\mu(t_n))$ and $\rho^*(\sigma(t_n)w, v) < \delta(\varepsilon)$, where $\delta(\varepsilon)$ is the one for Property (A), which implies

$$\nu(t) \in V_\varepsilon(\mu(t + t_n)) \quad \text{for} \quad t \geq 0.$$

Thus $\mu(t)$ satisfies Property (L) and hence it is asymptotically almost periodic.

Next, suppose that $\mu(t)$ is asymptotically almost periodic, but does not have Property (A). Then there exists an $\varepsilon > 0$ and sequences $(\nu^n, v^n) \in \Omega_\sigma(\mu, w), \tau_n \geq 0$ and $t_n > 0$ such that

$$\nu^n(t_n) \in \partial V_\varepsilon(\mu(t_n + \tau_n)), \tag{3.1}$$

$$\nu^n(0) \in V_{1/n}(\mu(\tau_n)) \tag{3.2}$$

and

$$\rho^*(v^n, \sigma(\tau_n)w) < \frac{1}{n}, \tag{3.3}$$

where ∂V_ε is the boundary of V_ε. We can assume that for a function $\nu(t)$

$$d(\nu^n(t), \nu(t)) \to 0 \text{ uniformly on } \mathbf{R} \text{ as } n \to \infty, \tag{3.4}$$

because $\nu^n \in \Omega(\mu)$ and $\mu(t)$ is asymptotically almost periodic.

First, we shall show that $\tau_n \to \infty$ as $n \to \infty$. Suppose not. Then we may assume that for a constant $\tau \geq 0, \tau_n \to \tau$ as $n \to \infty$. Since

$$\rho^*(\sigma(\tau)w, v^n) \leq \rho^*(\sigma(\tau)w, \sigma(\tau_n)w) + \rho^*(\sigma(\tau_n)w, v^n),$$

we have by (3.3)
$$\rho^*(\sigma(\tau)w, v^n) \to 0 \text{ as } n \to \infty.$$

Here we note that since

$$\begin{aligned}w(t, s + \tau, \nu(s)) &= (\sigma(\tau)w)(t, s, \nu(s)) \\ &= \lim_{n \to \infty} v^n(t, s, \nu^n(s)) \\ &= \lim_{n \to \infty} \nu^n(t + s) = \nu(t + s),\end{aligned}$$

we have $w(t, s + \tau, \nu(s)) = \nu(s + t)$. Since

$$d(\mu(\tau), \nu(0)) \leq d(\mu(\tau), \mu(\tau_n)) + d(\mu(\tau_n), \nu^n(0)) + d(\nu^n(0), \nu(0))$$

and
$$d(\mu(\tau), \mu(\tau_n)) \to 0 \text{ as } n \to \infty,$$

it follows from (3.2) and (3.4) that $d(\mu(\tau), \nu(0)) = 0$. Then $\mu(t+\tau) = w(t, \tau, \mu(\tau)) = w(t, \tau, \nu(0)) = \nu(t)$ for all $t \in \mathbf{R}^+$. In particular,

$$d(\mu(t_n + \tau), \nu(t_n)) = 0 \text{ for all } n. \tag{3.5}$$

On the other hand, for sufficiently large n we get

$$\begin{aligned} d(\mu(t_n + \tau), \nu(t_n)) &\geq d(\mu(t_n + \tau_n), \nu^n(t_n)) - d(\mu(t_n + \tau_n), \mu(t_n + \tau)) \\ &\quad - d(\nu^n(t_n), \nu(t_n)) \\ &\geq \varepsilon - \frac{\varepsilon}{4} - \frac{\varepsilon}{4} \end{aligned}$$

by (3.1), (3.4) and the uniform continuity of $\mu(t)$ on $\mathbf{R}^+$. This contradicts (3.5).

By virtue of the sequential compactness of $H_\sigma(w)$ and the asymptotic almost periodicity of $\mu(t)$, we can assume that

$$(\mu^{\tau_n}, \sigma(\tau_n)w) \to (\eta, v) \text{ compactly on } \mathbf{R} \tag{3.6}$$

for some $(\eta, v) \in \Omega_\sigma(\mu, w)$. Since

$$d(\eta(0), \nu(0)) \leq d(\eta(0), \mu(\tau_n)) + d(\mu(\tau_n), \nu^n(0)) + d(\nu^n(0), \nu(0)),$$

(3.2), (3.4) and (3.6) imply $\eta(0) = \nu(0)$, and therefore

$$\begin{aligned} \eta(t) &= v(t, 0, \eta(0)) = v(t, 0, \nu(0)) \\ &= \lim_{n \to \infty} (\sigma(\tau_n)w)(t, 0, \nu^n(0)) \\ &= \lim_{n \to \infty} v^n(t, 0, \nu^n(0)) \\ &= \lim_{n \to \infty} \nu^n(t) = \nu(t) \end{aligned}$$

by (3.3), (3.4) and (3.6). Hence we have

$$\begin{aligned} d(\mu(\tau_n + t), \nu^n(t)) &\leq d(\mu(\tau_n + t), \eta(t)) + d(\nu(t), \nu^n(t)) \\ &< \varepsilon \end{aligned}$$

for all sufficiently large n, which contradicts (3.1). This completes the proof of Theorem 3.3.

The following theorem is an extension of the well known result (cf., [231, Theorem 16.1]) for $\mathcal{X} = \mathbf{R}^n$ to the case of $\dim \mathcal{X} = \infty$.

Theorem 3.4 *If the integral $\mu(t)$ on $\mathbf{R}^+$ of the almost periodic process w is asymptotically almost periodic, then there exists an almost periodic integral of the process w.*

Proof. Since $w(t, s, x)$ is an almost periodic process, there exsits a sequence $\{t_n\}, t_n \to \infty$ as $n \to \infty$, such that $(\sigma(t_n)w)(t, s, x) \to w(t, s, x)$ as $n \to \infty$ uniformly with respect to t, x in bounded sets and $s \in \mathbf{R}$. The integral $\mu(t)$ has the decomposition $\mu(t) = \phi(t) + \psi(t)$, where $\phi(t)$ is almost periodic and $\psi(t) \to 0$ as $t \to \infty$. Hence we may assume $\phi(t + t_n) \to \phi^*(t)$ as $n \to \infty$ uniformly on $t \in \mathbf{R}$, where we note $\phi^*(t)$ is almost periodic. Since $(\phi^*, w) \in \Omega_\sigma(\mu, w)$, $\phi^*(t)$ is an almost periodic integral of w.

3.2.2. Uniform Asymptotic Stability and Existence of Almost Periodic Integrals

We shall show the existence of an almost periodic integral under the assumption of uniform asymptotic stability.

Lemma 3.1 *Let $T > 0$. Then for any $\varepsilon > 0$, there exists a $\delta(\varepsilon) > 0$ with the property that $d(\mu(s), \phi(0)) < \delta(\varepsilon)$ and $\rho^*(\sigma(s)w, v) < \delta(\varepsilon)$ imply $\phi(t) \in V_\varepsilon(\mu(s+t))$ for $t \in [0, T]$, whenever $s \in \mathbf{R}^+$ and $(\phi, v) \in \Omega_\sigma(\mu, w)$.*

Proof. Suppose the contrary. Then, for some $\varepsilon > 0$ there exists sequences $\{s_n\}$, $s_n \in \mathbf{R}^+$, $\{\tau_n\}$, $0 < \tau_n < T$, and $(\phi^n, v^n) \in \Omega_\sigma(\mu, w)$ such that

$$\rho^*(\sigma(s_n)w, v^n) < \frac{1}{n},$$

$$\phi^n(0) \in V_{1/n}(\mu(s_n)),$$

$$\phi^n(t) \in V_\varepsilon(\mu(s_n + t)) \text{ for } t \in [0, \tau_n)$$

and

$$\phi^n(\tau_n) \in \partial V_\varepsilon(\mu(s_n + \tau_n)).$$

Since $\tau_n \in [0, T]$, we can assume that τ_n converges to a $\tau \in [0, T]$ as $n \to \infty$. Since $\Omega_\sigma(\mu, w)$ is compact, we may assume $(\phi^n, v^n) \to (\phi, v) \in \Omega_\sigma(\mu, w)$ and $(\mu^{s_n}, \sigma(s_n)w) \to (\eta, v) \in \Omega_\sigma(\mu, w)$ as $n \to \infty$, respectively. Then $\phi(\tau) \in \partial V_\varepsilon(\eta(\tau))$. On the other hand, since $\phi(0) = \eta(0)$, we get $\phi(t) = \eta(t)$ on $\mathbf{R}^+$ by (p2). This is a contradiction.

Theorem 3.5 *Suppose that w is an almost periodic process on $\mathcal{X}$, and let $\mu(t)$ be an integral on $\mathbf{R}^+$ of w such that the set $O^+(\mu)$ is relatively compact in $\mathcal{X}$. If the integral $\mu(t)$ is UAS, then it has Property (A). Consequently, it is asymptotically almost periodic.*

Proof. Suppose that $\mu(t)$ has not Property (A). Then there are sequences $\{t_n\}, t_n \geq 0, \{r_n\}, r_n > 0, (\phi^n, v^n) \in \Omega_\sigma(\mu, w)$ and a constant $\delta_1, 0 < \delta_1 < \delta_0/2$, such that

$$\phi^n(0) \in V_{1/n}(\mu(t_n)) \text{ and } \rho^*(v^n, \sigma(t_n)w) < \frac{1}{n} \tag{3.7}$$

and

$$\phi^n(r_n) \in \partial V_{\delta_1}(\mu(t_n + r_n)) \text{ and } \phi^n(t) \in V_{\delta_1}(\mu(t + t_n)) \text{ on } [0, r_n), \tag{3.8}$$

where δ_0 is the one given for the UAS of $\mu(t)$. By Theorem 3.2, $\mu(t)$ is UAS in $\Omega_\sigma(w)$. Let $\delta(\cdot)$ be the one given for US of $\mu(t)$ in $\Omega_\sigma(w)$. There exists a sequence $\{q_n\}, 0 < q_n < r_n$, such that

$$\phi^n(q_n) \in \partial V_{\delta(\delta_1/2)/2}(\mu(t_n + q_n)) \tag{3.9}$$

and
$$\phi^n(t) \in \overline{V_{\delta_1}(\mu(t+t_n))} \setminus V_{\delta(\delta_1/2)/2}(\mu(t+t_n)) \text{ on } [q_n, r_n], \tag{3.10}$$

for a large n by (3.7) and (3.8). Suppose that there exists a subsequence of $\{q_n\}$, which we shall denote by $\{q_n\}$ again, such that q_n converges some $q \in \mathbf{R}^+$. It follows from (3.7) that there exists an $n_0 > 0$ such that for any $n \geq n_0$, $q + 1 \geq q_n \geq 0$ and $\phi^n(t) \in V_{\delta(\delta_1/2)/4}(\mu(t_n + t))$ for $t \in [0, q+1]$ by Lemma 3.1, which contraicts (3.9). Therefore, we can see that $q_n \to \infty$ as $n \to \infty$.

Put $p_n = r_n - q_n$ and suppose that $p_n \to \infty$ as $n \to \infty$. Set $s_n = q_n + (p_n/2)$. By (3.7) and the compactness of $\Omega_\sigma(\mu, w)$, we may assume that $((\phi^n)^{s_n}, \sigma(s_n)v^n)$ and $(\mu^{t_n+s_n}, \sigma(t_n + s_n)w)$ tend to some (ϕ, v) and $(\eta, v) \in \Omega_\sigma(\mu, w)$ compactly on $\mathbf{R}$ as $n \to \infty$, respectively. For any fixed $t > 0$, one can take an $n_1 > 0$ such that for every $n \geq n_1$, $r_n - s_n = p_n/2 > t$, because $p_n \to \infty$ as $n \to \infty$. Therefore for $n \geq n_1$, we have $q_n < t + s_n < r_n$, and

$$\phi^n(t + s_n) \notin V_{\delta(\delta_1/2)/2}(\mu(t + t_n + s_n)) \tag{3.11}$$

by (3.10). There exists an $n_2 \geq n_1$ such that for every $n \geq n_2$

$$\phi^n(t + s_n) \in \overline{V_{\delta(\delta_1/2)/8}(\phi(t))} \text{ and } \eta(t) \in \overline{V_{\delta(\delta_1/2)/8}(\mu(t + t_n + s_n))}. \tag{3.12}$$

It follows from (3.11) and (3.12) that for every $n \geq n_2$,

$$\begin{aligned} d(\phi(t), \eta(t)) &\geq d(\phi^n(t + s_n), \mu(t + t_n + s_n)) - d(\mu(t + t_n + s_n), \eta(t)) \\ &\quad - d(\phi^n(t + s_n), \phi(t)) \\ &\geq \delta(\delta_1/2)/4. \end{aligned} \tag{3.13}$$

However, since $\eta(0) \in V_{\delta_0/2}(\phi(0))$, the UAS of $\mu(t)$ in $\Omega_\sigma(w)$ implies $d(\phi(t), \eta(t)) \to 0$ as $t \to \infty$, which contradicts (3.13).

Now we may assume that p_n converges to some $p \in \mathbf{R}^+$ as $n \to \infty$, and that $0 \leq p_n < p + 1$ for all n. Moreover, we may assume that $((\phi^n)^{q_n}, \sigma(q_n)v^n)$ and $(\mu^{t_n+q_n}, \sigma(t_n + q_n)w)$ tend to some $(\psi, u), (\nu, u) \in \Omega_\sigma(\mu, w)$ as $n \to \infty$, respectively. Since $d(\psi(0), \nu(0)) = \delta(\delta_1/2)/2$ by (3.9), we have $\psi(p) \in V_{\delta_1/2}(\nu(p))$. However, we have a contradiction by (3.8), because $d(\psi(p), \nu(p)) \geq d(\mu(t_n + r_n), \phi^n(r_n)) - d(\psi(p), \phi^n(q_n + p)) - d(\phi^n(q_n + p_n), \phi^n(q_n + p)) - d(\phi^n(q_n + p_n), \phi^n(r_n)) - d(\mu(t_n + r_n), \nu(p_n)) - d(\nu(p_n), \nu(p)) \geq \delta_1/2$ for all large n. Thus the integral $\mu(t)$ must have Property (A).

3.2.3. Separation Condition and Existence of Almost Periodic Integrals

We shall establish an existence theorem of almost periodic integrals under a separation condition. When $(\nu, v) \in \Omega_\sigma(\mu, w)$, we often write $\nu \in \Omega_v(\mu)$.

Definition 3.3 $\Omega_\sigma(\mu, w)$ is said to satisfy a separation condition if for any $v \in \Omega_\sigma(w)$, $\Omega_v(\mu)$ is a finite set and if ϕ and ψ, $\phi, \psi \in \Omega_v(\mu)$, are distinct integrals of v, then there exists a constant $\lambda(v, \phi, \psi) > 0$ such that

$$d(\phi(t), \psi(t)) \geq \lambda(v, \phi, \psi) \text{ for all } t \in \mathbf{R}.$$

To make expressions simple, we shall use the following notations. For a sequence $\{\alpha_k\}$, we shall denote it by α and $\beta \subset \alpha$ means that β is a subsequence of α. For $\alpha = \{\alpha_k\}$ and $\beta = \{\beta_k\}$, $\alpha + \beta$ will denote the sequence $\{\alpha_k + \beta_k\}$. Moreover, $L_\alpha x$ will denote $\lim_{k \to \infty} x(t + \alpha_k)$, whenever $\alpha = \{\alpha_k\}$ and limit exists for each t.

Lemma 3.2 *Suppose that $\Omega_\sigma(\mu, w)$ satisfies the separation condition. Then one can choose a number λ_0 independent of $v \in \Omega_\sigma(w)$, ϕ and ψ for which $d(\phi(t), \psi(t)) \geq \lambda_0$ for all $t \in \mathbf{R}$. The number λ_0 is called the separation constant for $\Omega_\sigma(\mu, w)$.*

Proof. Obviously, we can assume that the number $\lambda(v, \phi, \psi)$ is independent of ϕ and ψ. Let v_1 and v_2 are in $\Omega_\sigma(w)$. Then there exists a sequence $r' = \{r'_k\}$ such that
$$v_2(t, s, x) = \lim_{k \to \infty} (\sigma(r'_k)v_1)(t, s, x)$$
uniformly on $\mathbf{R}^+ \times S$ for any bounded set S in $\mathbf{R}^+ \times \mathcal{X}$, that is, $L_{r'}v_1 = v_2$ uniformly on $\mathbf{R}^+ \times S$ for any bounded set S in $\mathbf{R} \times \mathcal{X}$. Let $\phi^1(t)$ and $\phi^2(t)$ be integrals in $\Omega_{v_1}(\mu)$. There exist a subsequence $r \subset r'$, $(\psi^1, v_2) \in H_\sigma(\phi^1, v_1)$ and $(\psi^2, v_2) \in H_\sigma(\phi^2, v_1)$ such that $L_r\phi^1 = \psi^1$ and $L_r\phi^2 = \psi^2$ in $\mathcal{X}$ compactly on $\mathbf{R}$. Since $H_\sigma(\phi^i, v_1) \subset \Omega_\sigma(\mu, w)$, $i = 1, 2$, ψ^1 and ψ^2 also are in $\Omega_{v_2}(\mu)$. Let ϕ^1 and ϕ^2 be distinct integrals. Then
$$\inf_{t \in \mathbf{bf} R} d(\phi^1(t + r_k), \phi^2(t + r_k)) = \inf_{t \in \mathbf{bf} R} d(\phi^1(t), \phi^2(t)) = \alpha_{12} > 0,$$
and hence
$$\inf_{t \in \mathbf{bf} R} d(\psi^1(t), \psi^2(t)) = \beta_{12} \geq \alpha_{12} > 0, \tag{3.14}$$

which means that ψ^1 and ψ^2 are distinct integrals of the process $v_2(t, s, x)$. Let $p_1 \geq 1$ and $p_2 \geq 1$ be the numbers of distinct integrals of processes $v_1(t, s, x)$ and $v_2(t, s, x)$, respectively. Clearly, $p_1 \leq p_2$. In the same way, we have $p_2 \leq p_1 =: p$.

Now, let $\alpha = \min\{\alpha_{ik} : i, k = 1, 2, \cdots, p, i \neq k\}$ and $\beta = \min\{\beta_{jm} : j, m = 1, 2, \cdots, p, j \neq m\}$. By (3.14), we have $\alpha \leq \beta$. In the same way, we have $\alpha \geq \beta$. Therfore $\alpha = \beta$, and we may set $\lambda_0 = \alpha = \beta$.

Theorem 3.6 *Assume that $\mu(t)$ is an integral on $\mathbf{R}^+$ of the almost periodic process w such that the set $O^+(\mu)$ is relatively compact in $\mathcal{X}$, and suppose that $\Omega_\sigma(\mu, w)$ satisfies the separation condition. Then $\mu(t)$ has Property (A). Consequently, $\mu(t)$ is asymptotically almost periodic.*

Proof. Suppose that $\mu(t)$ has not Property (A). Then there exists an $\varepsilon > 0$ and sequences $(\phi^k, v^k) \in \Omega_\sigma(\mu, w)$, $\tau_k \geq 0$ and $t_k \geq 0$ such that
$$d(\mu(t_k + \tau_k), \phi^k(t_k)) = \varepsilon(< \lambda_0/2), \tag{3.15}$$

$$d(\mu(\tau_k), \phi^k(0)) = 1/k \tag{3.16}$$

and
$$\rho^*(\sigma(\tau_k)w, v^k) = 1/k, \qquad (3.17)$$

where λ_0 is the separation constant for $\Omega_\sigma(\mu, w)$.

First, we shall show that $t_k + \tau_k \to \infty$ as $k \to \infty$. Suppose not. Then there exists a subsequence of $\{\tau_k\}$, which we shall denote by $\{\tau_k\}$ again, and a constant $\tau \geq 0$ and that $\tau_k \to \tau$ as $k \to \infty$. Since

$$\rho^*(\sigma(\tau)w, v^k) \leq \rho^*(\sigma(\tau)w, \sigma(\tau_k)w) + \rho^*(\sigma(\tau_k)w, v^k),$$

(3.17) implies that

$$\rho^*(\sigma(\tau)w, v^k) \to 0 \text{ as } k \to \infty. \qquad (3.18)$$

Moreover, we can assume that

$$(\phi^k, v^k) \to (\phi, v) \text{ compactly on } \mathbf{R}$$

for some $(\phi, v) \in \Omega_\sigma(\mu, w)$. From (3.18), it follows that $v = \sigma(\tau)w$. Since

$$d(\mu(\tau), \phi(0)) \leq d(\mu(\tau), \mu(\tau_k)) + d(\mu(\tau_k), \phi^k(0)) + d(\phi^k(0), \phi(0)) \to 0$$

as $k \to \infty$ by (3.16), we get $\mu(\tau) = \phi(0)$, and hence

$$\mu(t + \tau) = (\sigma(\tau)w)(t, 0, \mu(\tau)) = (\sigma(\tau)w)(t, 0, \phi(0)) = v(t, 0, \phi(0)) = \phi(t).$$

However, we have $d(\mu(t_k + \tau_k), \phi(t_k)) \geq \varepsilon/2$ for a sufficiently large k by (3.15), because

$$d(\mu(t_k + \tau_k), \phi^k(t_k)) \geq d(\mu(t_k + \tau_k), \phi^k(t_k)) - d(\phi^k(t_k), \phi(t_k)).$$

This is a contradiction. Thus we must have $t_k + \tau_k \to \infty$ as $k \to \infty$.

Now, set $q_k = t_k + \tau_k$ and $\nu^k(t) = \phi^k(t_k + t)$. Then

$$(\mu^{q_k}, \sigma(q_k)w) \in H_\sigma(\mu, w) \text{ and } (\nu^k, \sigma(t_k)v^k) \in \Omega_\sigma(\mu, w),$$

respectively. We may assume that $(\mu^{q_k}, \sigma(q_k)w) \to (\bar{\mu}, \bar{w})$ compactly on $\mathbf{R}$ for some $(\bar{\mu}, \bar{w}) \in \Omega_\sigma(\mu, w)$, because $q_k \to \infty$ as $k \to \infty$. Since

$$\begin{aligned}\rho^*(\bar{w}, \sigma(t_k)v^k) &\leq \rho^*(\bar{w}, \sigma(q_k)w) + \rho^*(\sigma(q_k)w, \sigma(t_k)v^k) \\ &= \rho^*(\bar{w}, \sigma(q_k)w) + \rho^*(\sigma(\tau_k)w, v^k),\end{aligned}$$

we see that $\rho^*(\bar{w}, \sigma(t_k)v^k) \to 0$ as $k \to \infty$ by (3.17). Hence, we can choose a subsequence $\{\nu^{k_j}\}$ of $\{\nu^k\}$ and a $\bar{\nu} \in \Omega_{\bar{w}}(\mu)$ such that

$$(\nu^{k_j}, \sigma(t_{k_j})v^{k_j}) \to (\bar{\nu}, \bar{w}) \text{ compactly on } \mathbf{R}.$$

Since

$$\lim_{j\to\infty}\{d(\mu(t_{k_j}+\tau_{k_j}),\phi^{k_j}(t_{k_j}))-d(\nu^{k_j}(0),\bar{\nu}(0))-d(\bar{\mu}(0),\mu(q_{k_j}))\}$$
$$\leq d(\bar{\mu}(0),\bar{\nu}(0))$$
$$\leq \lim_{j\to\infty}\{d(\mu(t_{k_j}+\tau_{k_j}),\phi^{k_j}(t_{k_j}))+d(\nu^{k_j}(0),\bar{\nu}(0))+d(\bar{\mu}(0),\mu(q_{k_j}))\},$$

it follows from (3.15) that $d(\bar{\mu}(0),\bar{\nu}(0))=\varepsilon$, which contradicts the separation condition of $\Omega_\sigma(\mu,w)$.

If for any $v\in\Omega_\sigma(w)$, $\Omega_v(\mu)$ consists of only one element, then $\Omega_\sigma(\mu,w)$ clearly satisfies the separation condition. Thus, the following result (cf.[54]) is an immediate consequence of Theorems 3.4 and 3.6.

Corollary 3.1 *If for any $v\in\Omega_\sigma(w)$, $\Omega_v(\mu)$ consists of only one element, then there exists an almost periodic integral of the process w.*

3.2.4. Relationship between the Uniform Asymptotic Stability and the Separation Condition

Theorem 3.7 *Assume that $\mu(t)$ is an integral on $\mathbf{R}^+$ of the almost periodic process w such that the set $O^+(\mu)$ is relatively compact in $\mathcal{X}$. Then the following statements are equivalent:*

i) $\Omega_\sigma(\mu,w)$ *satisfies the separation condition;*

ii) *there exists a number $\delta_0>0$ with the property that for any $\varepsilon>0$ there exists a $t_0(\varepsilon)>0$ such that $d(\phi(s),\psi(s))<\delta_0$ implies $d(\phi(t),\psi(t))<\varepsilon$ for $t\geq s+t_0(\varepsilon)$, whenever $s\in\mathbf{R}, v\in\Omega_\sigma(w)$ and $\phi,\psi\in\Omega_v(\mu)$.*

Consequently, the UAS of the integral $\mu(t)$ on $\mathbf{R}^+$ implies the separation condition on $\Omega_\sigma(\mu,w)$.

Proof. If we set $\delta_0=\lambda_0$, then Claim i) clearly implies Claim ii).

We shall show that Claim ii) implies Claim i). First of all, we shall verify that any distinct integrals $\phi(t),\psi(t)$ in $\Omega_v(\mu)$, $v\in\Omega_\sigma(w)$, satisfy

$$\liminf_{t\to-\infty}d(\phi(t),\psi(t))\geq\delta_0. \tag{3.19}$$

Suppose not. Then for some $v\in\Omega_\sigma(w)$, there exists two distinct integrals $\phi(t)$ and $\psi(t)$ in $\Omega_v(\mu)$ which satisfy

$$\liminf_{t\to-\infty}d(\phi(t),\psi(t))<\delta_0. \tag{3.20}$$

Since $\phi(t)$ and $\psi(t)$ are distinct integrals, we have $d(\phi(s),\psi(s))=\varepsilon$ at some s and for some $\varepsilon>0$. Then there is a t_1 such that $t_1<s-t_0(\varepsilon/2)$ and $d(\phi(t_1),\psi(t_1))<\delta_0$ by (3.20). Then $d(\phi(s),\psi(s))<\varepsilon/2$, which contradicts $d(\phi(s),\psi(s))=\varepsilon$. Thus we have (3.19). Since $O^+(\mu)$ is compact, there are a finite number of coverings which consists of m_0 balls with diameter $\delta_0/4$. We shall show that the number of integrals in $\Omega_v(\mu)$ is at most m_0. Suppose not. Then there are $m_0 + 1$ integrals in $\Omega_v(\mu), \phi^j(t), j = 1, 2, \cdots, m_0 + 1$, and a t_2 such that

$$d(\phi^j(t_2), \phi^i(t_2)) \geq \delta_0/2 \text{ for } i \neq j, \tag{3.21}$$

by (3.19). Since $\phi^j(t_2), j = 1, 2, \cdots, m_0 + 1$, are in $\overline{O^+(\mu)}$, some of these integrals, say $\phi^i(t), \phi^j(t) (i \neq j)$, are in one ball at time t_2, and hence $d(\phi^j(t_2), \phi^i(t_2)) < \delta_0/4$, which contradicts (3.21). Therefore the number of integrals in $\Omega_v(\mu)$ is $m \leq m_0$. Thus

$$\Omega_v(\mu) = \{\phi^1(t), \phi^2(t), \cdots, \phi^m(t)\} \tag{3.22}$$

and

$$\liminf_{t \to -\infty} d(\phi^j(t), \phi^i(t)) \geq \delta_0, i \neq j. \tag{3.23}$$

Consider a sequence $\{\tau_k\}$ such that $\tau_k \to -\infty$ as $k \to \infty$ and $\rho^*(\sigma(\tau_k)v, v) \to 0$ as $k \to \infty$. For each $j = 1, 2, \cdots, m$, set $\phi^{j,k}(t) = \phi^j(t + \tau_k)$. Since $(\phi^{j,k}, \sigma(\tau_k)v) \in H_\sigma(\phi^j, v)$, we can assume that

$$(\phi^{j,k}, \sigma(\tau_k)v) \to (\psi^j, v) \text{ compactly on } \mathbf{R}$$

for some $(\psi^j, v) \in H_\sigma(\phi^j, v) \subset \Omega_\sigma(\mu, w)$. Then it follows from (3.23) that

$$d(\psi^j(t), \psi^i(t)) \geq \delta_0, \text{ for all } t \in \mathbf{R} \text{ and } i \neq j. \tag{3.24}$$

Since the number of integrals in $\Omega_v(\mu)$ is m, $\Omega_v(\mu)$ consists of $\psi^1(t), \psi^2(t), \cdots, \psi^m(t)$ and we have (3.24). This shows that $\Omega_\sigma(\mu, w)$ satisfies the separation condition.

3.2.5. Existence of an Almost Periodic Integral of Almost Quasi- Processes

By using the same arguments as in the proof of Theorem 3.3, we can see that $\mu(t)$ on $\mathbf{R}^+$ of the quasi-process w is asymptotically almost periodic if and only if it has Property (A). Hence we have the following theorem by using the same arguments as in the proof of Theorem 3.5, if $O^+(\mu)$ is contained in $\mathcal{Y}$.

Theorem 3.8 *Let $\mathcal{Y}$ be a nonempty closed set in $\mathcal{X}$. Suppose that w is an almost periodic quasi-process on $\mathcal{X}$, and let $\mu(t)$ be an integral on $\mathbf{R}^+$ of w such that the set $O^+(\mu)$ is contained in $\mathcal{Y}$ and relatively compact in $\mathcal{X}$. If the integral $\mu(t)$ is $\mathcal{Y}$-UAS, then it has Property (A). Consequently, it is asymptotically almost periodic.*

3.3. PROCESSES AND QUASI-PROCESSES GENERATED BY ABSTRACT FUNCTIONAL DIFFERENTIAL EQUATIONS

3.3.1. Abstract Functional Differential Equations with Infinite Delay

We shall treat abstract functional differential equations on *a fading memory space* (resp. *uniform fading memory space*) and show that quasi-processes (resp. processes) are naturally generated by functional differential equations.

We first explain some notation and convention employed throughout this section. Let $\mathbf{X}$ be a Banach space with norm $\|\cdot\|_\mathbf{X}$. For any interval $J \subset \mathbf{R}$, we denote by $BC(J;\mathbf{X})$ the space of all bounded and continuous functions mapping J into $\mathbf{X}$. Clearly $BC(J;\mathbf{X})$ is a Banach space with the norm $\|\cdot\|_{BC(J;\mathbf{X})}$ defined by $\|\phi\|_{BC(J;\mathbf{X})} = \sup\{\|\phi(t)\|_\mathbf{X} : t \in J\}$. If $J = \mathbf{R}^-, \mathbf{R}^- = (-\infty, 0]$, then we simply write $BC(J;\mathbf{X})$ and $\|\cdot\|_{BC(J;\mathbf{X})}$ as BC and $\|\cdot\|_{BC}$, respectively. For any function $u : (-\infty, a) \mapsto \mathbf{X}$ and $t < a$, we define a function $u_t : \mathbf{R}^- \mapsto \mathbf{X}$ by $u_t(s) = u(t+s)$ for $s \in \mathbf{R}^-$. Let $\mathcal{B} = \mathcal{B}(\mathbf{R}^-;\mathbf{X})$ be a real Banach space of functions mapping $\mathbf{R}^-$ into $\mathbf{X}$ with a norm $\|\cdot\|_\mathcal{B}$. The space $\mathcal{B}$ is assumed to have the following properties:

(A1) There exist a positive constant N and locally bounded functions $K(\cdot)$ and $M(\cdot)$ on $\mathbf{R}^+$ with the property that if $u : (-\infty, a) \mapsto \mathbf{X}$ is continuous on $[\sigma, a)$ with $u_\sigma \in \mathcal{B}$ for some $\sigma < a$, then for all $t \in [\sigma, a)$,

(i) $u_t \in \mathcal{B}$,

(ii) u_t is continuous in t (w.r.t. $\|\cdot\|_\mathcal{B}$),

(iii) $N\|u(t)\|_\mathbf{X} \leq \|u_t\|_\mathcal{B} \leq K(t-\sigma)\sup_{\sigma \leq s \leq t}\|u(s)\|_\mathbf{X} + M(t-\sigma)\|u_\sigma\|_\mathcal{B}$.

(A2) If $\{\phi^n\}$ is a sequence in $\mathcal{B} \cap BC$ converging to a function ϕ compactly on $\mathbf{R}^-$ and $\sup_n \|\phi^n\|_{BC} < \infty$, then $\phi \in \mathcal{B}$ and $\|\phi^n - \phi\|_\mathcal{B} \to 0$ as $n \to \infty$.

It is known [107, Proposition 7.1.1] that the space $\mathcal{B}$ contains BC and that there is a constant $J > 0$ such that

$$\|\phi\|_\mathcal{B} \leq J\|\phi\|_{BC}, \quad \phi \in BC. \tag{3.25}$$

Set $\mathcal{B}_0 = \{\phi \in \mathcal{B} : \phi(0) = 0\}$ and define an operator $S_0(t) : \mathcal{B}_0 \mapsto \mathcal{B}_0$ by

$$[S_0(t)\phi](s) = \begin{cases} \phi(t+s) & \text{if } t+s \leq 0, \\ 0 & \text{if } t+s > 0 \end{cases}$$

for each $t \geq 0$. In virtue of (A1), one can see that the family $\{S_0(t)\}_{t \geq 0}$ is a strongly continuous semigroup of bounded linear operators on $\mathcal{B}_0$. We consider the following properties:

(A3) $\lim_{t \to \infty} \|S_0(t)\phi\|_\mathcal{B} = 0, \quad \phi \in \mathcal{B}_0$.

(A3') $\lim_{t \to \infty} \|S_0(t)\| = 0$.

Here and hereafter, we denote by $\|\cdot\|$ the operator norm of linear bounded operators. The space $\mathcal{B}$ is called *a fading memory space* (resp. *a uniform fading memory space*), if it satisfies (A3) (resp. (A3')) in addition to (A1) and (A2). It is obvious that $\mathcal{B}$ is a fading memory space whenever it is a uniform fading memory space. It is known [107, Proposition 7.1.5] that the functions $K(\cdot)$ and $M(\cdot)$ in (A1) can be chosen as $K(t) \equiv J$ and $M(t) \equiv (1+(J/N))\|S_0(t)\|$. Note that (A3) implies

$\sup_{t \geq 0} \|S_0(t)\| < \infty$ by the Banach-Steinhaus theorem. Therefore, whenever $\mathcal{B}$ is a fading memory space, we can assume that the functions $K(\cdot)$ and $M(\cdot)$ in (A1) satisfy $K(\cdot) \equiv K$ and $M(\cdot) \equiv M$, constants.

We provide a typical example of fading memory spaces. Let $g : \mathbf{R}^- \mapsto [1, \infty)$ be any continuous, nonincreasing function such that $g(0) = 1$ and $g(s) \to \infty$ as $s \to -\infty$. We set

$$C_g^0 := C_g^0(\mathbf{X}) = \{\phi \in C(\mathbf{R}^-, \mathbf{X}), \lim_{s \to -\infty} \|\phi(s)\|_\mathbf{X}/g(s) = 0\}.$$

Then the space C_g^0 equipped with the norm

$$\|\phi\|_g = \sup_{s \leq 0} \frac{\|\phi(s)\|_\mathbf{X}}{g(s)}, \quad \phi \in C_g^0,$$

is a separable Banach space and it satisfies (A1)-(A3). Moreover, one can see that (A3') holds if and only if $\sup\{g(s+t)/g(s) : s \leq -t\} \to 0$ as $t \to \infty$. Therefore, if $g(s) \equiv e^{-s}$, then the space C_g^0 is a uniform fading memory space. On the other hand, if $g(s) = 1 + \|s\|^k$ for some $k > 0$, then the space C_g^0 is a fading memory space, but not a uniform fading memory space.

Throughout the remainder of this section, we assume that $\mathcal{B}$ is a fading memory space or a uniform fading memory space which is separable.

Now we consider the following functional differential equation

$$\frac{du}{dt} = Au(t) + F(t, u_t), \tag{3.26}$$

where A is the infinitesimal generator of a compact semigroup $\{T(t)\}_{t \geq 0}$ of bounded linear operators on $\mathbf{X}$ and $F(t, \phi) \in C(\mathbf{R}^+ \times \mathcal{B}; \mathbf{X})$. In what follows, we shall show that (3.26) generates a quasi-process on an appropriate space under some conditions and deduce equivalence relationships between some stability properties of (3.26) and those of its limiting equations as an application of Theorem 3.1.

We assume the following conditions on F:

(H1) $F(t, \phi)$ is uniformly continuous on $\mathbf{R}^+ \times \mathcal{K}$ for any compact set $\mathcal{K}$ in $\mathcal{B}$, and $\{F(t, \phi) \mid t \in \mathbf{R}^+\}$ is a relative compact subset of $\mathbf{X}$ for each $\phi \in \mathcal{B}$.

(H2) For any $H > 0$, there is an $L(H) > 0$ such that $\|F(t, \phi)\|_\mathbf{X} \leq L(H)$ for all $t \in \mathbf{R}^+$ and $\phi \in \mathcal{B}$ such that $\|\phi\|_\mathcal{B} \leq H$.

For $\tau \in \mathbf{R}^+$, we denote the τ-translation F^τ of $F(t, \phi)$ by

$$F^\tau(t, \phi) = F(t + \tau, \phi), \quad (t, \phi) \in \mathbf{R}^+ \times \mathcal{B}.$$

Clearly, F^τ is in $C(\mathbf{R}^+ \times \mathcal{B}; \mathbf{X})$, too. Set

$$H(F) = \overline{\{F^\tau; \tau \in \mathbf{R}^+\}},$$

where $\overline{\{F^\tau; \tau \in \mathbf{R}^+\}}$ denotes the closure of $\{F^\tau; \tau \in \mathbf{R}^+\}$ in $C(\mathbf{R} \times \mathcal{B}; \mathbf{X})$. The subspace $H(F)$ of $C(\mathbf{R}^+ \times \mathcal{B}; \mathbf{X})$ is called the hull of F. It is known [107, Proposition 8.1.3] that $H(F)$ is metrizable. Clearly, the hull $H(F)$ is invariant with respect to the τ-translation; that is, $G^\tau \in H(F)$ whenever $G \in H(F)$ and $\tau \in \mathbf{R}^+$. Moreover, from (H1) and [107, Theorem 7.1.4] it follows that $H(F)$ is a compact set in $C(\mathbf{R}^+ \times \mathcal{B}; \mathbf{X})$. If $G \in H(F)$, one can choose a sequence $\{\tau_n\} \subset \mathbf{R}^+$ so that F^{τ_n} tends to G in $C(\mathbf{R}^+ \times \mathcal{B}; \mathbf{X})$, that is, $F(t + \tau_n, \phi) \to G(t, \phi)$ compactly on $\mathbf{R}^+ \times \mathcal{B}$. Assume (H1) and (H2), and let $\{t'_n\}$ with $t'_n \to \infty$ as $n \to \infty$. Then the family of functions $\{F(t + t'_n, \phi) : n = 1, 2, \ldots\}$ is equicontinuous on the set $[-J, J] \times \mathcal{B}$ for each $J > 0$. Indeed, if $(s_0, \phi_0) \in [-J, J] \times \mathcal{B}$ and $\delta := \|s - s_0\| + \|\phi - \phi_0\|_\mathcal{B}$ is small, then $s_0 + t'_n \in \mathbf{R}^+$ and $s + t'_n \in \mathbf{R}^+$ for sufficiently large n and

$$\begin{aligned}
&\|F(s + t'_n, \phi) - F(s_0 + t'_n, \phi_0)\|_\mathbf{X} \\
&\leq \|F(s + t'_n, \phi - \phi_0)\|_\mathbf{X} + \|F(s + t'_n, \phi_0) - F(s_0 + t'_n, \phi_0)\|_\mathbf{X} \\
&\leq L\delta + \sup\{\|F(t, \phi_0) - F(\tau, \phi_0)\|_\mathbf{X} : t, \tau \in \mathbf{R}^+, \|t - \tau\| \leq \delta\}
\end{aligned}$$

by (H2). Therefore the desired equicontinuity follows from (H1). For each $(t, \phi) \in \mathbf{R} \times \mathcal{B}$, the set $\{F(t + t'_n, \phi) : n \geq n_0\}$ is relatively compact in $\mathbf{X}$ by (H1); here n_0 is a positive integer such that $t'_n + t \in \mathbf{R}^+$ for $n \geq n_0$. Since $\mathcal{B}$ is separable, by the Ascoli-Arzéla theorem and the diagonalization procedure, one can select a subsequence of $\{F(t + t'_n, \phi) : n = 1, 2, \ldots\}$ which is uniformly convergent on any compact subset on $\mathbf{R} \times \mathcal{B}$. Hence there exist a subsequence $\{t_n\} \subset \{t'_n\}$ and a continuous function $G : \mathbf{R} \times \mathcal{B} \mapsto \mathbf{X}$ such that

$$F(t + t_n, \phi) \to G(t, \phi) \quad \text{compactly on } \mathbf{R} \times \mathcal{B}. \tag{3.27}$$

We denote by $\Omega(F)$ the set of all functions $H : \mathbf{R} \times \mathcal{B} \mapsto \mathbf{X}$ for which one can choose a sequence $\{t_n\} \subset \mathbf{R}$ so that $t_n \to \infty$ as $n \to \infty$ and $F(t + t_n, \phi) \to G(t, \phi)$ compactly on $\mathbf{R} \times \mathcal{B}$. Note that the conditions (H1) and (H2) are satisfied for $G(t, \phi)$ as well as $F(t, \phi)$. We denote by $\Omega(F)$ the set of all elements G in $H(F)$ for which one can choose a sequence $\{\tau_n\} \subset \mathbf{R}^+$ so that $\{\tau_n\} \to \infty$ as $n \to \infty$ and $F^{\tau_n} \to G$ in $C(\mathbf{R} \times \mathcal{B}; \mathbf{X})$. If $G \in H(F)$, the system

$$\frac{du}{dt} = Au(t) + G(t, u_t) \quad t \in \mathbf{R}^+, \tag{3.28}$$

is called *an equation in the hull* of System (3.26). In particular, if $G \in \Omega(F)$, then it is called *a limiting equation* of (3.26).

Under the conditions (H1) and (H2), it is known that for any $(\sigma, \phi) \in \mathbf{R}^+ \times \mathcal{B}$, there exists a function $u \in C((-\infty, t_1); \mathbf{X})$ such that $u_\sigma = \phi$ and the following relation holds:

$$u(t) = T(t - \sigma)\phi(0) + \int_\sigma^t T(t - s)G(s, u_s)ds, \quad \sigma \leq t \leq t_1,$$

(cf. [91, Theorem 1]). The function u is called the (mild) solution of (3.28) through (σ, ϕ) defined on $[\sigma, t_1]$ and denoted by $u(t) := u(t, \sigma, \phi, G)$. In the above, t_1 can be taken as $t_1 = \infty$ if $\sup_{t \leq t_1} \|u(t)\|_X < \infty$ (cf. [91, Corollary 2]). In the following, we always assume that (3.26) is *regular*, that is, the following condition is satisfied, too:

(H3) For any $G \in H(F)$ and $(\sigma, \phi) \in \mathbf{R}^+ \times \mathcal{B}$, Equation (3.28) has a unique solution through (σ, ϕ) which exists for all $t \geq \sigma$.

3.3.2. Processes and Quasi-Processes Generated by Abstract Functional Differential Equations with Infinite Delay

Consider a mapping $\Phi : \mathbf{R}^+ \times \mathbf{R}^+ \times \mathcal{B} \times H(F) \mapsto \mathcal{B}$ defined by

$$\Phi(t, s, \phi, G) = u_{t+s}(s, \phi, G) \in \mathcal{B}, \qquad (t, s, \phi, G) \in \mathbf{R}^+ \times \mathbf{R}^+ \times \mathcal{B} \times H(F).$$

Proposition 3.1 *Assume that $\mathcal{B}$ is a fading memory space and that conditions (H1)–(H3) are hold. Then the mapping Φ is continuous.*

Proof. Assume that the mapping Φ is not continuous. Then there exist an $\varepsilon > 0$, $(t_0, s_0, \phi^0, G) \in \mathbf{R}^+ \times \mathbf{R}^+ \times \mathcal{B} \times H(F)$ and sequences $\{t_k\} \subset \mathbf{R}^+, \{s_k\} \subset \mathbf{R}^+, \{\phi^k\} \subset \mathcal{B}$ and $\{G_k\} \subset H(F)$ such that $(t_k, s_k, \phi^k, G_k) \to (t_0, s_0, \phi^0, G)$ and $|\Phi(t_k, s_k, \phi^k, G_k) - \Phi(t_0, s_0, \phi^0, G)|_\mathcal{B} \geq 3\varepsilon$ for $k \in \mathbf{N}$ ($\mathbf{N}$ denotes the set of all positive integers). Since $\Phi(t, s_0, \phi^0, G) \in \mathcal{B}$ is continuous in $t \in \mathbf{R}^+$ by (A1-ii), we may assume that $|x^k_{t_k} - x_{t_k}|_\mathcal{B} \geq 2\varepsilon$ for all $k \in \mathbf{N}$, where $x(t) = u(t + s_0, s_0, \phi^0, G)$ and $x^k(t) = u(t + s_k, s_k, \phi^k, G_k)$ for $k \in \mathbf{N}$. There exist $\gamma > 0$, σ_k and τ_k, $0 < \sigma_k < \tau_k \leq t_k$, such that $\gamma < \min\{\varepsilon/(1+M), N\varepsilon/K\}$,

$$\|x^k_{\sigma_k} - x_{\sigma_k}\|_\mathcal{B} = \gamma, \qquad \|x^k_{\tau_k} - x_{\tau_k}\|_\mathcal{B} = 2\varepsilon,$$

$$\|x^k_t - x_t\|_\mathcal{B} < \gamma \quad (0 \leq t < \sigma_k)$$

and

$$\|x^k_t - x_t\|_\mathcal{B} < 2\varepsilon \quad (0 \leq t < \tau_k)$$

for $k \in \mathbf{N}$; here the functions $M(\cdot)$ and $K(\cdot)$ in (A1) may be chosen as positive constants M and K, respectively, because $\mathcal{B}$ is a fading memory space. By choosing a subsequence if necessarily, we may assume that $\sigma_k \to \sigma_0 \in [0, t_0]$. We claim that

$$\sigma_0 > 0. \tag{3.29}$$

Indeed, if (3.29) is false, then we have, for any $0 \leq t \leq \min\{\sigma_k, 1\}$,

$$\begin{aligned}\|x^k(t) - x(t)\|_X &= |T(t)x^k(0) + \int_0^t T(t-\tau)G_k(s_k + \tau, x^k_\tau)d\tau \\ &\quad - T(t)x(0) - \int_0^t T(t-\tau)G(s_0 + \tau, x_\tau)d\tau|_X \\ &\leq C_1\{(1/N)|\phi^k - \phi^0|_\mathcal{B} + 2\int_0^t L(H)d\tau\},\end{aligned}$$

where $H = \sup\{|x_t|_\mathcal{B}, |x_t^k|_\mathcal{B} : 0 \leq t \leq \tau_k, k \in \mathbf{N}\}$ and $C_1 = \sup_{0 \leq s \leq 1} ||T(s)||$; hence

$$\begin{aligned} \gamma &= |x_{\sigma_k}^k - x_{\sigma_k}|_\mathcal{B} \\ &\leq K \sup_{0 \leq t \leq \sigma_k} |x^k(t) - x(t)|_\mathbf{X} + M|\phi^k - \phi^0|_\mathcal{B} \\ &\leq KC_1\{(1/N)|\phi^k - \phi^0|_\mathcal{B} + 2\sigma_k L(H)\} + M|\phi^k - \phi^0|_\mathcal{B} \to 0 \end{aligned}$$

as $k \to \infty$, a contradiction.

Next we prove that the set $O := \{x^k(t) : 0 \leq t \leq \tau_k, \; k \in \mathbf{N}\}$ is relatively compact in $\mathbf{X}$. To do this, we consider the sets $O_\eta = \{x^k(t) : \eta \leq t \leq \tau_k, \; k \in \mathbf{N}\}$ and $\tilde{O}_\eta = \{x^k(t) : 0 \leq t \leq \eta, \; k \in \mathbf{N}\}$ for any $\eta > 0$ such that $\eta < \inf_k \tau_k$. Then $\alpha(O) = \max\{\alpha(O_\eta), \alpha(\tilde{O}_\eta)\}$, where $\alpha(\cdot)$ is *Kuratowski's measure of noncompactness* of sets in $\mathbf{X}$. For the details of the properties of $\alpha(\cdot)$, see [133, Section 1.4]. Let $0 < \nu < \min\{1, \eta\}$. Since $x^k(t)$ is a mild solution of $\frac{du}{dt} = Au(t) + G_k(t + s_k, u_t)$ through $(0, \phi^k)$, we get

$$\begin{aligned} x^k(t) &= T(t)\phi^k(0) + \int_0^t T(t-s)h^k(s)ds \\ &= T(\eta)[T(t-\eta)\phi^k(0) + \int_0^{t-\eta} T(t-\eta-s)h^k(s)ds] + \\ &\quad + \int_{t-\eta}^t T(t-s)h^k(s)ds \\ &= T(\eta)x^k(t-\eta) + T(\nu)\int_{t-\eta}^{t-\nu} T(t-s-\nu)h^k(s)ds + \\ &\quad + \int_{t-\nu}^t T(t-s)h^k(s)ds \end{aligned}$$

for $t \geq \eta$, where $h^k(t) = G_k(t + s_k, x_t^k)$. The set $\{\int_{t-\eta}^{t-\nu} T(t-s-\nu)h^k(s)ds : \eta \leq t \leq \tau_k, \; k \in \mathbf{N}\}$ is bounded in $\mathbf{X}$, and hence $T(\nu)\{\int_{t-\eta}^{t-\nu} T(t-s-\nu)h^k(s)ds : \eta \leq t \leq \tau_k, \; k \in \mathbf{N}\}$ is relatively compact in $\mathbf{X}$ by the compactness of the semigroup $\{T(t)\}_{t \geq 0}$. Similarly, one can get the relative compactness of the set $T(\eta)\{x^k(t-\eta) : \eta \leq t \leq \tau_k, \; k \in \mathbf{N}\}$. Since

$$\begin{aligned} \alpha(O_\eta) &= \alpha(\{\int_{t-\nu}^t T(t-s)h^k(s)ds : \eta \leq t \leq \tau_k, \; k \in \mathbf{N}\}) \\ &\leq C_1 L(H)\nu, \end{aligned}$$

letting $\nu \to 0$ in the above, we get $\alpha(O_\eta) = 0$. Hence

$$\begin{aligned} \alpha(O) = \alpha(\tilde{O}_\eta) &= \alpha(\{T(t)\phi^k(0) + \int_0^t T(t-s)h^k(s)ds : 0 \leq t \leq \eta, \; k \in \mathbf{N}\}) \\ &= \alpha(\{\int_0^t T(t-s)h^k(s)ds : 0 \leq t \leq \eta, \; k \in \mathbf{N}\}) \\ &\leq C_1 L(H)\eta \end{aligned}$$

for all $0 < \eta < \inf_k \tau_k$, which shows $\alpha(O) = 0$; consequently, O must be relatively compact in $\mathbf{X}$.

Since the set $\{x^k(\sigma_k), x(\sigma_k) : k \in \mathbf{N}\}$ is relatively compact in $\mathbf{X}$, $\|[T(t-\sigma_k) - I](x^k(\sigma_k) - x(\sigma_k))\|_\mathbf{X} \to 0$ as $|t - \sigma_k| \to 0$. Therefore, repeating almost the same arguments as in the proof of (3.29), we obtain $\inf\{\tau_k - \sigma_k : k \in \mathbf{N}\} =: 2a > 0$, because of the inequality $|x^k_{\tau_k} - x_{\tau_k}|_\mathcal{B} \leq K \sup_{\sigma_k \leq t \leq \tau_k} |x^k(t) - x(t)|_\mathbf{X} + M|x^k_{\sigma_k} - x_{\sigma_k}|_\mathcal{B}$ or $\varepsilon < K \sup_{\sigma_k \leq t \leq \tau_k} |x^k(t) - x(t)|_\mathbf{X}$. Noting that $|x^k(t) - x^k(s)|_\mathbf{X} \leq \sup\{|T(t-s)z - z|_\mathbf{X} : z \in O\} + C_1 L(H)|t - s|$ when $0 \leq s \leq t \leq \sigma_0 + a$ and $t \leq s + 1$, we have that $x^k(t)$ is equicontinuous on $[0, \sigma_0 + a]$. Hence, one may assume that $x^k(t)$ converges to some continuous function $y(t)$ uniformly on $[0, t_0 + a]$ as $k \to \infty$. Putting $y_0 = \phi^0$, we have $x^k_t \to y_t$ in $\mathcal{B}$ uniformly on $[0, \sigma_0 + a]$, because of $x^k_0 = \phi^k \to \phi^0$ in $\mathcal{B}$. Letting $k \to \infty$ in the relation

$$x^k(t) = T(t)\phi^k(0) + \int_0^t T(t-\tau)G_k(s_k + \tau, x^k_\tau)d\tau$$

for $t \in [0, \sigma_k + a]$, we have

$$y(t) = T(t)y(0) + \int_0^t T(t-\tau)G(s_0 + \tau, y_\tau)d\tau$$

for $t \in [0, \sigma_0 + (a/2)]$; hence $y(t - s_0) \equiv u(t, s_0, \phi^0, G) = x(t - s_0)$ on $[s_0, s_0 + \sigma_0 + (a/2)]$ by (H3). Consequently $|y_{\sigma_0} - x_{\sigma_0}|_\mathcal{B} = 0$. This is a contradiction, because we must get $|y_{\sigma_0} - x_{\sigma_0}|_\mathcal{B} = \gamma$ by letting $k \to \infty$ in $|x^k_{\sigma_k} - x_{\sigma_k}|_\mathcal{B} = \gamma$. This completes the proof of the proposition.

Now we take $\mathcal{X} = \mathcal{Y} = \mathcal{B}$ and consider a function $w^G_\mathcal{B} : \mathbf{R}^+ \times \mathbf{R}^+ \times \mathcal{B} \mapsto \mathcal{B}$ defined by

$$w^G_\mathcal{B}(t, s, \phi) = u_{t+s}(s, \phi, G), \qquad (t, s, \phi) \in \mathbf{R}^+ \times \mathbf{R}^+ \times \mathcal{B}.$$

By virtue of (H3), we see that the mapping $w^G_\mathcal{B}$ satisfies (p1) and (p2) with $\mathcal{X} = \mathcal{B}$. Moreover from Proposition 3.1 it follows that $w^G_\mathcal{B}$ satisfies (p3). Thus the mapping $w^G_\mathcal{B}$ is a $\mathcal{B}$-quasi-process on $\mathcal{B}$. In fact, $w^G_\mathcal{B}$ is precisely a process on $\mathcal{B}$ in a sense of [53], [54], [77], [78]. We call $w^G_\mathcal{B}$ a process on $\mathcal{B}$ generated by (3.28).

Now we consider the process $w^F_\mathcal{B}$ on $\mathcal{B}$ generated by (3.26). (H3) yields the relation $u(t + \tau, s + \tau, \phi, F) = u(t, s, \phi, F^\tau)$ for $t \in \mathbf{R}$ and $(s, \tau, \phi) \in \mathbf{R}^+ \times \mathbf{R}^+ \times \mathcal{B}$. Hence $(\sigma(\tau)w^F_\mathcal{B})(t, s, \phi) = w^F_\mathcal{B}(t, \tau + s, \phi) = u_{t+\tau+s}(\tau + s, \phi, F) = u_{t+s}(s, \phi, F^\tau) = w^{F^\tau}_\mathcal{B}(t, s, \phi)$ for $(t, s, \phi) \in \mathbf{R}^+ \times \mathbf{R}^+ \times \mathcal{B}$; in other words,

$$\sigma(\tau)w^F_\mathcal{B} \equiv w^{F^\tau}_\mathcal{B}.$$

Therefore $H_\sigma(w^F_\mathcal{B}) = \{w^G_\mathcal{B} : G \in H(F)\}$ and $\Omega_\sigma(w^F_\mathcal{B}) = \{w^G_\mathcal{B} : G \in \Omega(F)\}$. In particular, the process $w^F_\mathcal{B}$ satisfies the condition (p4). Moreover, we see that $H_\sigma(w^F_\mathcal{B})$ is sequentially compact and that a mapping $\pi_\mathcal{B}(t) : \mathcal{B} \times H_\sigma(w^F_\mathcal{B}) \mapsto \mathcal{B} \times H_\sigma(w^F_\mathcal{B})$ defined by

$$\pi_\mathcal{B}(t)(\phi, w^G_\mathcal{B}) = (u_t(0, \phi, G), \sigma(t)w^G_\mathcal{B}), \qquad (\phi, w^G_\mathcal{B}) \in \mathcal{B} \times H_\sigma(w^F_\mathcal{B}),$$

satisfies the condition (p5), and hence $\pi_\mathcal{B}(t)$ is the skew product flow of $w^F_\mathcal{B}$.

Lemma 3.3 *The skew product flow $\pi_B(t)$ is $\mathcal{B}$-strongly asymptotically smooth whenever $\mathcal{B}$ is a uniform fading memory space.*

Proof. Let B be any bounded closed set in $\mathcal{B} \times H_\sigma(w_{\mathcal{B}}^F)$. Then there is an $H > 0$ such that $B \subset \mathcal{B}_H \times H_\sigma(w_{\mathcal{B}}^F)$, where $\mathcal{B}_H = \{\phi \in \mathcal{B} : |\phi|_{\mathcal{B}} \leq H\}$. Let $L := L(H)$ be a constant in (H2), and set

$$Q_1 = \{\int_0^1 T(\tau)h(\tau)d\tau : \; h \in BC([0,1];\mathbf{X}) \text{ with } |h|_{\mathrm{BC}([0,1];\mathbf{X})} \leq L\}.$$

By the same reason as the one for the set O in the proof of Proposition 3.1, we see that the set Q_1 is relativly compact in $\mathbf{X}$. Then there exists a compact set O_B in $\mathbf{X}$ satisfying

$$T(1)\mathbf{X}_{H/N} + Q_1 \subset O_B, \tag{3.30}$$

where N is the constant in (A1), and $X_{H/N} = \{x \in \mathbf{X} : |x|_{\mathbf{X}} \leq H/N\}$. Denote by J_B the set of all elements ϕ in BC with the property that $\phi(\theta) \in O_B$ for $\theta \in \mathbf{R}^-$ and

$$|\phi(t) - \phi(s)|_{\mathbf{X}} \leq \sup\{|T(t-s)z - z|_{\mathbf{X}} : z \in O_B\} + C_1 L |t-s|$$

for all s,t satisfying $s - 1 \leq t - 1 \leq s \leq 0$, where $C_1 = \sup_{0 \leq s \leq 1} \|T(s)\|$. From (A2) and Ascoli-Arzéla's theorem, we see that J_B is a compact set in $\mathcal{B}$.

Now, let $\{t_n\} \subset \mathbf{R}^+$ and $\{(\phi^n, w_{\mathcal{B}}^{G_n})\} \subset B$ be sequences with the property that $\lim_{n \to \infty} t_n = \infty$ and $\pi_B(t)(\phi^n, w_{\mathcal{B}}^{G_n}) = (u_t(0, \phi^n, G_n), w_{\mathcal{B}}^{G_n^t}) \in B$ for all $t \in [0, t_n]$. We shall show that $\{\pi_B(t_n)(\phi^n, w_{\mathcal{B}}^{G_n})\}$ has a subsequence which approaches to the compact set $J_B \times H_\sigma(w_{\mathcal{B}}^F)$. We may assume that $t_n > 2$ for $n = 1, 2, \cdots$. Set $x^n(t) = u(t, 0, \phi^n, G_n), n = 1, 2, \cdots$. Since $|x_t^n|_{\mathcal{B}} \leq H$ for $t \in [0, t_n]$, we get

$$\begin{aligned} x^n(t) &= T(1)x^n(t-1) + \int_{t-1}^t T(t-s) G_n(s, x_s^n) ds \\ &= T(1)x^n(t-1) + \int_0^1 T(\tau) h^{n,t}(\tau) d\tau \end{aligned}$$

for any $t \in [1, t_n]$, where $h^{n,t}(\tau) = G_n(t - \tau, x_{t-\tau}^n)$. Note that $h^{n,t} \in BC([0,1];\mathbf{X})$ with $|h^{n,t}|_{\mathrm{BC}([0,1];\mathbf{X})} \leq L$ and that $|x^n(t-1)|_{\mathbf{X}} \leq (1/N)|x_{t-1}^n|_{\mathcal{B}} \leq H/N$. It follows from (3.30) that the set $\{x^n(t) : 1 \leq t \leq t_n, \; n = 1, 2, \cdots\} \subset O_B$. Moreover, if $1 \leq s \leq t \leq t_n$ and $|t - s| \leq 1$, then

$$\begin{aligned} |x^n(t) - x^n(s)|_{\mathbf{X}} &\leq |T(t-s)x^n(s) - x^n(s)|_{\mathbf{X}} + |\int_s^t T(t-\tau) G_n(\tau, x_\tau^n) d\tau|_{\mathbf{X}} \\ &\leq \sup\{|T(t-s)z - z|_{\mathbf{X}} : z \in O_B\} + C_1 L |t - s|. \end{aligned}$$

Therefore, if we consider a function y^n defined by $y^n(t) = x^n(t)$ if $1 \leq t \leq t_n$, and $y^n(t) = x^n(1)$ if $t \leq 1$, then $y_{t_n}^n \in J_B$. Observe that

$$u_{t_n}(0, \phi_n, G_n) = y_{t_n}^n + S_0(t_n - 1)[x_1^n - x^n(1)\xi],$$

where $\xi(\theta) = 1$ for $\theta \leq 0$. Since $|x_1^n - x^n(1)\xi|_\mathcal{B} \leq H + JH/N$ by (3.25), we get $|u_{t_n}(0, \phi^n, G_n) - y_{t_n}^n|_\mathcal{B} \leq \|S_0(t_n - 1)\|(1 + J/N)H \to 0$ as $n \to \infty$, because $\mathcal{B}$ is a uniform fading memory space. Thus $\{\pi_\mathcal{B}(t_n)(\phi^n, w_\mathcal{B}^{G_n})\}$ approaches to the compact set $J_B \times H_\sigma(w_\mathcal{B}^F)$.

Suppose the following condition:

(H4) Equation (3.26) has a bounded solution $\bar{u}(t)$ defined on $\mathbf{R}^+$ such that $\bar{u}_0 \in$ BC.

A subset $\mathcal{F}$ of $C(\mathbf{R}^+; \mathbf{X})$ is said to be uniformly equicontinuous on $\mathbf{R}^+$, if $\sup\{\|x(t + \delta) - x(t)\|_\mathbf{X} : t \in \mathbf{R}^+, x \in \mathcal{F}\} \to 0$ as $\delta \to 0^+$. For any set $\mathcal{F}$ in $C(\mathbf{R}^+; \mathbf{X})$ and any set S in $\mathcal{B}$, we set

$$R(\mathcal{F}) = \{x(t) : t \in \mathbf{R}^+, \ x \in \mathcal{F}\}$$

$$W(S, \mathcal{F}) = \{x(\cdot) : \mathbf{R} \mapsto \mathbf{X} \mid x_0 \in S, x|_{\mathbf{R}^+} \in \mathcal{F}\}$$

and

$$V(S, \mathcal{F}) = \{x_t \mid t \in \mathbf{R}^+, x \in W(S, \mathcal{F})\}.$$

Lemma 3.4 *Let $\mathcal{B}$ be a fading memory space. If S is a compact subset in $\mathcal{B}$ and if $\mathcal{F}$ is a uniformly equicontinuous set in $C(\mathbf{R}^+, \mathbf{X})$ such that the set $R(\mathcal{F})$ is relatively compact in $\mathbf{X}$, then the set $V(S, \mathcal{F})$ is relatively compact in $\mathcal{B}$.*

Proof. We shall prove that any sequence $\{x_{t_k}^k\}, t_k \geq 0, x_{t_k}^k \in V(S, \mathcal{F})$, contains a convergent subsequence. Taking a subsequence if necessary, we may assume that $t_k \to t_0 \leq \infty$ and $x_0^k := \phi^k \to \phi$ in S as $k \to \infty$, because S is compact. Let

$$x_{t_k}^k = y_{t_k}^k + S_0(t_k)\psi^k,$$

where

$$y^k(s) = \begin{cases} x^k(s) & s \geq 0 \\ x^k(0) & s \leq 0, \end{cases}$$

$$\psi^k = x_0^k - x^k(0)\chi$$

and

$$\chi(\theta) = 1 \qquad \theta \leq 0.$$

Then $\psi^k \to \psi := \phi - \phi(0)\chi$ in $\mathcal{B}$ as $k \to \infty$. Clearly $\xi^k := y_{t_k}^k$ lies in BC, and the sequence $\{\xi^k\}$ is equicontinuous on $\mathbf{R}^-$. Moreover, for each $\theta \leq 0$ the set $\{\xi^k(\theta) : k = 1, 2, \ldots\}$ is relatively compact in $\mathbf{X}$, because it is contained in the set $R(\mathcal{F})$ which is relatively compact in $\mathbf{X}$. By applying the Ascoli-Arzéla theorem and (A2), we can assume that $\{\xi^k\}$ is a convergent sequence in $\mathcal{B}$. On the other hand, since $\sup_{t \geq 0} \|S_0(t)\| < \infty$, we have $\|S_0(t_k)\psi^k - S_0(t_k)\psi\|_\mathcal{B} \leq \sup_{t \geq 0} \|S_0(t)\|\|\psi^k - \psi\|_\mathcal{B} \to 0$ as $k \to \infty$. If $t_0 < \infty$, then $S_0(t_k)\psi \to S_0(t_0)\psi$ as $k \to \infty$, if $t_0 = \infty$, then

$S_0(t_k)\psi \to 0$ as $k \to \infty$ by (A3). As a result, $\{S_0(t_k)\psi^k\}$ is a convergent sequence in $\mathcal{B}$. Therefore, the sequence $\{x_{t_k}^k\}$ has the desired property.

By noting Lemma 3.4, for the solution $\bar{u}(t)$ of (3.26) whose existence is assumed in (H4), we have the following lemma by using the same arguments as in the proof of Proposition 3.1.

Lemma 3.5 $O_{\bar{u},\mathbf{R}^+} := \overline{\{\bar{u}(t) \mid 0 \leq t\}}$ *is compact in* $\mathbf{X}$, $\bar{u}(t)$ *is uniformly continuous on* $\mathbf{R}^+$ *and* $\mathbf{X}_{\bar{u},\mathbf{R}^+} := \overline{\{\bar{u}_t \mid 0 \leq t\}}$ *is compact in* $\mathcal{B}$.

Therefore, for any sequence $\{\tau_n'\} \subset \mathbf{R}^+$ one can choose a subsequence $\{\tau_n\} \subset \{\tau_n'\}$, $\bar{v} \in C(\mathbf{R}; \mathbf{X})$ and $G \in H(F)$ such that $\lim_{n\to\infty} F^{\tau_n} = G$ in $C(\mathbf{R}^+ \times \mathcal{B}; \mathbf{X})$ and $\lim_{n\to\infty} |\bar{u}_{t+\tau_n} - \bar{v}_t|_\mathcal{B} = 0$ uniformlu on any compact set in $\mathbf{R}^+$, that is,

$$(\bar{u}^{\tau_n}, F^{\tau_n}) \to (\bar{v}, G) \text{ compactly.}$$

Denote by $H(\bar{u}, F)$ the set of all $(\bar{v}, G) \in C(\mathbf{R}; \mathbf{X}) \times H(F)$ such that $(\bar{u}^{\tau_n}, F^{\tau_n}) \to (\bar{v}, G)$ compactly for some sequence $\{\tau_n\} \subset \mathbf{R}^+$. In particular, we denote by $\Omega(\bar{u}, F)$ the set of all elements $(\bar{v}, G)$ in $H(\bar{u}, F)$ for which one can choose a sequence $\{\tau_n\} \subset \mathbf{R}^+$ so that $\lim_{n\to\infty} \tau_n = \infty$ and $(\bar{u}^{\tau_n}, F^{\tau_n}) \to (\bar{v}, G)$ compactly. We can easily see that $\bar{v}$ is a solution of (3.28) whenever $(\bar{v}, G) \in H(\bar{u}, F)$.

For any function $\xi : \mathbf{R} \mapsto \mathbf{X}$ such that $\xi_0 \in \mathcal{B}$ and ξ is continuous on $\mathbf{R}^+$, we define a continuous function $\mu_\mathcal{B}^\xi : \mathbf{R}^+ \mapsto \mathcal{B}$ by the formula

$$\mu_\mathcal{B}^\xi(t) = \xi_t, \quad t \in \mathbf{R}^+.$$

It is clear that $\mu_\mathcal{B}^{\bar{u}}$ is an integral of the process $w_\mathcal{B}^F$ on $\mathbf{R}^+$. Also, we get $H_\sigma(\mu_\mathcal{B}^{\bar{u}}, w_\mathcal{B}^F) = \{(\mu_\mathcal{B}^{\bar{v}}, w_\mathcal{B}^G) : (\bar{v}, G) \in H(\bar{u}, F)\}$ and $\Omega_\sigma(\mu_\mathcal{B}^{\bar{u}}, w_\mathcal{B}^F) = \{(\mu_\mathcal{B}^{\bar{v}}, w_\mathcal{B}^G) : (\bar{v}, G) \in \Omega(\bar{u}, F)\}$.

The $\mathcal{B}$-*stabilities* for the solution $\bar{u}(t)$ of (3.26) is defined via those of the integral $\mu_\mathcal{B}^{\bar{u}}$ of the process $w_\mathcal{B}^F$ in Definition 3.1 with $\mathcal{X} = \mathcal{Y} = \mathcal{B}$. For example, the solution $\bar{u}(t)$ of (3.26) is $\mathcal{B}$-uniformly stable in $\Omega(F)$($\mathcal{B}$-US in $\Omega(F)$), if for any $\varepsilon > 0$ there exists a $\delta(\varepsilon) > 0$ such that $|u_t(s, \phi, G) - \bar{v}_t|_\mathcal{B} < \varepsilon$ for $t \geq s \geq 0$ whenever $(\bar{v}, G) \in \Omega(\bar{u}, F)$ and $|\phi - \bar{v}_s|_\mathcal{B} < \delta(\varepsilon)$. The other $\mathcal{B}$-stabilities for $\bar{u}(t)$ are given in a similar way; we omit the details.

3.3.3. Stability Properties for Abstract Functional Differential Equations with Infinite Delay

Combining the above observation with Theorem 3.1 and Lemma 3.3, we get the following result on $\mathcal{B}$-stabilities (cf. [101]). We emphasize that the additional condition that $\mathcal{B}$ is a uniform fading memory space cannot be removed because a fading memory space $\mathcal{B}$ must be a uniform fading memory space whenever there is a functional differential equation on $\mathcal{B}$ which has a $\mathcal{B}$-UAS solution ([107, Theorem 7.2.6], [159]).

Theorem 3.9 *Let $\mathcal{B}$ be a uniform fading memory space which is separable, and suppose that the conditions (H1)–(H4) are satisfied. Then the following statements are equivalent:*

i) *The solution $\bar{u}(t)$ of (3.26) is $\mathcal{B}$-UAS.*

ii) *The solution $\bar{u}(t)$ of (3.26) is $\mathcal{B}$-US and $\mathcal{B}$-attractive in $\Omega(F)$.*

iii) *The solution $\bar{u}(t)$ of (3.26) is $\mathcal{B}$-UAS in $\Omega(F)$.*

iv) *The solution $\bar{u}(t)$ of (3.26) is $\mathcal{B}$-WUAS in $\Omega(F)$.*

Next we shall construct a quasi-process with $\mathcal{X} = \mathrm{BC}_\rho$ associated with (3.28); here and hereafter, BC_ρ denotes the space BC which is equipped with the metric ρ defined by

$$\rho(\phi,\psi) = \sum_{n=0}^{\infty} \frac{1}{2^n} \frac{|\phi-\psi|_n}{1+|\phi-\psi|_n}, \qquad \phi, \psi \in \mathrm{BC}_\rho,$$

where $|\phi-\psi|_n = \sup_{-n \leq \theta \leq 0} |\phi(\theta) - \psi(\theta)|_{\mathbf{X}}$. Then (BC,ρ) is a metric space. Furthermore, it is clear that $\rho(\phi^k,\phi) \to 0$ as $k \to \infty$ if and only if $\phi^k \to \phi$ compactly on $\mathbf{R}^-$.

We first provide an example which shows that a process on BC_ρ cannot be always constructed for functional differential equations with infinite delay.

Example 3.1 *Consider a scalar delay equation*

$$\dot{x}(t) = \sum_{n=1}^{\infty} (1/n^3) x(t-n), \qquad (3.31)$$

which is a special case of (3.26) with $\mathcal{B} = C_g^0(\mathbf{R})$ ($g(s) = s+1$), $A = 0$ and $F(t,\phi) = \sum_{n=1}^{\infty} (1/n^3) \phi(-n)$.

It is clear that the conditions (H1)–(H3) are satisfied for this equation. Consider a sequence $\{\phi^k\} \subset \mathrm{BC}$ defined by $\phi^k(\theta) = 0$ if $-k \leq \theta \leq 0$, k^4 if $\theta \leq -k-1$ and linear if $-k-1 \leq \theta \leq -k$. Clearly $\phi^k \to 0$ in BC_ρ. Let denote by $x(t,s,\phi)$ the solution of (3.31) through (s,ϕ). Then

$$\begin{aligned}
x(1,0,\phi^k) &= \int_0^1 \sum_{n=1}^{\infty} (1/n^3) \phi^k(s-n) ds \\
&\geq 1/(k+2)^3 \int_0^1 \phi^k(s-k-2) ds \\
&\geq k^4/(k+2)^3 \geq 1
\end{aligned}$$

for $k \geq 10$. Note that $x(t,0,0) \equiv 0$ and $x_1(0,\phi^k) \not\to x_1(0,0)$ in BC_ρ. Hence the associated mapping $w : \mathbf{R}^+ \times \mathbf{R}^+ \times \mathrm{BC}_\rho \mapsto \mathrm{BC}_\rho$ defined by $w(t,s,\phi) = x_{t+s}(s,\phi)$, $(t,s,\phi) \in \mathbf{R}^+ \times \mathbf{R}^+ \times \mathrm{BC}_\rho$, is not continuous on $\mathbf{R}^+ \times \mathbf{R}^+ \times \mathrm{BC}_\rho$.

From the above example, we see that the concept of processes does not fit in with the study of the ρ-stabilities in functional differential equations. In what follows, we shall consider a subset $\mathcal{Y}$ of BC_ρ and construct a $\mathcal{Y}$-quasi-process on BC_ρ associated with (3.26) to overcome the above difficulty.

Let U be a closed and bounded subset of $\mathbf{X}$ whose interior U^i contains the closure of the set $\{\bar{u}(t) : t \in \mathbf{R}\}$, where $\bar{u}$ is the one in (H4). Set

$$BC_\rho^U = \{\phi \in BC_\rho : \phi(\theta) \in U \text{ for all } \theta \in \mathbf{R}^-\}.$$

It is clear that BC_ρ^U is a nonempty closed subset of BC_ρ. With $\mathcal{X} = BC_\rho$ and $\mathcal{Y} = BC_\rho^U$, we shall construct the quasi-process associated with (3.28). Consider a function $w_\rho^G : \mathbf{R}^+ \times \mathbf{R}^+ \times BC_\rho \mapsto BC_\rho$ defined by

$$w_\rho^G(t,s,\phi) = u_{t+s}(s,\phi,G), \qquad (t,s,\phi) \in \mathbf{R}^+ \times \mathbf{R}^+ \times BC_\rho,$$

which is the restriction of $w_\mathcal{B}^G$ to $\mathbf{R}^+ \times \mathbf{R}^+ \times BC_\rho$.

Lemma 3.6 w_ρ^G is a BC_ρ^U-quasi-process on BC_ρ.

Proof. From (H3) we easily see that w_ρ^G satisfies (p1) and (p2) with $\mathcal{X} = BC_\rho$. We shall show that w_ρ^G satisfies (p3) with $\mathcal{X} = BC_\rho$ and $\mathcal{Y} = BC_\rho^U$. Suppose the condition (p3) is not satisfied for w_ρ^G. Then there exist a point $(\bar{t}, \bar{s}, \bar{\phi}) \in \mathbf{R}^+ \times \mathbf{R}^+ \times BC_\rho$ and sequences $\{t_n\} \subset \mathbf{R}^+$, $\{s_n\} \subset \mathbf{R}^+$ and $\{\phi^n\} \subset BC_\rho^U$ such that $(t_n, s_n, \phi^n) \to (\bar{t}, \bar{s}, \bar{\phi})$ in $\mathbf{R}^+ \times \mathbf{R}^+ \times BC_\rho$ as $n \to \infty$ and that $\inf_n \rho(u_{t_n+s_n}(s_n, \phi^n, G), u_{\bar{t}+\bar{s}}(\bar{s}, \bar{\phi}, G)) > 0$. Then there exists an integer $l > 0$ such that $\inf_n |u_{t_n+s_n}(s_n, \phi^n, G) - u_{\bar{t}+\bar{s}}(\bar{s}, \bar{\phi}, G)|_{BC([-l,0];\mathbf{X})} > 0$, and hence there exists a sequence $\{\tau_n\} \subset [-l, 0]$ such that

$$\inf_n |u(t_n + s_n + \tau_n, s_n, \phi^n, G) - u(\bar{t} + \bar{s} + \tau_n, \bar{s}, \bar{\phi}, G)|_\mathbf{X} > 0. \qquad (3.32)$$

Since $u(t, \bar{s}, \bar{\phi}, G)$ is continuous in $t \in \mathbf{R}$, we get $\inf_n |u(t_n+s_n+\tau_n, s_n, \phi^n, G) - u(t_n + \bar{s}+\tau_n, \bar{s}, \bar{\phi}, G)|_\mathbf{X} > 0$. Therefore it must hold that $t_n + \tau_n \geq 0$ for all sufficiently large n, because of $\rho(\phi^n, \bar{\phi}) \to 0$ as $n \to \infty$. Thus we can assume that $\lim_{n\to\infty} \tau_n = \bar{\tau}$ for some $\bar{\tau} \in [-l, 0]$ with $\bar{t} + \bar{\tau} \geq 0$. Since $\lim_{n\to\infty} |\phi^n - \phi|_\mathcal{B} = 0$ by (A2), it follows from Proposition 3.1 that $\lim_{n\to\infty} |u_{t_n+s_n+\tau_n}(s_n, \phi^n, G) - u_{\bar{t}+\bar{s}+\bar{\tau}}(\bar{s}, \bar{\phi}, G)|_\mathcal{B} = 0$; which implies that $\lim_{n\to\infty} |u(t_n + s_n + \tau_n, s_n, \phi^n, G) - u(\bar{t} + \bar{s} + \bar{\tau}, \bar{s}, \bar{\phi}, G)|_\mathbf{X} = 0$ by (A1-iii). Therefore $\lim_{n\to\infty} |u(t_n + s_n + \tau_n, s_n, \phi^n, G) - u(\bar{t} + \bar{s} + \tau_n, \bar{s}, \bar{\phi}, G)|_\mathbf{X} = 0$, which is a contradiction to (3.32).

The mapping w_ρ^G constructed above is called the BC_ρ^U-quasi-process on BC_ρ generated by (3.28).

Now we consider the BC_ρ^U-quasi-process w_ρ^F on BC_ρ generated by (3.26). By the same calculation as for $w_\mathcal{B}^F$, we see that $\sigma(\tau)w_\rho^F = w_\rho^{F^\tau}$, $H_\sigma(w_\rho^F) = \{w_\rho^G : G \in H(F)\}$ and $\Omega_\sigma(w_\rho^F) = \{w_\rho^G : G \in \Omega(F)\}$. Moreover, we see that $H_\sigma(w_\rho^F)$ is sequentially compact and the BC_ρ^U- quasi-process $w_\mathcal{B}^F$ satisfies the condition (p4). For $t \in \mathbf{R}^+$, consider a mapping $\pi_\rho(t) : BC_\rho \times H_\sigma(w_\rho^F) \mapsto BC_\rho \times H_\sigma(w_\rho^F)$ defined by

$$\pi_\rho(t)(\phi, w_\rho^G) = (u_t(0, \phi, G), \sigma(t)w_\rho^G), \qquad (\phi, w_\rho^G) \in \mathrm{BC}_\rho \times H_\sigma(w_\rho^F).$$

Notice that $\lim_{n\to\infty} |\phi^n - \phi|_\mathcal{B} = 0$ whenever $\{\phi^n\} \subset \mathrm{BC}_\rho^U$ satisfies the condition $\lim_{n\to\infty} \rho(\phi^n, \phi) = 0$. Therefore, repeating almost the same argument as in the proof of Lemma 3.6, one can see that $\pi_\rho(t)$ satisfies the condition (p5) with $\mathcal{Y} = \mathrm{BC}_\rho^U$, and hence $\pi_\rho(t)$ is the skew product flow of w_ρ^F.

Lemma 3.7 *The skew product flow $\pi_\rho(t)$ is BC_ρ^U-strongly asymptotically smooth.*

Proof. It suffices to show that for the set $\mathrm{BC}_\rho^U \times H_\sigma(w_\rho^F)$ there exists a compact set $J \subset \mathrm{BC}_\rho^U \times H_\sigma(w_\sigma^F)$ with the property that $\{\pi_\rho(t_n)(\phi^n, w_\rho^{G_n})\}$ has a subsequence which approaches to J whenever sequences $\{t_n\} \subset \mathbf{R}^+$ and $\{(\phi^n, w_\rho^{G_n})\} \subset \mathrm{BC}_\rho^U \times H_\sigma(w_\rho^F)$ satisfy $\lim_{n\to\infty} t_n = \infty$ and $\pi_\rho(t)(\phi^n, w_\rho^{G_n}) \subset \mathrm{BC}_\rho^U \times H_\sigma(w_\rho^F)$ for all $t \in [0, t_n]$. This can be done by the same arguments as in the proof of Lemma 3.3. Indeed, since the set U is bounded in $\mathbf{X}$, putting $B = \mathrm{BC}_\rho^U \times H_\sigma(w_\rho^F)$ we can construct the set J_B as in the proof of Lemma 3.3. Then the set $J := (J_B \cap \mathrm{BC}_\rho^U) \times H_\sigma(w_\rho^F)$ is a compact set in $\mathrm{BC}_\rho^U \times H_\sigma(w_\rho^F)$. By virtue of (A2) and Ascoli-Arzéla's theorem, J has the desired property, because the function $x^n(t)$ in the proof of Lemma 3.3 satisfies $x^n(t) \in O_B$ and $|x^n(t) - x^n(s)|_\mathbf{X} \leq \sup\{|T(t-s)z - z|_\mathbf{X} : z \in O_B\} + C_1 L|t-s|$ for any s, t with $1 \leq s \leq t \leq t_n$ and $|t-s| \leq 1$.

For any function $\xi : \mathbf{R} \mapsto \mathbf{X}$ such that $\xi_0 \in \mathrm{BC}_\rho^U$ and ξ is continuous on $\mathbf{R}^+$, we define a continuous function $\mu_\rho^\xi : \mathbf{R}^+ \mapsto \mathrm{BC}_\rho$ by

$$\mu_\rho^\xi(t) = \xi_t, \qquad t \in \mathbf{R}^+.$$

It follows from (H4) that $\mu_\rho^{\bar{u}}$ is an integral of the quasi-process w_ρ^F on $\mathbf{R}^+$. Let $\eta > 0$ be chosen so that the interior of U contains the η-neighborhood of the set $\{\bar{u}(t) : t \in \mathbf{R}\}$. Then we easily see that (p6) is satisfied with $\delta_1 := \eta$ as $\mathcal{Y} = \mathrm{BC}_\rho^U$, $w = w_\rho^F$ and $\mu = \mu_\rho^{\bar{u}}$ because of the inequality $|u(t+s, s, \phi, F) - \bar{u}(t+s)|_\mathbf{X} \leq \rho(w_\rho^F(t,s,\phi), \mu_\rho^{\bar{u}}(t))$. Also, we get $H_\sigma(\mu_\rho^{\bar{u}}, w_\rho^F) = \{(\mu_\rho^{\bar{v}}, w_\rho^G) : (\bar{v}, G) \in H(\bar{u}, F)\}$ and $\Omega_\sigma(\mu_\rho^{\bar{u}}, w_\rho^F) = \{(\mu_\rho^{\bar{v}}, w_\rho^G) : (\bar{v}, G) \in \Omega(\bar{u}, F)\}$.

The BC_ρ^U-stabilities of the integral $\mu_\rho^{\bar{u}}$ of the quasi-process w_ρ^F yield the ρ-stabilities with respect to U for the solution $\bar{u}(t)$ of (3.26). For example, the solution $\bar{u}(t)$ of (3.26) is ρ-uniformly stable with respect to U in $\Omega(F)$(ρ-US with respect to U in $\Omega(F)$), if for any $\varepsilon > 0$ there exists a $\delta(\varepsilon) > 0$ such that $\rho(\bar{u}_t(s, \phi, G), \bar{v}_t) < \varepsilon$ for $t \geq s \geq 0$ whenever $(\bar{v}, G) \in \Omega(\bar{u}, F)$, $\rho(\phi, \bar{v}_s) < \delta(\varepsilon)$ and $\phi(s) \in U$ for all $s \in \mathbf{R}^-$. The other ρ-stabilities with respect to U for $\bar{u}(t)$ are given in a similar way; we omit the details.

The following result is a direct consequence of Theorem 3.9 and Lemmas 3.6 and 3.7.

Theorem 3.10 *Let $\mathcal{B}$ be a fading memory space which is separable and suppose that the conditions (H1)-(H4) are satisfied. Also, let U be a closed and bounded subset of $\mathbf{X}$ whose interior contains the closure of the set $\{\bar{u}(t) : t \in \mathbf{R}\}$. Then the following statements are equivalent:*

CHAPTER 3. STABILITY METHODS

i) *The solution $\bar{u}(t)$ of (3.26) is ρ-UAS with respect to U.*

ii) *The solution $\bar{u}(t)$ of (3.26) is ρ-US with respect to U and ρ-attractive with respect to U in $\Omega(F)$.*

iii) *The solution $\bar{u}(t)$ of (3.26) is ρ-UAS with respect to U in $\Omega(F)$.*

iv) *The solution $\bar{u}(t)$ of (3.26) is ρ-WUAS with respect to U in $\Omega(F)$.*

In the above, if the terms $\rho(u_t(\sigma, \phi, F), \bar{u}_t)$ and $\rho(u_t(\sigma, \phi, G), \bar{v}_t)$ are replaced by $|u(t, \sigma, F) - \bar{u}(t)|_{\mathbf{X}}$ and $|u(t, \sigma, G) - \bar{v}(t)|_{\mathbf{X}}$ respectively, then we have another concept of ρ-stability, which will be referred to as the $(\rho, \mathbf{X})$-*stability*. The equivalence of these two concepts of ρ-stabilities are given by the following proposition.

Proposition 3.2 *Let U be a closed and bounded subset of $\mathbf{X}$ whose interior U^i contains the closure of the set $\{\bar{u}(t) : t \in \mathbf{R}\}$. Then the solution $\bar{u}(t)$ of (3.26) is ρ-US if and only if it is $(\rho, \mathbf{X})$-US.*

Proof. The proof of the "only if" part is obvious. We shall establish the "if" parts. Take any $\varepsilon > 0, (\sigma, \phi) \in \mathbf{R}^+ \times BC$ with $\phi(s) \in U$, for all $s \in U$, $\rho(\phi, \bar{u}_\sigma) < \delta(\varepsilon)$, where $\delta(\cdot)$ is the one for $(\rho, \mathbf{X})$-US of the solution $\bar{u}(t)$ of (3.26). Then $v(t) = v(t, \sigma, \phi, F)$ satisfies

$$|v(t) - \bar{u}(t)|_{\mathbf{X}} < \varepsilon \text{ for } t \geq \sigma. \tag{3.33}$$

To estimate $\rho(v_t, \bar{u}_t)$, we first estimate $|\bar{u}_t - v_t|_j$. Let $t \geq \sigma$, and denote by k the largest integer which does not exceed $t - \sigma$. If $j \leq k$, then $j \leq t - \sigma$; hence $|\bar{u}_t - v_t|_j = \sup_{-j \leq s \leq 0} |\bar{u}(t+s) - v(t+s)|_{\mathbf{X}} < \varepsilon$ by (3.33). On the other hand, if $j \geq k+1$, then $j \geq t - \sigma$, hence

$$\begin{aligned}|\bar{u}_t - v_t|_j &= \max\{\sup_{-j \leq s \leq \sigma - t} |\bar{u}(t+s) - v(t+s)|_{\mathbf{X}}, \\ &\qquad \sup_{\sigma - t \leq s \leq 0} |\bar{u}(t+s) - v(t+s)|_{\mathbf{X}}\} \\ &\leq \max\{\sup_{-j \leq \theta \leq 0} |\phi(\theta) - \bar{u}(\sigma + \theta)|_{\mathbf{X}}, \sup_{\sigma \leq \theta} |v(\theta) - \bar{u}(\theta)|_{\mathbf{X}}\} \\ &< |\phi - \bar{u}(\sigma)|_j + \varepsilon\end{aligned}$$

by (3.33). Then

$$\begin{aligned}\rho(v_t, \bar{u}_t) &= (\sum_{j=1}^{k} + \sum_{j=k+1}^{\infty}) 2^{-j} |v_t - \bar{u}_t|_j / [1 + |v_t - \bar{u}_t|_j] + \varepsilon \\ &< \sum_{j=1}^{k} 2^{-j} \varepsilon / (1+\varepsilon) + \sum_{j=k+1}^{\infty} 2^{-j} [|\phi - \bar{u}_\sigma|_j + \varepsilon] / [1 + |\phi - \bar{u}_\sigma|_j + \varepsilon] \\ &\leq \sum_{j=1}^{k} 2^{-j} \varepsilon / (1+\varepsilon) + \sum_{j=k+1}^{\infty} 2^{-j} |\phi - \bar{u}_\sigma|_j / [1 + |\phi - \bar{u}_\sigma|_j] \\ &< \varepsilon + \delta(\varepsilon) \leq 2\varepsilon,\end{aligned}$$

which shows that the solution $\bar{u}(t)$ of (3.26) is ρ-US with $\delta(\cdot/2)$.

3.4. EQUIVALENT RELATIONSHIPS BETWEEN BC-STABILITIES AND ρ-STABILITIES

3.4.1. BC-Stabilities in Abstract Functional Differential Equations with Infinite Delay

Let $\mathcal{B}$ be a fading memory space which is separable. We shall discuss some relationships between BC-stabilities and ρ-stabilities, and extend some results due to Murakami and Yoshizawa [161] for $\mathbf{X} = \mathbf{R}^n$ to the case where dim$\mathbf{X} = \infty$.

The solution $\bar{u}(t)$ of (3.26) in (H4) is said to be *BC-totally stable* (BC-TS) if for any $\varepsilon > 0$ there exists a $\delta(\varepsilon) > 0$ with the property that $\sigma \in \mathbf{R}^+, \phi \in \mathrm{BC}$ with $|\bar{u}_\sigma - \phi|_{\mathrm{BC}} < \delta(\varepsilon)$ and $h \in \mathrm{BC}([\sigma, \infty); \mathbf{X})$ with $\sup_{t \in [\sigma, \infty)} |h(t)|_{\mathbf{X}} < \delta(\varepsilon)$ imply $|\bar{u}(t) - u(t, \sigma, \phi, F + h)|_{\mathbf{X}} < \varepsilon$ for $t \geq \sigma$, where $u(\cdot, \sigma, \phi, F + h)$ denotes the solution of

$$\frac{du}{dt} = Au(t) + F(t, u_t) + h(t), \quad t \geq \sigma, \tag{3.34}$$

through (σ, ϕ).

The other BC-stabilities for $\bar{u}(t)$ are given in a similar way; we omit the details.

Theorem 3.11 *Assume that conditions* (H1)–(H4) *hold. If $\bar{u}(t)$ of (3.26) is BC-UAS, then it is BC-UAS in $\Omega(F)$ and BC-TS.*

Proof. First, we shall show that $\bar{u}(t)$ is BC-UAS in $\Omega(F)$.

Let $(\delta(\cdot), \delta_0, t_0(\cdot))$ be the triple for BC-UAS of $\bar{u}(t)$, where we may assume $\delta_0 < \delta(1)$. We first establish that

$$\sigma \in \mathbf{R}^+, \ (\bar{v}, G) \in \Omega(\bar{u}, F) \text{ and } |\phi - \bar{v}_\sigma|_{BC} < \delta(\frac{\eta}{2})$$
$$\text{imply } |x(t, \sigma, \phi, G) - v(t)|_{\mathbf{X}} < \eta \text{ for } t \geq \sigma. \tag{3.35}$$

Select a sequence $\{t_n\}$ with $t_n \to \infty$ as $n \to \infty$ such that $(\bar{u}^{t_n}, F^{t_n}) \to (\bar{v}(t), G(t, \phi))$ compactly, and consider any solution $x(\cdot, \sigma + t_n, \phi - \bar{v}_\sigma + \bar{u}_{\sigma + t_n}, F)$. For any $n \in \mathbf{N}$, set $x^n(t) = x(t + t_n, \sigma + t_n, \phi - \bar{v}_\sigma + \bar{u}_{\sigma + t_n}, F)$, $t \in \mathbf{R}$. Since the solution $\bar{u}(t)$ of (3.26) is BC-UAS, from the fact that $|x_\sigma^n - \bar{u}_{\sigma + t_n}|_{BC} = |\phi - v_\sigma|_{BC} < \delta(\eta/2)$ it follows that

$$|x^n(t) - \bar{u}(t + t_n)|_{\mathbf{X}} < \frac{\eta}{2} \text{ for all } t \geq \sigma \text{ and } n \in \mathbf{N}. \tag{3.36}$$

Observe that the set $\{x^n(\sigma) : n \in \mathbf{N}\}$ is relatively compact in $\mathbf{X}$. In virtue of this fact and (3.36), repeating almost the same argument as in the proof of Lemma 3.5 we can see that the set $\{x^n(t) : t \geq \sigma, \ n \in \mathbf{N}\}$ is relatively compact in $\mathbf{X}$ and that the sequence $\{x^n\}$ is uniformly equicontinuous on $[\sigma, \infty)$. Thus we may assume that $x^n(t) \to y(t)$ compactly on $[\sigma, \infty)$ for some function $y(t) : [\sigma, \infty) \mapsto \mathbf{X}$. Since $x^n(\sigma) = \phi(0) - \bar{v}(\sigma) + \bar{u}(\sigma + t_n)$, we obtain $y(\sigma) = \phi(0)$. Hence, if we extend the function y by setting $y_\sigma = \phi$, then $y \in C(\mathbf{R}, \mathbf{X})$ and $|x_t^n - y_t|_{\mathcal{B}} \to 0$ compactly on $[\sigma, \infty)$. Letting $n \to \infty$ in the relation

$$x^n(t) = T(t-\sigma)\{\phi(0) - \bar{v}(\sigma) + \bar{u}(\sigma + t_n)\} + \int_\sigma^t T(t-s)F(s+t_n, x_s^n)ds, \quad t \geq \sigma,$$

we obtain

$$y(t) = T(t-\sigma)\phi(0) + \int_\sigma^t T(t-s)G(s, y_s)ds, \quad t \geq \sigma,$$

which means that $y(t) \equiv x(t, \sigma, \phi, G)$ for $t \geq \sigma$ by the regularity assumption (H3). Then (3.35) follows from (3.36) by letting $n \to \infty$.

Repeating the above arguments with $\eta = 2$, we can establish that

$$\sigma \in \mathbf{R}^+, \ (\bar{v}, G) \in \Omega(\bar{u}, F) \text{ and } |\phi - v_\sigma|_{BC} < \delta_0$$
$$\text{imply } |x(t, \sigma, \phi, G) - \bar{v}(t)|_\mathbf{X} < \varepsilon \text{ for } t \geq \sigma + t_0(\frac{\varepsilon}{2}). \quad (3.37)$$

Next, we suppose that the solution $\bar{u}(t)$ is not BC-TS. Then there exist an $\varepsilon, 0 < \varepsilon < \delta_0$, sequences $\{\tau_n\} \subset \mathbf{R}^+, \{r_n\}, r_n > 0, \{\phi_n\} \subset BC, \{h_n\}, h_n \in BC([\tau_n, \infty); \mathbf{X})$, and solutions $\{x(\cdot, \tau_n, \phi_n, F + h_n)\}$ such that, for all $n \in \mathbf{N}$,

$$|\phi_n - \bar{u}_{\tau_n}|_{BC} < \frac{1}{n} \text{ and } \sup_{t \geq \tau_n} |h_n(t)| < \frac{1}{n} \quad (3.38)$$

and

$$|z^n(\tau_n + r_n) - \bar{u}(\tau_n + r_n)|_\mathbf{X} = \varepsilon \text{ and } |z^n(t) - \bar{u}(t)|_\mathbf{X} < \varepsilon \ \forall t \in (-\infty, \tau_n + r_n), \quad (3.39)$$

here $z^n(t) := x(t, \tau_n, \phi_n, F + h_n)$. We first consider the case that $\{r_n\}$ is unbounded. Without loss of generality, we may assume that

$$(\bar{u}^{\tau_n + r_n - t_0}, F^{\tau_n + r_n - t_0}) \to (\bar{v}, G)$$

compactly, for some $(\bar{v}, G) \in \Omega(\bar{u}, F)$ and that $z^n(t + \tau_n + r_n - t_0) \to z(t)$ compactly on $(-\infty, t_0]$ for some function z, where $t_0 = t_0(\varepsilon/2)$. Repeating almost the same argument as in the proof of the claim (3.35), we see by (3.38) that z satisfies (3.28) on $[0, t_0]$. Let $n \to \infty$ in (3.39) to obtain $|z(t) - \bar{v}(t)|_\mathbf{X} \leq \varepsilon$ on $(-\infty, t_0]$ and $|z(t_0) - \bar{v}(t_0)|_\mathbf{X} = \varepsilon$. This is a contradiction, because $|z_0 - \bar{v}_0|_{BC} \leq \varepsilon < \delta_0$ implies $|z(t_0) - \bar{v}(t_0)|_\mathbf{X} < \varepsilon$ by (3.37). Therefore the sequence $\{r_n\}$ must be bounded. Thus we may assume that $\{r_n\}$ converges to some $r, 0 \leq r < \infty$. Moreover, we may assume that $\{z^n(\tau_n + t)\} \to \xi$ compactly on $(-\infty, r]$. Consider the case where the sequence $\{\tau_n\}$ is unbounded; hence we may assume that $(\bar{u}^{\tau_n}, F^{\tau_n}) \to (w, H)$ compactly, for some $(w, H) \in \Omega(\bar{u}, F)$. Then $\xi(t)$ satisfies

$$\xi(t) = T(t)\xi(0) + \int_0^t T(t-s)H(s, \xi_s)ds$$

on $[0, r]$, and moreover we have $|\xi_0 - w_0|_{BC} = 0$ and $|\xi(r) - w(r)|_\mathbf{X} = \varepsilon$ by letting $n \to \infty$ in (3.38) and (3.39). This is a contradiction, because we must have $\xi \equiv w$ on $[0, r]$ by (H3). Thus the sequence $\{\tau_n\}$ must be bounded, too. Hence we may assume that $\lim_{n \to \infty} \tau_n = \tau$ for some $\tau < \infty$. Then $\xi(t - \tau)$ satisfies (3.26) on $[\tau, \tau + r]$, and moreover we have $|\xi_0 - \bar{u}_\tau|_{BC} = 0$ and $|\xi(r) - \bar{u}(\tau + r)|_\mathbf{X} = \varepsilon$ by (3.38) and (3.39). This again contradicts the fact that the solution $\bar{u}(t)$ of (3.26) is BC-UAS.

3.4.2. Equivalent Relationship between BC-Uniform Asymptotic Stability and ρ-Uniform Asymptotic Stability

Theorem 3.12 *Assume that conditions* (H1)–(H4) *hold. If the solution $\bar{u}(t)$ of* (3.26) *is BC-UAS, then it is ρ-US w.r.t. U for any closed bounded set U in $\mathbf{X}$ such that $U^i \supset O_{\bar{u}}$.*

Proof. We assume that the solution $\bar{u}(t)$ of (3.26) is BC-UAS but not ρ-US w.r.t. U; here $U \subset \{x \in \mathbf{X} : |x|_{\mathbf{X}} \leq c\}$ for some $c > 0$. Since the solution $\bar{u}(t)$ of (3.26) is not $(\rho, \mathbf{X})$-US w.r.t. U as noted in Proposition 3.2 there exist an $\varepsilon \in (0,1)$, sequences $\{\tau_m\} \subset \mathbf{R}^+$, $\{t_m\}$ $(t_m > \tau_m)$, $\{\phi^m\} \subset$ BC with $\phi^m(s) \in U$ for $s \in \mathbf{R}^-$ and solutions $\{u(t, \tau_m, \phi^m, F) =: \hat{u}^m(t)\}$ of (3.26) such that

$$\rho(\phi^m, \bar{u}_{\tau_m}) < 1/m \tag{3.40}$$

and that

$$|\hat{u}^m(t_m) - \bar{u}(t_m)|_{\mathbf{X}} = \varepsilon \text{ and } |\hat{u}^m(t) - \bar{u}(t)|_{\mathbf{X}} < \varepsilon \text{ on } [\tau_m, t_m) \tag{3.41}$$

for $m \in \mathbf{N}$. For each $m \in \mathbf{N}$ and $r \in \mathbf{R}^+$, we define $\phi^{m,r} \in$ BC by

$$\phi^{m,r}(\theta) = \begin{cases} \phi^m(\theta) & \text{if } -r \leq \theta \leq 0, \\ \phi^m(-r) + \bar{u}(\tau_m + \theta) - \bar{u}(\tau_m - r) & \text{if } \theta < -r. \end{cases}$$

We note that $|\phi^{m,r} - \bar{u}_{\tau_m}|_{\text{BC}} = |\phi^m - \bar{u}_{\tau_m}|_{[-r,0]}$ and

$$\sup\{|\phi^{m,r} - \phi^m|_{\mathcal{B}} : m \in \mathbf{N}\} \to 0 \text{ as } r \to \infty. \tag{3.42}$$

For, if (3.42) is false, then there exist an $\varepsilon > 0$ and sequences $\{m_k\} \subset \mathbf{N}$ and $\{r_k\}$, $r_k \to \infty$ as $k \to \infty$, such that $|\phi^{m_k, r_k} - \phi^{m_k}|_{\mathcal{B}} \geq \varepsilon$ for $k = 1, 2, \cdots$. Put $\psi^k := \phi^{m_k, r_k} - \phi^{m_k}$. Clearly, $\{\psi^k\}$ is a sequence in BC which converges to zero function compactly on $\mathbf{R}^-$ and $\sup_k |\psi^k|_{\text{BC}} < \infty$. Then Axiom (A2) yields that $|\psi^k|_{\mathcal{B}} \to 0$ as $k \to \infty$, a contradiction. Furthermore, the set $\{\phi^m, \phi^{m,r} : m \in \mathbf{N}, r \in \mathbf{R}^+\}$ is relatively compact in $\mathcal{B}$. Indeed, since the set $\{\bar{u}_t : t \in \mathbf{R}^+\}$ is relatively compact in $\mathcal{B}$, (3.40) and Axiom (A2) yield that any sequence $\{\phi^{m_j}\}_{j=1}^{\infty}$ $(m_j \in \mathbf{N})$ has a convergent subsequence in $\mathcal{B}$.

Therefore, it sufficies to show that any sequence $\{\phi^{m_j, r_j}\}_{j=1}^{\infty} (m_j \in \mathbf{N}, r_j \in \mathbf{R}^+)$ has a convergent subsequence in $\mathcal{B}$. We assert that the sequence of functions $\{\phi^{m_j, r_j}(\theta)\}_{j=1}^{\infty}$ contains a subsequence which is equicontinuous on any compact set in $\mathbf{R}^-$. If this is the case, then the sequence $\{\phi^{m_j, r_j}\}_{j=1}^{\infty}$ would have a convergent subsequence in $\mathcal{B}$ by the Ascoli-Arzéla theorem and Axiom (A2), as required. Now, notice that the sequence of functions $\{\bar{u}(\tau_{m_j} + \theta)\}$ is equicontinuous on any compact set in $\mathbf{R}^-$. Then the assertion obviously holds true when the sequence $\{m_j\}$ is bounded. Taking a subsequence if necessary, it is thus sufficient to consider the case $m_j \to \infty$ as $j \to \infty$. In this case, it follows from (3.40) that $\phi^{m_j}(\theta) - \bar{u}(\tau_{m_j} + \theta) =: w^j(\theta) \to 0$ compactly on $\mathbf{R}^-$. Concequently, $\{w^j(\theta)\}$ is equicontinuous on any compact set in $\mathbf{R}^-$, and so is $\{\phi^{m_j}(\theta)\}$. Therefore the assertion immediately follows from this observation.

Now, for any $m \in \mathbf{N}$, set $u^m(t) = \hat{u}^m(t+\tau_m)$ if $t \le t_m - \tau_m$ and $u^m(t) = u^m(t_m - \tau_m)$ if $t > t_m - \tau_m$. Moreover, set $u^{m,r}(t) = \phi^{m,r}(t)$ if $t \in \mathbf{R}^-$ and $u^{m,r}(t) = u^m(t)$ if $t \in \mathbf{R}^+$. Since $|u_t^m|_\mathcal{B} \le K\{1+|\bar{u}|_{[0,\infty)}\}+M|\phi^m|_\mathcal{B} \le K\{1+|\bar{u}|_{[0,\infty)}\}+MJc$ by (3.25) and (A1-iii), by using the same arguments as in the proof of Lemma 3.5 we see that $\{u^m(t)\}$ is uniformly equicontinuous in $C(\mathbf{R}^+, \mathbf{X})$. Then it follows from Lemma 3.4 and the relative compactness of the set $\{\phi^m, \phi^{m,r} : m \in \mathbf{N}, r \in \mathbf{R}^+\}$ that the set $W := \overline{\{u_t^{m,r}, u_t^m : m \in \mathbf{N}, t \in \mathbf{R}^+, r \in \mathbf{R}^+\}}$ is compact in $\mathcal{B}$. Hence $F(t, \phi)$ is uniformly continuous on $\mathbf{R}^+ \times W$ by (H1). Define a continuous function $q_{m,r}$ on $\mathbf{R}^+$ by $q_{m,r}(t) = F(t+\tau_m, u_t^m) - F(t+\tau_m, u_t^{m,r})$ if $0 \le t \le t_m - \tau_m$, and $q_{m,r}(t) = q_{m,r}(t_m - \tau_m)$ if $t > t_m - \tau_m$. Since $|u_t^{m,r} - u_t^m|_\mathcal{B} \le M|\phi^{m,r} - \phi^m|_\mathcal{B}$ ($t \in \mathbf{R}^+, m \in \mathbf{N}$) by (A1-iii), it follows from (3.42) that $\sup\{|u_t^{m,r} - u_t^m|_\mathcal{B} : t \in \mathbf{R}^+, m \in \mathbf{N}\} \to 0$ as $r \to \infty$; hence one can choose $r = r(\varepsilon) \in \mathbf{N}$ in such a way that

$$\sup\{|q_{m,r}(t)|_\mathbf{X} : m \in \mathbf{N}, t \in \mathbf{R}^+\} < \delta(\varepsilon/2)/2,$$

where $\delta(\cdot)$ is the one for the BC-TS of $\bar{u}(t)$. Moreover, for this r, select an $m \in \mathbf{N}$ such that $m > 2^r(1+\delta(\varepsilon/2))/\delta(\varepsilon/2)$. Then $2^{-r}|\phi^m - \bar{u}_{\tau_m}|_r/[1+|\phi^m - \bar{u}_{\tau_m}|_r] \le \rho(\phi^m, \bar{u}_{\tau_m}) < 2^{-r}\delta(\varepsilon/2)/[1+\delta(\varepsilon/2)]$ by (3.40), which implies that

$$|\phi^m - \bar{u}_{\tau_m}|_r < \delta(\varepsilon/2) \text{ or } |\phi^{m,r} - \bar{u}_{\tau_m}|_{\text{BC}} < \delta(\varepsilon/2).$$

The function $u^{m,r}$ satisfies $u_0^{m,r} = \phi^{m,r}$ and

$$\begin{aligned} u^{m,r}(t) &= u^m(t) \\ &= T(t)\phi^m(0) + \int_0^t T(t-s)\{F(s+\tau_m, u_s^m)\}ds \\ &= T(t)\phi^{m,r}(0) + \int_0^t T(t-s)\{F(s+\tau_m, u_s^{m,r}) + q_{m,r}(s)\}ds \end{aligned}$$

for $t \in [0, t_m - \tau_m)$. Since $\bar{u}^m(t) = \bar{u}(t+\tau_m)$ is a solution of

$$\frac{du}{dt} = Au + F(t+\tau_m, u_t)$$

with the same $\delta(\cdot)$ as the one for $\bar{u}(t)$, from the fact that $\sup_{t \ge 0}|q_{m,r}(t)|_\mathbf{X} < \delta(\varepsilon/2)/2 < \delta(\varepsilon/2)$ it follows that $|u^{m,r}(t) - \bar{u}(t+\tau_m)|_\mathbf{X} < \varepsilon/2$ on $[0, t_m - \tau_m)$. In particular, we have $|u^{m,r}(t_m - \tau_m) - \bar{u}(t_m)|_\mathbf{X} < \varepsilon$ or $|\hat{u}^m(t_m) - \bar{u}(t_m)|_\mathbf{X} < \varepsilon$, which contradicts (3.41).

Theorem 3.13 *Assume that conditions* (H1)–(H4) *are hold. Then the solution $\bar{u}(t)$ of* (3.26) *is* BC-UAS *if and only if it is ρ-UAS w.r.t. U for any closed bounded set U in $\mathbf{X}$ such that $U^i \supset O_{\bar{u}} := \{\bar{u}(t) : t \in \mathbf{R}\}$.*

Proof. The "if" part is easily shown by noting that $\rho(\phi, \xi) \le |\phi - \xi|_{\text{BC}}$ for $\phi, \xi \in$ BC. We shall establish the "only if" part. The solution $\bar{u}(t)$ of (3.26) is ρ-US w.r.t. U by Theorem 3.12. Thus it suffices to establish the following assertion:

(*) For any $\varepsilon > 0$ there exists a $t^*(\varepsilon) > 0$ such that $\rho(\phi, \bar{u}_\tau) < \delta_1 := \delta^*(\delta_0/4)$ with $\phi(s) \in U, s \in \mathbf{R}^-$, implies $\rho(u_t(\tau, \phi, F), \bar{u}_t) < \varepsilon$ for all $t \geq \tau + t^*(\varepsilon)$, where $\delta^*(\cdot)$ is the number for ρ-US of $\bar{u}(t)$ and δ_0 is the number for the BC-UAS in $\Omega(F)$ of $\bar{u}(t)$.

If this assertion is not true, then there exist an $\varepsilon > 0$ and sequences $\{\tau_k\} \subset \mathbf{R}^+$, $\{t_k\}$, $t_k \geq \tau_k + 2k$, $\{\phi^k\} \subset \mathrm{BC}$, and solutions $\{u(t, \tau_k, \phi^k, F)\}$ such that

$$\rho(\phi^k, \bar{u}_{\tau_k}) < \delta_1, \quad \phi^k(s) \in U, s \in \mathbf{R}^- \tag{3.43}$$

and

$$\rho(u_{t_k}(\tau_k, \phi^k, F), \bar{u}_{t_k}) \geq \varepsilon \tag{3.44}$$

for all $k \in \mathbf{N}$. Since $\bar{u}(t)$ is ρ-US, (3.43) and (3.44) imply that

$$\rho(u_t(\tau_k, \phi^k, F), \bar{u}_t) < \frac{\delta_0}{4} \text{ for all } t \geq \tau_k \tag{3.45}$$

and $\rho(u_t(\tau_k, \phi^k, F), \bar{u}_t) \geq \delta^*(\varepsilon)$ for all $t \in [\tau_k, \tau_k + 2k]$ or

$$\rho(u_{t+\tau_k+k}(\tau_k, \phi^k, F), \bar{u}_{t+\tau_k+k}) \geq \delta^*(\varepsilon) \text{ for all } t \in [-k, k], \tag{3.46}$$

respectively. Set $u^k(t) = u(t + \tau_k + k, \tau_k, \phi^k)$ for $t \in \mathbf{R}$. Since $\rho(\phi, \psi) \geq 2^{-1}|\phi - \psi|_1/[1 + |\phi - \psi|_1] \geq 2^{-1}|\phi(0) - \psi(0)|_\mathbf{X}/[1 + |\phi(0) - \psi(0)|_\mathbf{X}]$, we have $|\phi(0) - \psi(0)|_\mathbf{X} \geq 2\rho(\phi, \psi)/[1 - 2\rho(\phi, \psi)]$ whenever $\rho(\phi, \psi) \leq 1/2$; hence (3.45) implies that

$$|u^k(t) - \bar{u}(t + \tau_k + k)|_\mathbf{X} \leq \frac{\delta_0}{2 - \delta_0} \text{ for all } t \in [-k, k]. \tag{3.47}$$

We shall show that the set $O := \{u^k(t) : -k+1 \leq t \leq k, k \in \mathbf{N}\}$ is relatively compact in $\mathbf{X}$ and that $\{u^k(t)\}$ is a family of equicontinuous on any bounded interval in $\mathbf{R}$. To certify the assertion, we use the Kuratovski's measure $\alpha(\cdot)$ of noncompactness of set in $\mathbf{X}$. Let $0 < \nu < 1$. Since $u^k(t)$ is a mild solution of $\frac{du(t)}{dt} = Au(t) + F(t + \tau_k + k, u_t)$ through $(-k, \phi^k)$, we get

$$\begin{aligned}
u^k(t) &= T(t+k)\phi^k(0) + \int_{-k}^t T(t-s)F(s+\tau_k+k, u_s^k)ds \\
&= T(1)[T(t+k-1)\phi^k(0) + \int_{-k}^{t-1} T(t-1-s)F(s+\tau_k+k, u_s^k)ds] \\
&\quad + \int_{t-1}^t T(t-s)F(s+\tau_k+k, u_s^k)ds \\
&= T(1)[u^k(t-1)] + T(\nu)\int_{t-1}^{t-\nu} T(t-s-\nu)F(s+\tau_k+k, u_s^k)ds \\
&\quad + \int_{t-\nu}^t T(t-s)F(s+\tau_k+k, u_s^k)ds
\end{aligned}$$

for $t \geq -k+1$. It follows from Axiom (A1-iii), (H2) and (3.47) that $\sup_{s \in [-k, \infty)} |F(s + \tau_k + k, u_s^k)|_{\mathbf{X}} =: C_2 < \infty$. Then the set $\{\int_{t-1}^{t-\nu} T(t-s-\nu) F(s+\tau_k+k, u_s^k) ds : t \geq -k+1\}$ is bounded in $\mathbf{X}$, and hence $T(\nu)\{\int_{t-1}^{t-\nu} T(t-s-\nu) F(s+\tau_k+k, u_s^k) ds : t \geq -k+1\}$ is relatively compact in $\mathbf{X}$ because of the compactness of the semigroup $\{T(t)\}_{t \geq 0}$. Similarly, one can get the relative compactness of the set $T(1)\{u^k(t-1) : t \geq -k+1\}$. Thus we obtain

$$\begin{aligned}\alpha(O) &\leq \alpha(\{\int_{t-\nu}^{t} T(t-s) F(s + \tau_k + k, u_s^k) ds : t \geq -k+1\}) \\ &\leq C_2 C_3 \nu,\end{aligned}$$

where $C_3 = \sup_{0 \leq \tau \leq 1} \|T(\tau)\|$. Letting $\nu \to 0$ in the above, we get $\alpha(O) = 0$; consequently, O must be relatively compact in $\mathbf{X}$. To certify the assertion on the equicontinuity, let $-k+1 \leq s \leq t \leq s+1$. Then

$$u^k(t) = T(t-s)u^k(s) + \int_s^t T(t-\tau) F(\tau + \tau_k + k, u_\tau^k) d\tau. \tag{3.48}$$

Hence

$$\begin{aligned}|u^k(t) - u^k(s)|_{\mathbf{X}} &\leq |T(t-s)u^k(s) - u^k(s)|_{\mathbf{X}} \\ &\quad + |\int_s^t T(t-s) F(\tau + \tau_k + k, u_{\tau_k}^k) d\tau|_{\mathbf{X}} \\ &\leq \sup\{|T(t-s)z - z|_{\mathbf{X}} : z \in O\} + C_2 C_3 |t-s|.\end{aligned}$$

Since the set O is relatively compact in $\mathbf{X}$, $T(\tau)z$ is uniformly continuous in $\tau \in [0, 1]$ uniformly for $z \in O$. This observation proves the assertion.

Now, applying the Ascoli-Arzéla theorem and the diagonalization procedure, one may assume that $u^k(t) \to w(t)$ compactly on $\mathbf{R}$ for some $w(t) \in \mathrm{BC}(\mathbf{R}; \mathbf{X})$. Then $u_t^k \to w_t$ compactly on $\mathbf{R}$ by Axiom (A1-iii). Also, we may assume that $(\bar{u}^{\tau_k + k}, F^{\tau_k + k}) \to (\bar{v}, G)$ compactly on $\mathbf{R}^+ \times \mathcal{B}$ for some $(\bar{v}, G) \in \Omega(\bar{u}, F)$. Letting $k \to \infty$ in (3.47) and (3.48), $w(t) = u(t, 0, w_0, G)$ on $\mathbf{R}$ and $|w_0 - \bar{v}_0|_{\mathrm{BC}} \leq \delta_0/[2 - \delta_0] < \delta_0$. Then $|w(t) - \bar{v}(t)|_{\mathbf{X}} \to 0$ as $t \to \infty$, by Theorem 3.11. On the other hand, letting $k \to \infty$ in (3.46), we have $\rho(w_t, \bar{v}_t) \geq \delta^*(\varepsilon)$ for all $t \in \mathbf{R}$, a contradiction. This completes the proof of Theorem 3.13.

3.4.3. Equivalent Relationship Between BC-Total Stability and ρ-Total Stability

The bounded solution $\bar{u}(t)$ of (3.26) is said to be ρ-totally stable with respect to U (ρ-TS w.r.t.U) if for any $\varepsilon > 0$ there exists a $\delta(\varepsilon) > 0$ with the property that $\sigma \in \mathbf{R}^+, \phi(s) \in U$ for $s \in \mathbf{R}^-$, with $\rho(\bar{u}_\sigma, \phi) < \delta(\varepsilon)$ and $h \in \mathrm{BC}([\sigma, \infty); \mathbf{X})$ with $\sup_{t \in [\sigma, \infty)} |h(t)|_{\mathbf{X}} < \delta(\varepsilon)$ imply $\rho(\bar{u}_t, u_t(\sigma, \phi, F + h)) < \varepsilon$ for $t \geq \sigma$.

In the above, if the term $\rho(u_t(\sigma, \phi, F + h), \bar{u}_t)$ is replaced by $|u(t, \sigma, F + h) - \bar{u}(t)|_{\mathbf{X}}$, then we have another concept of ρ-total stability, which will be referred to as the $(\rho, \mathbf{X})$-total stability $((\rho, \mathbf{X})$-TS). As was shown in Proposition 3.2, these two concepts of ρ-TS are equivalent.

We shall discuss the equivalence between the BC-TS and the ρ-TS, and extend a result due to Murakami and Yoshizawa [161, Theorems 1]) for $\mathbf{X} = \mathbf{R}^n$ to the case of $\dim \mathbf{X} = \infty$.

Theorem 3.14 *Assume that conditions* (H1), (H2) *and* (H4) *hold. Then the solution* $\bar{u}(t)$ *of* (3.26) *is BC-TS if and only if it is ρ-TS w.r.t. U for any bounded set U in* $\mathbf{X}$ *such that* $U^i \supset O_{\bar{u}} := \overline{\{\bar{u}(t) : t \in \mathbf{R}\}}$.

Proof. The "if" part is easily shown by noting that $\rho(\phi, \psi) \leq |\phi - \psi|_{\mathrm{BC}}$ for $\phi, \psi \in$ BC. We shall establish the "only if" part. We assume that the solution $\bar{u}(t)$ of (3.26) is BC-TS but not $(\rho, \mathbf{X})$-TS w.r.t.U; here $U \subset \{x \in \mathbf{X} : |x|_{\mathbf{X}} \leq c\}$ for some $c > 0$. Since the solution $\bar{u}(t)$ of (3.26) is not $(\rho, \mathbf{X})$-TS w.r.t.U, there exist an $\varepsilon \in (0, 1)$, sequences $\{\tau_m\} \subset \mathbf{R}^+, \{t_m\}(t_m > \tau_m), \{\phi^m\} \subset$ BC with $\phi^m(s) \in U, s \in \mathbf{R}^-, \{h_m\}$ with $h_m \in \mathrm{BC}([\tau_m, \infty))$, and solutions $\{u(t, \tau_m, \phi^m, F + h_m) := \hat{u}^m(t)\}$ of

$$\frac{du}{dt} = Au(t) + F(t, u_t) + h_m(t)$$

such that

$$\rho(\phi^m, \bar{u}_{\tau_m}) < 1/m \text{ and } |h_m|_{[\tau_m, \infty)} < 1/m$$

and that

$$|\hat{u}^m(t_m) - \bar{u}(t_m)|_{\mathbf{X}} = \varepsilon \text{ and } |\hat{u}^m(t) - \bar{u}(t)|_{\mathbf{X}} < \varepsilon \text{ on } [\tau_m, t_m] \quad (3.49)$$

for $m \in \mathbf{N}$. Thus if we can show that $\{u^m(t)\}$ is a family of uniformly equicontinuous on $\mathbf{R}^+$, where $\{u^m(t)\}$ is given by the parallel style as in the proof of Theorem 3.12, we have the conclusion by using the same arguments as in the proof of Theorem 3.12.

For any $m \in \mathbf{N}$, set $u^m(t) = \hat{u}^m(t + \tau_m)$ if $t \leq t_m - \tau_m$ and $u^m(t) = u^m(t - \tau_m)$ if $t > t_m - \tau_m$. Moreover, set $u^{m,r}(t) = \phi^{m,r}(t)$ if $t \in \mathbf{R}^-$ and $u^{m,r}(t) = u^m(t)$ if $t \in \mathbf{R}^+$. In what follows, we shall show that $\{u^m(t)\}$ is a family of uniformly equicontinuous functions on $\mathbf{R}^+$. To do this, we first prove that

$$\inf_m (t_m - \tau_m) > 0. \quad (3.50)$$

Assume that (3.50) is false. By taking a subsequence if necessary, we may assume that $\lim_{m \to \infty} (t_m - \tau_m) = 0$. If $m \geq 3$ and $0 \leq t \leq \min\{t_m - \tau_m, 1\}$, then

$$|u^m(t) - \bar{u}(t + \tau_m)|_{\mathbf{X}} =$$

$$= |T(t)\phi^m(0) + \int_0^t T(t-s)\{F(s + \tau_m, u_s^m) + h_m(s + \tau_m)\}ds$$

$$- T(t)\bar{u}(\tau_m) - \int_0^t T(t-s)F(s + \tau_m, \bar{u}_{s+\tau_m})ds|_{\mathbf{X}}$$

$$\leq C_2\{|\phi^m - \bar{u}_{\tau_m}|_1 + \int_0^t (2L(H) + 1)ds\}$$

$$\leq C_2\{2/(m-2) + t(2L(H) + 1)\},$$

where $H = \sup\{|\bar{u}_s|_\mathcal{B}, |u_s^m|_\mathcal{B} : 0 \leq s \leq t_m - \tau_m, m \in \mathbf{N}\}$ and $C_2 = \sup_{0 \leq s \leq 1} \|T(s)\|$. Then (3.49) yields that

$$\varepsilon \leq C_2\{2/(m-2) + (t_m - \tau_m)(2L(H) + 1)\} \to 0$$

as $m \to \infty$, a contradiction. We next prove that the set $O := \{u^m(t) : t \in \mathbf{R}^+, m \in \mathbf{N}\}$ is relatively compact in $\mathbf{X}$. To do this, for each η such that $0 < \eta < \inf_m(t_m - \tau_m)$ we consider the sets $O_\eta = \{u^m(t) : t \geq \eta, m \in \mathbf{N}\}$ and $\tilde{O}_\eta = \{u^m(t) : 0 \leq t \leq \eta, m \in \mathbf{N}\}$. Then $\alpha(O) = \max\{\alpha(O_\eta), \alpha(\tilde{O}_\eta)\}$, where $\alpha(\cdot)$ is the Kuratowski's measure of noncompactness of sets in $\mathbf{X}$. Let $0 < \nu < \min\{1, \eta\}$. If $\eta \leq t \leq t_m - \tau_m$, then

$$\begin{aligned} u^m(t) &= T(t)\phi^m(0) + \int_0^t T(t-s)\{F(s+\tau_m, u_s^m) + h_m(s+\tau_m)\}ds \\ &= T(\nu)[T(t-\nu)u^m(0) + \\ &\quad + \int_0^{t-\nu} T(t-\nu-s)\{F(s+\tau_m, u_s^m) + h(s+\tau_m)\}ds] \\ &\quad + \int_{t-\nu}^t T(t-s)\{F(s+\tau_m, u_s^m) + h(s+\tau_m)\}ds \\ &= T(\nu)u^m(t-\nu) + \int_{t-\nu}^t T(t-s)\{F(s+\tau_m, u_s^m) + h(s+\tau_m)\}ds. \end{aligned}$$

Since the set $T(\nu)\{u^m(t-\nu) : t \geq \eta, m \in \mathbf{N}\}$ is relatively compact in $\mathbf{X}$ because of the compactness of the semigroup $\{T(t)\}_{t \geq 0}$, it follows that

$$\alpha(O_\eta) \leq C_2\{L(H) + 1\}\nu.$$

Letting $\nu \to 0$ in the above, we get $\alpha(O_\eta) = 0$ for all η such that $0 < \eta < \inf_m(t_m - \tau_m)$. Observe that the set $\{T(t)\phi^m(0) : 0 \leq t \leq \eta, m \in \mathbf{N}\}$ is relatively compact in $\mathbf{X}$. Then

$$\begin{aligned} \alpha(O) &= \alpha(\tilde{O}_\eta) \\ &= \alpha(\{T(t)\phi^m(0) \\ &\quad + \int_0^t T(t-s)\{F(s+\tau_m, u_s^m) + h(s+\tau_m)\}ds : 0 \leq t \leq \eta, m \in \mathbf{N}\}) \\ &= \alpha(\{\int_0^t T(t-s)\{F(s+\tau_m, u_s^m) + h(s+\tau_m)\}ds : 0 \leq t \leq \eta, m \in \mathbf{N}\}) \\ &= C_2(L(H) + 1)\eta \end{aligned}$$

for all η such that $0 < \eta < \inf_m(t_m - \tau_m)$, which shows $\alpha(O) = 0$; consequently O must be relatively compact in $\mathbf{X}$.

Now, in order to establish the uniform equicontinuity of the family $\{u^m(t)\}$ on $\mathbf{R}^+$, let $\sigma \leq s \leq t \leq s + 1$ and $t \leq t_m - \tau_m$. Then

$$|u^m(t) - u^m(s)|_{\mathbf{X}} \leq |T(t-s)u^m(s) - u^m(s)|_{\mathbf{X}} +$$
$$+ |\int_s^t T(t-\tau)\{F(\tau + \tau_m, u_\tau^m) + h(\tau + \tau_m)\}d\tau|_{\mathbf{X}}$$
$$\leq \sup\{|T(t-s)z - z|_{\mathbf{X}} : z \in O\} + C_2\{L(H) + 1\}|t - s|.$$

Since the set O is relatively compact in $\mathbf{X}$, $T(\tau)z$ is uniformly continuous in $\tau \in [0, 1]$ uniformly for $z \in O$. This leads to $\sup\{|u^m(t) - u^m(s)|_{\mathbf{X}} : 0 \leq s \leq t \leq s+1, m \in \mathbf{N}\} \to 0$ as $|t - s| \to 0$, which proves the uniform equicontinuity of $\{u^m\}$ on $\mathbf{R}^+$.

3.4.4. Equivalent Relationships of Stabilities for Linear Abstract Functional Differential Equations with Infinite Delay

Next, we consider the case where (3.26) is linear, that is,

(H5) $F(t, \phi)$ is linear in ϕ with $\sup_{t \geq 0} |F(t, \cdot)|_{\mathbf{X}} \leq L$.

Theorem 3.15 *Assume that the conditions* (H2) *and* (H5) *hold. Let $\mathcal{B}$ be a fading memory space. Then the following statements hold:*

i) *If the null solution of* (3.26) *is BC-TS, then it is BC-UAS.*

ii) *Assume that $\mathcal{B}$ is a uniform fading memory space. If the null solution of* (3.26) *is $\mathcal{B}$-TS then it is $\mathcal{B}$- UAS.*

Proof.   Claim i). Let $\sigma \in \mathbf{R}^+$ and $\phi \in BC$ with $|\phi|_{BC} < \min(1, \delta(1))$, where $\delta(\cdot)$ is the one given for the BC-TS of the null solution of (3.26). Then $u(t) := u(t, \sigma, \phi)$ satisfies $|u(t)|_{\mathbf{X}} < 1$ for all $t \in \mathbf{R}$. Now, for any $\varepsilon > 0$, $0 < \varepsilon < 1$, and $\alpha > 0$, we set

$$a(t) := a(t, \alpha, \varepsilon) = \begin{cases} (1 + 2\alpha t)(1 + \varepsilon \alpha t), & t \geq 0 \\ 1, & t < 0, \end{cases}$$

and define $v(t)$ and $h(t)$ by

$$v(t) = a(t - \sigma)u(t), \quad t \in \mathbf{R}$$

and

$$h(t) = \dot{a}(t - \sigma)u(t) + a(t - \sigma)F(t, u_t) - F(t, v_t), \quad t \geq \sigma, \tag{3.51}$$

respectively, where $\dot{a}$ denotes the right hand derivative of a. Clearly $h(t)$ is continuous in $t \geq \sigma$ and it satisfies

$$|h(t)|_{\mathbf{X}} \leq 2\alpha|u(t)|_{\mathbf{X}} + |F(t, a(t-\sigma)u_t - v_t)|_{\mathbf{X}}$$
$$\leq 2\alpha + L|a(t-\sigma)u_t - v_t|_{\mathcal{B}}$$

for $t \geq \sigma$ by (H2), because of $|\dot{a}(t)| \leq 2\alpha$ for $t \geq 0$. We first assert that $v(t)$ is the (mild) solution of (3.34) with $h(t)$ given by (3.51); that is, $v(t)$ satisfies the relation

$$v(t) = T(t-\sigma)v(\sigma) + \int_\sigma^t T(t-s)\{\dot{a}(s-\sigma)u(s) + a(s-\sigma)F(s,u_s)\}ds, \quad (3.52)$$

for all $\tau \geq t > \sigma$. Indeed, one can take sequences $\{x_n\} \subset \mathcal{D}(A)$ and $\{h^n\} \subset BC([\sigma,\tau];\mathbf{X})$ such that h^n is continuously differentiable on $[\sigma,\tau]$ and $|x_n - \phi(0)|_\mathbf{X} + \sup_{\sigma \leq t \leq \tau} |h^n(t) - F(t,u_t)|_\mathbf{X} \to 0$ as $n \to \infty$. Set $u^n(t) = T(t-\sigma)x_n + \int_\sigma^t T(t-s)h^n(s)ds$ and $v^n(t) = a(t-\sigma)u^n(t)$ for $t \in [\sigma,\tau]$. Then $u^n(t) \to u(t)$ and $v^n(t) \to v(t)$ in $\mathbf{X}$ as $n \to \infty$ uniformly for $t \in [\sigma,\tau]$. From [179, Theorem 4.2.4] it follows that u^n is continuously differentiable on (σ,τ) with $(du^n(t))(dt) = Au^n(t) + h^n(t)$. Hence v^n is continuously differentiable on (σ,τ) with $(dv^n(t))(dt) = Av^n(t) + \dot{a}(t-\sigma)u^n(t) + a(t-\sigma)h^n(t)$. Then, from [179, Corollary 4.2.2] it follows that

$$v^n(t) = T(t-\sigma)v^n(\sigma) + \int_\sigma^t T(t-s)\{\dot{a}(s-\sigma)u^n(s) + a(s-\sigma)h^n(s)\}ds$$

for $t \in [\sigma,\tau]$. Letting $n \to \infty$ in the above, we get (3.52) as required.

Now, for each positive integer n we set $S_n := \sup\{|\phi|_B : \phi \in BC, |\phi|_{BC} \leq 1$ and supp $\phi \subset (-\infty,-n]\}$, where supp ϕ denotes the support of ϕ. Observing that $S_n \to 0$, by (A2), as $n \to \infty$, we take so large $n := n(\varepsilon)$ that $S_n < \varepsilon \delta(1)/(8L)$, and then choose $\alpha := \alpha(\varepsilon)$ such that $2\alpha(1+LJn) < \delta(1)/2$. We claim that $|h|_{[\sigma,\infty)} < \delta(1)$. To see this, for any $t \geq \sigma$ we consider functions q, q^n and w^n in BC defined by

$$q(\theta) = a(t-\sigma)u(t+\theta) - v(t+\theta)$$
$$= [a(t-\sigma) - a(t+\theta-\sigma)]u(t+\theta), \quad \theta \leq 0,$$

$$q^n(\theta) = \begin{cases} q(\theta) & \text{if } -n \leq \theta \leq 0, \\ \text{linear} & \text{if } -n-1 \leq \theta \leq -n, \\ 0 & \text{if } \theta \leq -n-1, \end{cases}$$

and $w^n = q - q^n$. Then the support of w^n is contained in $(-\infty,-n]$, and $|w^n|_{BC} \leq 2\sup_{s \in \mathbf{R}} |a(s)| \leq 4/\varepsilon$. Consequently, $|w^n|_B \leq (4/\varepsilon)S_n < \delta(1)/(2L)$. Also, since $|q^n|_{BC} \leq |q|_{[-n,0]} \leq 2\alpha n$, we get $|q^n|_B \leq J|q^n|_{BC} \leq 2\alpha Jn$. Then $|h(t)|_\mathbf{X} \leq 2\alpha + L|q|_B < \delta(1)$ or $|h|_{[\sigma,\infty)} < \delta(1)$ as required. Since the null solution of (3.26) is BC-TS, we get $|v(t)|_\mathbf{X} < 1$ for all $t \geq \sigma$. Hence, if $t \geq \sigma + (1-\varepsilon)/(\varepsilon\alpha(\varepsilon))$, then $|u(t)|_\mathbf{X} < 1/a(t-\sigma) < \varepsilon$, which proves the first claim of the theorem.

Claim ii). Let $\sigma \in \mathbf{R}^+$ and $\phi \in \mathcal{B}$ with $|\phi|_B < \delta(1)$, where $\delta(\cdot)$ is the one given for the $\mathcal{B}$-total stability of the null solution of (3.26). In what follows, we employ the same notation as in the proof of the first claim. Since $\mathcal{B}$ is a uniform fading memory space, we may assume that the functions $K(\cdot)$ and $M(\cdot)$ in (A1) satisfy $\sup_{t \geq 0} K(t) =: K < \infty$, $\sup_{t \geq 0} M(t) =: M < \infty$ and $M(t) \to 0$ as $t \to \infty$. In virtue of (A1-iii), we obtain $|u_t|_B \leq K\sup_{\sigma \leq s \leq t} |u(s)|_\mathbf{X} + M|u_\sigma|_B \leq K + M\delta(1)$ for $t \geq \sigma$. Similarly, we can get $\sup_{t \geq \sigma} |v_t|_B \leq (2/\varepsilon)K + M\delta(1)$. Take an $t_0 := t_0(\varepsilon) > 0$ so that

$$L(2/\varepsilon + 1)(2K/\varepsilon + M\delta(1))M(t_0) < \delta(1)/2,$$

and then choose an $\alpha := \alpha(\varepsilon) > 0$ so that

$$2\alpha\{1 + Lt_0(K + M\delta(1))\} < \delta(1)/2.$$

If $t \geq t_0 + \sigma$, then

$$\begin{aligned}
|h(t)|_{\mathbf{X}} &\leq 2\alpha + L|a(t-\sigma)u_t - v_t|_{\mathcal{B}} \\
&\leq 2\alpha + L\{K \sup_{-t_0 \leq \theta \leq 0} |a(t-\sigma) - a(t-\sigma+\theta)|\|u(t+\theta)|_{\mathbf{X}} \\
&\quad + M(t_0)|a(t-\sigma)u_{t-t_0} - v_{t-t_0}|_{\mathcal{B}}\} \\
&\leq 2\alpha + L\{2\alpha Kt_0 + (2/\varepsilon + 1)(2K/\varepsilon + M\delta(1))M(t_0)\} \\
&< \delta(1)
\end{aligned}$$

by (A1-iii). On one hand, if $\sigma \leq t \leq t_0 + \sigma$, then

$$\begin{aligned}
|h(t)|_{\mathbf{X}} &\leq 2\alpha + L\{K \sup_{\sigma-t \leq \theta \leq 0} |a(t-\sigma) - a(t-\sigma+\theta)|\|u(t+\theta)|_{\mathbf{X}} \\
&\quad + M|a(t-\sigma) - 1|\|\phi|_{\mathcal{B}}\} \\
&\leq 2\alpha + L\{2K\alpha(t-\sigma) + 2M\alpha(t-\sigma)\delta(1)\} \\
&< \delta(1)/2.
\end{aligned}$$

We thus obtain $|h|_{[\sigma,\infty)} < \delta(1)$. Then the $\mathcal{B}$-UAS of the null solution of (3.26) follows from the same reasoning as in the proof of the first claim.

It is natural to ask if the additional assumption that $\mathcal{B}$ is a uniform fading memory space can be removed in the second claim of Theorem 3.15. As the following example shows, however, one cannot remove the assumption, in general.

Example 3.2 *For $\mathbf{X} = \mathbf{R}$ and $g(s) = 1 - s$, we consider the space C_g^0 constructed in Section 3.3, and define a functional $G : \mathbf{R}^+ \times C_g^0 \mapsto \mathbf{R}$ by*

$$G(t, \phi) = \frac{\phi(-t)}{g(-t)}, \quad (t, \phi) \in \mathbf{R}^+ \times C_g^0.$$

For each $t \in \mathbf{R}^+$, $G(t, \cdot) : C_g^0 \mapsto \mathbf{R}$ is a bounded linear operator with $\|G(t, \cdot)\| \leq 1$. Moreover, one can see that G is continuous on $\mathbf{R}^+ \times C_g^0$. We now consider the linear functional differential equation

$$\frac{du}{dt} = -u(t) + G(t, u_t), \quad t \geq 0. \tag{3.53}$$

We first show that the null solution of (3.53) is C_g^0-TS. Indeed, if $\sigma \in \mathbf{R}^+$, $\phi \in C_g^0$ with $|\phi|_g < \varepsilon/3$ and $h \in BC([\sigma, \infty); \mathbf{R})$ with $|h|_{[\sigma,\infty)} < \varepsilon/3$, then the solution $v(t)$ of $(d/dt)u = -u(t) + G(t, u_t) + h(t)$, $t \geq \sigma$, through (σ, ϕ) satisfies

$$|v(t)| = |e^{-(t-\sigma)}\phi(0) + \int_\sigma^t e^{-(t-s)}(G(s, v_s) + h(s))ds|$$

$$\leq |\phi(0)| + \int_\sigma^t e^{-(t-s)}(|\phi(-\sigma)|/(1+s) + |h(s)|)ds$$

$$\leq |\phi|_g + \int_\sigma^t e^{-(t-s)}(|\phi|_g + |h|_{[\sigma,\infty)})ds$$

$$\leq 2|\phi|_g + |h|_{[\sigma,\infty)} < \varepsilon$$

for $t \geq \sigma$, which shows the C_g^0-TS of the null solution of (3.53). We claim that the null solution of (3.53) is not C_g^0-UAS. Indeed, if this is not true, then for any $\varepsilon > 0$ there exists a $t_0 := t_0(\varepsilon) > 1$ such that $\sup_{\sigma \geq 0} |u(\sigma + t_0, \sigma, \phi)| < \varepsilon$ whenever $\phi \in C_g^0$ with $|\phi|_g \leq 1$. For each $\sigma \geq 0$, choose a nonnegative function ϕ^σ in C_g^0 so that $\phi^\sigma(0) = 0$ and $|\phi^\sigma|_g = \phi^\sigma(-\sigma)/g(-\sigma) = 1$. Then

$$\sup_{\sigma \geq 0} |u(\sigma + t_0, \sigma, \phi^\sigma)| < \varepsilon. \tag{3.54}$$

On the other hand,

$$\begin{aligned}
u(\sigma + t_0, \sigma, \phi^\sigma) &= e^{-t_0}\phi^\sigma(0) + \int_\sigma^{\sigma+t_0} e^{-(\sigma+t_0-s)}G(s, u_s)ds \\
&= \int_0^{t_0} e^{\theta-t_0}G(\theta+\sigma, u_{\theta+\sigma})d\theta \\
&= \int_0^{t_0} e^{\theta-t_0}(1+\sigma)/(1+\theta+\sigma)d\theta \\
&\geq \int_0^{t_0} e^{\theta-t_0}d\theta \cdot (1+\sigma)/(1+t_0+\sigma) \\
&\geq (1-e^{-t_0})(1+\sigma)/(1+t_0+\sigma).
\end{aligned}$$

Hence we get $\sup_{\sigma \geq 0} |u(\sigma+t_0, \sigma, \phi^\sigma)| \geq 1 - e^{-t_0} > 1 - e^{-1}$, which is a contradiction to (3.54). Consequently, the null solution of (3.53) cannot be C_g^0-UAS.

As a direct consequence of Theorems 3.11 and 3.15, one can obtain the following result which is an extension of [93, Theorem] and [98, Theorem 3] with $\mathbf{X} = \mathbf{R}^n$ to the case where $\dim \mathbf{X} = \infty$.

Theorem 3.16 *Let $\mathcal{B}$ be a fading memory space, and assume (H1)-(H2) and (H5). Then the following statements hold.*

i) *The null solution of (3.26) is BC-TS if and only if it is BC-UAS.*

ii) *Assume that $\mathcal{B}$ is a uniform fading memory space. Then the null solution of (3.26) is $\mathcal{B}$-TS if and only if it is $\mathcal{B}$-UAS.*

We note that in Claim ii) of Theorem 3.16, the additional condition that $\mathcal{B}$ is a uniform fading memory space cannot necessarily be removed as Example 3.2 shows.

3.5. EXISTENCE OF ALMOST PERIODIC SOLUTIONS

3.5.1. Almost Periodic Abstract Functional Differential Equations with Infinite Delay

In this section, we assume that $\mathcal{B}$ is a fading memory space which is separable.

Now we shall consider the following functional differential equation

$$\frac{du}{dt} = Au(t) + F(t, u_t), \tag{3.55}$$

where A is the infinitesimal generator of a compact semigroup $\{T(t)\}_{t \geq 0}$ of bounded linear operators on $\mathbf{X}$ and $F(t, \phi) \in C(\mathbf{R} \times \mathcal{B}; \mathbf{X})$.

We always impose the following conditions on (3.55) in addition to (H1)–(H2):

(H6) $F(t, \phi)$ is almost periodic in t uniformly for $\phi \in \mathcal{B}$, where $F(t, \phi)$ is said to be almost periodic in t uniformly for $\phi \in \mathcal{B}$, if for any $\varepsilon > 0$ and any compact set W in $\mathcal{B}$, there exists a positive number $l(\varepsilon, W)$ such that any interval of length $l(\varepsilon, W)$ contains a τ for which

$$|F(t + \tau, \phi) - F(t, \phi)|_\mathbf{X} \leq \varepsilon$$

for all $t \in \mathbf{R}$ and all $\phi \in W$.

A sequence $\{F^k\}$ in $C(\mathbf{R} \times \mathcal{B}; \mathbf{X})$ is said to converge to G Bohr-uniformly on $\mathbf{R} \times \mathcal{B}$ if F^k converges to G uniformly on $\mathbf{R} \times W$ for any compact set W in $\mathcal{B}$ as $k \to \infty$. It is known (e.g. [121], [231, Theorems 2.2 and 2.3]) that $F(t, \phi)$ is almost periodic in t uniformly for $\phi \in \mathcal{B}$ if and only if for any sequence $\{t_k\}$ in $\mathbf{R}$, the sequence $\{F(t + t_k, \phi)\}$ contains a Bohr-uniformly convergent subsequence.

We denote by $H(F)$ the set of all functions $G(t, \phi)$ such that $\{F(t + t_k, \phi)\}$ converges to $G(t, \phi)$ Bohr-uniformly for some sequence $\{t_k\}$. In particular, $\Omega(F)$ is the subset of $H(F)$ for $\{t_k\}$ which tends to ∞ as $k \to \infty$. Clearly $G(t, \phi)$ is almost periodic in t uniformly for $\phi \in \mathcal{B}$ if $G \in H(F)$. For $\bar{u}(t)$ assured in (H4), we shall denote by $\Omega(\bar{u}, F)$ the set of all $(\bar{v}, g) \in H(\bar{u}, F)$ for which there exists a sequence $\{t_k\}, t_k \to \infty$ as $k \to \infty$, such that $F(t + t_k, \phi) \to G(t, \phi) \in H(F)$ Bohr-uniformly and $\bar{u}(t + t_k) \to \bar{v}(t)$ compactly on $\mathbf{R}$. We can see that $\Omega(\bar{u}, F)$ is nonempty and that $\bar{v}(t)$ is the solution of

$$\frac{dv}{dt} = Av(t) + G(t, v_t), \tag{3.56}$$

whenever $(\bar{v}, G) \in \Omega(\bar{u}, F)$.

The following proposition follows from Theorem 3.4, immediately.

Proposition 3.3 *Assume that conditions* (H1), (H2), (H4) *and* (H6) *hold. If the solution $\bar{u}(t)$ of* (3.55) *is an asymptotically almost periodic solution, then* (3.55) *has an almost periodic solution.*

3.5.2. Existence Theorems of Almost Periodic Solutions for Nonlinear Systems

We shall discuss the existence of an almost periodic solution of an almost periodic system (3.55).

Theorem 3.17 *Assume that conditions* (H1), (H2), (H4) *and* (H6) *hold. If the solution $\bar{u}(t)$ of* (3.55) *is BC-TS, then it is asymptotically almost periodic in t. Consequently,* (3.55) *has an almost periodic solution.*

Proof. By Theorem 3.12, the solution $\bar{u}$ of (3.55) is ρ-TS w.r.t. U for any bounded set U in $\mathbf{X}$ such that $O_{\bar{u},\mathbf{R}^+}\overline{\{\bar{u}(t) : t \in \mathbf{R}\}} \subset U^i$.

For any sequence $\{\tau_k'\}$ such that $\tau_k' \to \infty$ as $k \to \infty$, there is a subsequence $\{\tau_k\}$ of $\{\tau_k'\}$ and a $(\bar{v}, G) \in \Omega(\bar{u}, F)$ such that $\bar{u}(t + \tau_k) \to \bar{v}(t)$ compactly on $\mathbf{R}$ and $F(t + \tau_k, \phi)$ converges to $G(t, \phi)$ Bohr-uniformly on $\mathbf{R} \times \mathcal{B}$. We shall show that $\bar{u}(t + \tau_k)$ is convergent uniformly on $\mathbf{R}^+$.

Suppose that $\bar{u}(t + \tau_k)$ is not convergent uniformly on $\mathbf{R}^+$. Then, for some $\varepsilon > 0$ there are sequence $\{t_j\}, \{k_j\}$ and $\{m_j\}$ such that

$$k_j \to \infty, m_j \to \infty \text{ as } j \to \infty,$$

$$|\bar{u}(\tau_{k_j} + t_j) - \bar{u}(\tau_{m_j} + t_j)|_{\mathbf{X}} = \varepsilon \tag{3.57}$$

and

$$|\bar{u}(\tau_{k_j} + t) - \bar{u}(\tau_{m_j} + t)|_{\mathbf{X}} < \varepsilon \text{ on } [0, t_j). \tag{3.58}$$

Put $v^j(t) = \bar{u}(\tau_{k_j} + t)$ and $w^j(t) = \bar{u}(\tau_{m_j} + t)$. Since the sequences $\{v^j(t)\}$ and $\{w^j(t)\}$ converge to $\bar{v}(t)$ uniformly on any compact interval in $\mathbf{R}$, we can assume that

$$\rho(v_0^j, w_0^j) := \sum_{J=1}^{\infty} 2^{-J} |v_0^j - w_0^j|_J / \{1 + |v_0^j - w_0^j|_J\} < \frac{1}{j}, \ j = 1, 2, \cdots. \tag{3.59}$$

The set $\{v_0^j, w_0^j : j \in \mathbf{N}\}$ is relatively compact in $\mathcal{B}$, because the set $X_{\bar{u},\mathbf{R}^+}$ is compact in $\mathcal{B}$ by Lemma 3.5 and $v_0^j \in X_{\bar{u},\mathbf{R}^+}$. Moreover, the set $\{v^j(t), w^j(t) : j \in \mathbf{N}, t \in \mathbf{R}^+\}$ is contained in the compact set $O_{\bar{u},\mathbf{R}^+}$. From these observations and Lemma 3.4 it follows that the set $W := \{v_t^j, w_t^j : j \in \mathbf{N}, t \in \mathbf{R}^+\}$ is relatively compact in $\mathcal{B}$. Consequently,

$$\sup\{|F(t + \tau_k, \phi) - G(t, \phi)|_{\mathbf{X}} : t \in \mathbf{R}, \phi \in W\} \to 0 \text{ as } k \to \infty. \tag{3.60}$$

Define a continuous function q^j on $\mathbf{R}^+$ by

$$q^j(t) = \begin{cases} F(t + \tau_{k_j}, v_t^j) - F(t + \tau_{m_j}, w_t^j) & 0 \le t \le t_j \\ q^{j,r}(t_j) & t_j < t. \end{cases}$$

By (3.60), we can choose $j_0 = j_0(\varepsilon) \in \mathbf{N}$ in such a way that

$$\sup\{|q^j(t)|_\mathbf{X} : j \geq j_0, \ t \in \mathbf{R}^+\} < \delta(\varepsilon/2)/2,$$

where $\delta(\cdot)$ is the one for BC-TS of the solution $\bar{u}(t)$ of (3.55). Since the function v^j is the solution of

$$\frac{dx}{dt} = Ax(t) + F(t + \tau_{m_j}, x_t) + q^{j,r}(t)$$

for $t \in [0, t_j]$, and since $w^j(t)$ is a *BC-TS* solution of

$$\frac{dx}{dt} = Ax(t) + F(t + \tau_{m_j}, x_t)$$

with the same $\delta(\cdot)$ as the one for $\bar{u}(t)$, from the fact that $\sup_{t \geq 0} |q^j(t)|_\mathbf{X} < \delta(\varepsilon/2)$ it follows that $|v^j(t) - w^j(t)|_\mathbf{X} < \varepsilon/2$ on $[0, t_j]$. In particular, we have $|v^j(t_j) - w^j(t_j)|_\mathbf{X} < \varepsilon$, which contradicts (3.58).

The following theorem follows immediately from Theorems 3.11 and 3.17.

Theorem 3.18 *Assume that conditions* (H1)–(H4) *and* (H6) *hold. If the solution $u(t)$ is* BC-UAS, *then it is asymptotically almost periodic in t. Consequently,* (3.55) *has an almost periodic solution.*

Question. Assume that conditions (H1), (H2) and (H4) hold and $F(t, \phi)$ is periodic (we do not assume the regularity assumption on (3.55)).

i) Does there exist an almost periodic solution of (3.55) if $\bar{u}$ is BC-US ?

ii) Does there exist a harmonic solution of (3.55) if $\bar{u}$ is BC-UAS ?

The $\mathcal{B}$-stability implies the BC-stability. Therefore, the following results are direct consequences of Theorem 3.18.

Corollary 3.2 *Assume that conditions* (H1)–(H4) *and* (H6) *hold. If the solution $\bar{u}(t)$ of* (3.55) *is $\mathcal{B}$-TS or $\mathcal{B}$-UAS, then it is asymptotically almost periodic in t. Consequently,* (3.55) *has an almost periodic solution.*

3.5.3. Existence Theorems of Almost Periodic Solutions for Linear Systems

We consider the case where (3.55) is linear. First, we shall give the existence theorem of a bounded solution of

$$\frac{du}{dt} = Au(t) + F(t, u_t) + f(t). \tag{3.61}$$

Proposition 3.4 *Assume that conditions* (H1), (H5) *and* (H6). *If the null solution of* (3.55) *is* BC-TS, *then for any bounded and continuous function $f(t)$ defined on $\mathbf{R}$,* (3.61) *has a bounded solution defined on $\mathbf{R}^+$ which is* BC-TS.

Proof. The solution $w(t)$, $w(t) = 0$ on $\mathbf{R}^-$, of

$$\frac{dw}{dt} = Aw(t) + F(t, w_t) + \frac{\delta(1)}{2Q} f(t)$$

satisfies $|w(t)|_{\mathbf{X}} < 1$ for $t \geq 0$, where $Q = \sup_{t \in \mathbf{R}} |f(t)|_{\mathbf{X}}$ and $\delta(1)$ is the one given for the BC-TS of the null solution of (3.61). Putting

$$\bar{u}(t) = \frac{2Q}{\delta(1)} w(t),$$

$\bar{u}(t)$ is a bounded solution of (3.61) and satisfies

$$|\bar{u}(t)|_{\mathbf{X}} < \frac{2Q}{\delta(1)} < \infty$$

on $\mathbf{R}^+$.

This $\bar{u}(t)$ also is BC-TS with same pair $(\varepsilon, \delta(\varepsilon))$ as the one for BC-TS of the null solution of (3.55). Because, if $|\bar{u}_\sigma - \phi|_{\mathrm{BC}} < \delta(\varepsilon)$ and $h \in BC([\sigma, \infty); \mathbf{X})$ with $\sup_{[0,\infty)} |h(t)|_{\mathbf{X}} < \delta(\varepsilon)$, then $y(t) := u(t, \sigma, \phi, F + f + h) - \bar{u}(t)$ is a solution of (3.55) through $(\sigma, \phi - \bar{u}_\sigma)$ and $|y(t)|_{\mathbf{X}} = |u(t, \sigma, \phi, F + f + h) - \bar{u}(t)|_{\mathbf{X}} < \varepsilon$ for all $t \geq \sigma$, where $u(t, \sigma, \phi, F + f + h)$ denotes the solution of

$$\frac{du}{dt} = Au(t) + F(t, u_t) + f(t) + h(t), \quad t \geq \sigma,$$

through (σ, ϕ).

We have the following theorem from Proposition 3.4 and Theorems 3.14 and 3.18, directly.

Theorem 3.19 *Assume that conditions (H1), (H5) and (H6) hold. If the null solution of (3.55) is BC-TS(equivalently, BC-UAS), then for any almost periodic function $f(t)$, (3.61) has an almost periodic solution.*

Question. Assume that conditions (H1), (H4)–(H6) hold. Does there exist an almost periodic solution of (3.61) if $\bar{u}$ is BC-US ?

Under some additional conditions, we can derive the differentiability of the almost periodic solution ensured in Theorem 3.19.

Theorem 3.20 *Assume that conditions (H1), (H5) and (H6). Suppose that the semigroup $\{T(t)\}_{t \geq 0}$ generated by A is analytic and compact, and that $F(t, \phi)$ and $f(t)$ are locally Hölder continuous in t uniformly for ϕ in bounded sets. Then the almost periodic solution ensured in Theorem 3.19 is continuously differentiable, and it satisfies the equation (3.61).*

Proof. By considering $A-\mu I$ and $F(t,\phi)+\mu\phi(0)$ with a sufficiently large constant μ instead of A and $F(t,\phi)$ respectively, it suffices to establish the theorem under the assumption that

$$\|T(t)\| \leq ce^{-\lambda t}, \quad t \geq 0; \tag{3.62}$$

here c and λ are some positive constants. Let p be the almost periodic solution of (3.61) ensured in Theorem 3.19, and set $Q(t) = F(t,p_t) + f(t)$. We claim that the function $Q : \mathbf{R} \mapsto \mathbf{X}$ is locally Hölder continuous. If the claim holds true, then the conclusion of the theorem follows from [179, Theorem 4.3.2]. Now, let r be a constant satisfying $r \geq \sup\{|p_t|_\mathcal{B} : t \in \mathbf{R}\}$, and let $\vartheta \in (0,1)$ be the Hölder exponent for $F(t,\phi)$ and $f(t)$. Then there is a constant C such that

$$|F(t_1,\phi) - F(t_2,\phi)|_\mathbf{X} + |f(t_1) - f(t_2)|_\mathbf{X} \leq C|t_1 - t_2|^\vartheta \tag{3.63}$$

for $(t_1,\phi), (t_2,\phi) \in \mathbf{R} \times \mathcal{B}$ with $|t_1| \leq r, |t_2| \leq r$ and $|\phi|_\mathcal{B} \leq r$. If $s \geq 1$ and $0 < h < 1$, then

$$\begin{aligned}
|p(s+h) - p(s)|_\mathbf{X} &\leq |T(s+h)p(0) - T(s)p(0)|_\mathbf{X} \\
&\quad + |\int_0^{s+h} T(s+h-\tau)Q(\tau)d\tau - \int_0^s T(s-\tau)Q(\tau)d\tau|_\mathbf{X} \\
&= |\int_s^{s+h} AT(\tau)p(0)d\tau|_\mathbf{X} + \int_s^{s+h} |T(s+h-\tau)Q(\tau)|_\mathbf{X} d\tau \\
&\quad + |\int_0^s \int_0^h (-A)^{1-\vartheta}T(\theta)(-A)^\vartheta T(s-\tau)Q(\tau)d\theta d\tau|_\mathbf{X};
\end{aligned}$$

here we used the relations

$$[T(h) - I]x = \int_0^h AT(\tau)xd\tau, \quad x \in \mathbf{X},$$

and

$$-AT(\theta + \tau)x = (-A)^{1-\alpha}T(\theta)(-A)^\alpha T(\tau)x,$$

where $x \in \mathbf{X}$, $\theta > 0$, $\tau > 0$, $0 < \alpha \leq 1$, $(-A)^\alpha$, the fractional power of $-A$, is well-defined in virtue of (3.62) (cf. [179, Section 2.6]). Using the inequality $\|(-A)^\alpha T(\tau)\| \leq c_\alpha \tau^{-\alpha} e^{-\lambda \tau}$ for $\tau > 0$ (cf [179, Theorem 2.6.13 (c)]), one can easily deduce that

$$|p(s+h) - p(s)|_\mathbf{X} \leq C_1 h^\vartheta, \quad s \geq 1, \ 0 < h < 1,$$

for some constant $C_1 > 0$. Since $p(t)$ is almost periodic, the above inequality must hold for all $s \in \mathbf{R}$. We thus get $|p_{t+h} - p_t|_{BC} \leq C_1 h^\vartheta$ for $t \in \mathbf{R}$ and $0 < h < 1$, and consequently

$$|p_{t+h} - p_t|_\mathcal{B} \leq C_2 h^\vartheta, \quad t \in \mathbf{R}, \ 0 < h < 1, \tag{3.64}$$

by (3.1), where $C_2 = C_1 J$. Then, if $|t| \leq r$ and $0 < h < 1$, it follows from (3.63) and (3.64) that

$$|Q(t+h) - Q(t)|_\mathbf{X} \leq |F(t+h, p_{t+h} - p_t)|_\mathbf{X} + |F(t+h, p_t) - F(t, p_t)|_\mathbf{X}$$
$$+ |f(t+h) - f(t)|_\mathbf{X}$$
$$\leq \{C_2 \sup_{|\tau| \leq r+1} \|F(\tau, \cdot)\| + C\} h^\vartheta,$$

which shows the local Hölder continuity of $Q(t)$, as required.

3.6. APPLICATIONS

3.6.1. Damped Wave Equation

Consider the equation

$$\begin{cases} u_{tt} = u_{xx} - u_t + f(t, x, u), & t > 0, \ 0 < x < 1, \\ u(0, t) = u(1, t) = 0, & t > 0, \end{cases} \tag{3.65}$$

where $f(t, x, u)$ is continuous in $(t, x, u) \in \mathbf{R} \times (0, 1) \times \mathbf{R}$ and it is an almost periodic function in t uniformly with respect to x and u which satisfies

$$|f(t, x, u)| \leq \frac{1}{6}|u| + J$$

and

$$|f(t, x, u_1) - f(t, x, u_2)| \leq \frac{1}{6}|u_1 - u_2|$$

for all $(t, x, u), (t, x, u_1), (t, x, u_2) \in \mathbf{R} \times (0, 1) \times \mathbf{R}$ and some constant $J > 0$. We consider a Banach space $\mathbf{X}$ given by $\mathbf{X} = H_0^1(0, 1) \times L^2(0, 1)$ equipped with the norm $\|(u, v)\| = \{\|u_x\|_{L^2}^2 + \|v\|_{L^2}^2\}^{1/2} = \{\int_0^1 (u_x^2 + v^2) dx\}^{1/2}$. Then (3.65) can be considered as an abstract equation

$$\frac{d}{dt}\begin{pmatrix} u \\ v \end{pmatrix} = A\begin{pmatrix} u \\ v \end{pmatrix} + \begin{pmatrix} 0 \\ f(t, x, u) \end{pmatrix} \tag{3.66}$$

in $\mathbf{X}$, where A is a (unbounded) linear operator in $\mathbf{X}$ defined by

$$A\begin{pmatrix} u \\ v \end{pmatrix} = \begin{pmatrix} v \\ u_{xx} - v \end{pmatrix}$$

for $(u, v) \in H^2(0, 1) \times H_0^1(0, 1)$. It is well known that A generates a C_0-semigroup of bounded linear operator on $\mathbf{X}$. In the following, we show that each (mild) solution of (3.66) is bounded in the future, and that it is UAS. Moreover, we show that each solution of (3.66) has a compact orbit in $\mathbf{X}$. Consequently, one can apply Theorem 3.5 or Corollary 3.2 to the process generated by solutions of (3.66) and its integrals to conclude that (3.65) has an almost periodic solution which is UAS.

Now, for any solution $\begin{pmatrix} u(t) \\ v(t) \end{pmatrix}$ of (3.66), we consider a function $V(t)$ defined by

$$V(t) = \int_0^1 (u_x^2 + 2\lambda uv + v^2) dx$$

with $\lambda = \frac{1}{12}$. Since $\|u\|_{L^2} \leq \|u_x\|_{L^2}$ for $u \in H_0^1(0,1)$, we get the inequality

$$\frac{11}{12}(\|u_x\|_{L^2}^2 + \|v\|_{L^2}^2) \leq V(t) \leq \frac{13}{12}(\|u_x\|_{L^2}^2 + \|v\|_{L^2}^2). \quad (3.67)$$

Moreover, we get

$$\frac{d}{dt}V(t) \leq 2\int_0^1 \{(\lambda-1)v^2 - \lambda u_x^2 - \lambda uv + (\lambda|u|+|v|)(\frac{|u|}{6}+J)\}dx$$
$$\leq 2(\lambda-1)\|v\|_{L^2}^2 - 2\lambda\|u_x\|_{L^2}^2 + \lambda(\varepsilon_1\|u\|_{L^2}^2 + \varepsilon_1^{-1}\|v\|_{L^2}^2)$$
$$+\frac{\lambda}{3}\|u\|_{L^2}^2 + \lambda(\varepsilon_2\|u\|_{L^2}^2 + \varepsilon_2^{-1}J^2) + \frac{1}{6}(\varepsilon_3\|u\|_{L^2}^2 + \varepsilon_3^{-1}\|v\|_{L^2}^2)$$
$$+(\varepsilon_4\|v\|_{L^2}^2 + \varepsilon_4^{-1}J^2)$$

for any positive constants $\varepsilon_1, \varepsilon_2, \varepsilon_3$ and ε_4. We set $\varepsilon_1 = \frac{1}{10}, \varepsilon_2 = \frac{1}{10}, \varepsilon_3 = \frac{1}{2}$ and $\varepsilon_4 = \frac{1}{3}$ to get

$$\frac{d}{dt}V(t) \leq -\frac{1}{3}\|v\|_{L^2}^2 - \frac{7}{180}\|u_x\|_{L^2}^2 + 4J^2. \quad (3.68)$$

By (3.67) and (3.68), we get

$$\frac{d}{dt}V(t) \leq -C_1 V(t) + C_2$$

for some positive constants C_1 and C_2. Then $V(t) \leq e^{-C_1(t-t_0)}V(t_0) + \frac{C_2}{C_1}$ for all $t \geq t_0$, and hence $\sup_{t \geq t_0} \|(u(t), v(t))\|_{\mathbf{X}} < \infty$. Thus each solution of (3.66) is bounded in the future.

Next, for any solutions $\begin{pmatrix} u(t) \\ v(t) \end{pmatrix}$ and $\begin{pmatrix} \bar{u}(t) \\ \bar{v}(t) \end{pmatrix}$ of (3.66), we consider a function $\bar{V}(t)$ defined by

$$\bar{V}(t) = \int_0^1 \{(u_x - \bar{u}_x)^2 + 2\lambda(u-\bar{u})(v-\bar{v}) + (v-\bar{v})^2\}dx$$

with $\lambda = \frac{1}{12}$. By almost the same arguments as for the function $V(t)$, we get

$$\frac{11}{12}\{\|u_x - \bar{u}_x\|_{L^2}^2 + \|v - \bar{v}\|_{L^2}^2\} \leq \bar{V}(t) \leq$$
$$\frac{13}{12}\{\|u_x - \bar{u}_x\|_{L^2}^2 + \|v - \bar{v}\|_{L^2}^2\} \quad (3.69)$$

and

$$\frac{d}{dt}\bar{V}(t) \leq -C_3 \bar{V}(t) \quad (3.70)$$

for some positive constant C_3. Then, by (3.69) and (3.70) we get

$$\begin{aligned}
\|u_x(t) - \bar{u}_x(t)\|_{L^2}^2 + \|v(t) - \bar{v}(t)\|_{L^2}^2 &\leq \frac{11}{12} V(t) \\
&\leq \frac{11}{12} e^{-C_3(t-t_0)} V(t_0) \\
&\leq e^{-C_3(t-t_0)} \{\|u_x(t_0) - \bar{u}_x(t_0)\|_{L^2}^2 \\
&\quad + \|v(t_0) - \bar{v}(t_0)\|_{L^2}^2\}
\end{aligned}$$

for any $t \geq t_0$, which shows that the solution $\begin{pmatrix} u(t) \\ v(t) \end{pmatrix}$ of (3.66) is UAS.

Finally, we shall show that each bounded solution $(u(t), v(t))$ of (3.66) has a compact orbit. To do this, it suffices to show that each increasing sequence $\{\tau_n\} \subset \mathbf{R}^+$ such that $\tau_n \to \infty$ as $n \to \infty$ has a subsequence $\{\tau_n'\}$ such that $\{u(\tau_n'), v(\tau_n')\}$ is a Cauchy sequence in $\mathbf{X}$. Taking a subsequence if necessary, we may assume that $\{f(t+\tau_n, x, u)\}$ converges uniformly in (x, u) in bounded sets and $t \in \mathbf{R}$. Therefore, for any $\gamma > 0$ there exists an $n_0 > 0$ such that if $n > m \geq n_0$, then

$$\begin{aligned}
&\sup_{(t,x,\xi) \in \mathbf{R} \times (0,1) \times O^+(u)} |f(t+\tau_n, x, \xi) - f(t+\tau_m, x, \xi)| \\
= &\sup_{(t,x,\xi) \in \mathbf{R} \times (0,1) \times O^+(u)} |f(t+\tau_n - \tau_m, x, \xi) - f(t, x, \xi)| < \gamma.
\end{aligned}$$

Fix n and m such that $n > m \geq n_0$, and set $u^\tau(t) = u(t+\tau), v^\tau(t) = v(t+\tau)$ and $f^\tau(t, x, \xi) = f(t+\tau, x, \xi)$ with $\tau = \tau_n - \tau_m$. Clearly, (u^τ, v^τ) is a solution of (3.66) with f^τ instead of f.

Consider a function U defined by

$$U(t) = \int_0^1 \{(u_x - u_x^\tau)^2 + 2\lambda(u - u^\tau)(v - v^\tau) + (v - v^\tau)^2\} dx$$

with $\lambda = \frac{1}{12}$. Then

$$\frac{11}{12}\{\|u_x - u_x^\tau\|_{L^2}^2 + \|v - v^\tau\|_{L^2}^2\} \leq U(t) \leq \frac{13}{12}\{\|u_x - u_x^\tau\|_{L^2}^2 + \|v - v^\tau\|_{L^2}^2\}$$

and

$$\begin{aligned}
\frac{d}{dt} U(t) \leq\ & (2\lambda - 2)\|v - v^\tau\|_{L^2}^2 - 2\lambda \|u_x - u_x^\tau\|_{L^2}^2 \\
& + 2\lambda \int_0^1 |u - u^\tau|\{|v - v^\tau| + |f(t, x, u) - f^\tau(t, x, u^\tau)|\} dx \\
& + 2\int_0^1 |v - v^\tau| |f(t, x, u) - f^\tau(t, x, u^\tau)| dx.
\end{aligned}$$

Since

$$|f(t,x,u) - f^\tau(t,x,u^\tau)| \leq |f(t,x,u) - f(t,x,u^\tau)| + $$
$$+ |f(t,x,u^\tau) - f^\tau(t,x,u^\tau)|$$
$$\leq \frac{1}{6}|u - u^\tau| + \gamma,$$

we get

$$\frac{d}{dt}U(t) \leq -\frac{11}{6}\|v - v^\tau\|_{L^2}^2 - \frac{1}{6}\|u_x - u_x^\tau\|_{L^2}^2$$
$$+ \frac{1}{12}(\frac{1}{10}\|u - u^\tau\|_{L^2}^2 + 10\|v - v^\tau\|_{L^2}^2) + \frac{1}{36}\|u - u^\tau\|_{L^2}^2$$
$$+ \frac{1}{12}(\frac{1}{2}\|u - u^\tau\|_{L^2}^2 + 2\gamma^2)$$
$$+ \frac{1}{6}(\varepsilon_5 \|v - v^\tau\|_{L^2}^2 + \frac{1}{\varepsilon_5}\|u - u^\tau\|_{L^2}^2) + (\varepsilon_6 \|v - v^\tau\|_{L^2}^2 + \frac{1}{\varepsilon_6}\gamma^2)$$
$$\leq \|v - v^\tau\|_{L^2}^2 (-1 + \frac{\varepsilon_5}{6} + \varepsilon_6)$$
$$+ \|u_x - u_x^\tau\|_{L^2}^2 (-\frac{1}{6} + \frac{1}{120} + \frac{1}{36} + \frac{1}{24} + \frac{1}{36\varepsilon_5}) + (\frac{1}{6} + \frac{1}{\varepsilon_6})\gamma^2.$$

Putting $\varepsilon_5 = 1$ and $\varepsilon_6 = \frac{1}{3}$, we have

$$\frac{d}{dt}U(t) \leq -C(\|u_x - u_x^\tau\|_{L^2}^2 + \|v - v_\tau\|_{L^2}^2) + 4\gamma^2$$
$$\leq -C_1 U(t) + 4\gamma^2,$$

where C and C_1 are some constants independent of γ. Then

$$U(\tau_m) \leq e^{-C_1 \tau_m} U(0) + \frac{4\gamma^2}{C_1}$$

or

$$\|u_x(\tau_m) - u_x(\tau_n)\|_{L^2}^2 + \|v(\tau_m) - v(\tau_n)\|_{L^2}^2 \leq K e^{-C_1 \tau_m} + \frac{8\gamma^2}{C_1},$$

where $K = 2U(0)$. Take $n_1 \geq n_0$ so that $Ke^{-C_1 \tau_{n_1}} < \frac{\gamma^2}{C_1}$. Then

$$\|u_x(\tau_m) - u_x(\tau_n)\|_{L^2}^2 + \|v(\tau_n) - v(\tau_m)\|_{L^2}^2 < \frac{9}{C_1}\gamma^2$$

if $n \geq m \geq n_1$, which shows that $\{(u(\tau_n), v(\tau_n))\}$ is a Cauchy sequence in **X**, as required.

3.6.2. Integrodifferential Equation with Diffusion

We consider the following integrodifferential equation with diffusion

$$\frac{\partial u}{\partial t}(t,x) = \frac{\partial^2 u}{\partial x^2}(t,x) - u^3(t,x)$$

$$+ \int_{-\infty}^{t} k(t,s,x)u(s,x)ds + h(t,x), \quad t > 0, \ 0 < x < \pi. \tag{3.71}$$

In [90, p.85], the equation (3.71) with $k \equiv 0$ and $h \equiv 0$ was treated under the Dirichlet boundary condition, and the uniform asymptotic stability of the zero solution was derived. In what follows, we shall treat the equation (3.71) under the Neumann boundary condition

$$\frac{\partial u}{\partial x}(t,0) = \frac{\partial u}{\partial x}(t,\pi) = 0, \quad t > 0, \tag{3.72}$$

and give a sufficient condition under which (3.71) possesses a bounded solution which is BC-TS. Consequently, one can apply Theorem 3.17 to assure the existence of an almost periodic solution of (3.71)–(3.72).

We assume that the functions $h(t,x)$ and $k(t,s,x)$ are continuous functions satisfying $2 < h(t,x) < 7$ and $0 \leq k(t,s,x) \leq K(t-s)$ for some continuous function $K(\tau)$ with $\int_0^\infty K(\tau)d\tau < 1/4$. Note that one can choose a continuous nonincreasing function $g : \mathbf{R}^- \mapsto [1,\infty)$ so that $g(0) = 1$, $\lim_{s \to -\infty} g(s) = \infty$ and $\int_0^\infty K(\tau)g(-\tau)d\tau < \infty$ (cf. [15]). We now consider the Banach space $\mathbf{X} = C([0,\pi];\mathbf{R})$, and define the linear operator A in $\mathbf{X}$ by

$$(A\xi)(x) = \frac{d^2\xi}{dx^2}(x), \quad 0 < x < \pi,$$

for

$$\xi \in D(A) := \{\xi \in C^2[0,\pi] : \xi'(0) = \xi'(\pi) = 0\}.$$

Then the operator A generates a compact semigroup $T(t)$ on $\mathbf{X}$, and (3.71)–(3.72) is represented as the functional differential equation (5.1) on $\mathbf{X}$ with

$$F(t,\phi)(x) = h(t,x) - \phi^3(0,x) + \int_{-\infty}^{0} k(t,t+s,x)\phi(s,x)ds$$

for $\phi \in C_g^0(\mathbf{X})$. We refer to a (mild) solution of (5.1) as a (mild) solution of (3.71)–(3.72).

Lemma 3.8 *Let $\phi(\theta,x) \equiv 3/2$ for all $(\theta,x) \in \mathbf{R}^- \times [0,\pi]$. Then the solution $u(t,x)$ of (3.71)–(3.72) through $(0,\phi)$ satisfies the inequality*

$$1 < u(t,x) < 2 \quad \text{on } [0,\infty) \times [0,\pi].$$

Proof. It is clear that there exists a (unique) local solution $u(t,x) \in C([0,a) \times [0,\pi])$ of (3.71)–(3.72) through $(0,\phi)$ for some $a > 0$. We shall prove that

$$1 < u(t,x) < 2 \quad \text{on } [0,a) \times [0,\pi]. \tag{3.73}$$

Then $a = \infty$ and the conclusion of the Lemma 3.8 must hold. We will prove (3.73) by a contradiction. Assume that (3.73) is false. Then there exists a $t_1 \in (0, a)$ such that $1 \leq u(t, x) \leq 2$ on $[0, t_1] \times [0, \pi]$ and $u(t_1, x_1) = 2$ or $u(t_1, x_1) = 1$ for some $x_1 \in [0, \pi]$. Consider a function $p(t, x) \in C([0, t_1] \times [0, \pi])$ defined by $p(t, x) = \int_{-\infty}^{t} k(t, s, x) u(s, x) ds + h(t, x)$, and choose a sequence $\{p_n(t, x)\} \in C^1([0, t_1] \times [0, \pi])$ such that $2 \leq p_n(t, x) \leq 7.5$ on $[0, t_1] \times [0, \pi]$ and that $p_n(t, x) \to p(t, x)$ uniformly on $[0, t_1] \times [0, \pi]$ as $n \to \infty$. There exists a (classical) solution $v_n(t, x)$ of

$$\frac{\partial v}{\partial t}(t, x) = \frac{\partial^2 v}{\partial x^2}(t, x) - v^3(t, x) + p_n(t, x), \quad 0 < t \leq t_1, \ 0 < x < \pi,$$

$$\frac{\partial v}{\partial x}(t, 0) = \frac{\partial v}{\partial x}(t, 1) = 0, \quad 0 < t \leq t_1,$$

$$v(0, x) = 3/2, \quad 0 < x < \pi.$$

Clearly, $v_n(t, x) \to u(t, x)$ uniformly on $[0, t_1] \times [0, \pi]$. We now assert that

$$\sqrt[3]{1.9} < v_n(t, x) < \sqrt[3]{7.6} \quad \text{on } [0, t_1] \times [0, \pi]. \tag{3.74}$$

If this assertion holds true, letting $n \to \infty$ in (3.74) we obtain $\sqrt[3]{1.9} \leq u(t, x) \leq \sqrt[3]{7.6}$ on $[0, t_1] \times [0, \pi]$, which contradicts $u(t_1, x_1) = 1$ or $u(t_1, x_1) = 2$. Consider the case (3.74) does not hold true. Then there exists a $(t_2, x_2) \in [0, t_1] \times [0, \pi]$ such that $v_n(t_2, x_2) = \sqrt[3]{7.6}$ (or $v_n(t_2, x_2) = \sqrt[3]{1.9}$) and that $\sqrt[3]{1.9} < v_n(t, x) < \sqrt[3]{7.6}$ on $[0, t_2) \times [0, \pi]$. If $x_2 \in (0, \pi)$, then $\partial^2 v_n / \partial x^2 \leq 0$ and $\partial v_n / \partial t \geq 0$ (or $\partial^2 v_n / \partial x^2 \geq 0$ and $\partial v_n / \partial t \leq 0$) at (t_2, x_2); consequently $7.5 \geq p_n = -\partial^2 v_n / \partial x^2 + \partial v_n / \partial t + v_n^3 \geq v_n^3 = 7.6$ (or $2 \leq p_n = -\partial^2 v_n / \partial x^2 + \partial v_n / \partial t + v_n^3 \leq v_n^3 = 1.9$) at (t_2, x_2), a contradiction. Thus we must get $\sqrt[3]{1.9} < v_n(t, x) < \sqrt[3]{7.6}$ for all $(t, x) \in [0, t_2] \times (0, \pi)$ and $x_2 = 0$ or $x_2 = \pi$; say $x_2 = \pi$. Hence, by the strong maximum principle [183, Theorem 3.7] we get $\partial v_n / \partial x(t_2, \pi) > 0$ (or $\partial v_n / \partial x(t_2, \pi) < 0$). But this is a contradiction, because of $\partial v_n \partial x(t_2, \pi) = 0$. Therefore, we must have the assertion (3.74).

Proposition 3.5 *Let $u(t, x)$ be the solution of (3.71)–(3.72) ensured in Lemma 3.8. Then $u(t, x)$ is BC-TS. Hence, if $h(t, x)$ and $k(t, t + s, x)$ are almost periodic in t uniformly for $(s, x) \in \mathbf{R}^- \times [0, \pi]$, then $u(t, x)$ is asymptotically almost periodic in t uniformly for $x \in [0, \pi]$.*

Proof. In order to show that $u(t, x)$ is BC-TS, it is sufficient to certify that

$$|u(t, x) - v(t, x)| < \varepsilon, \quad t \geq \sigma, \ x \in [0, \pi], \tag{3.75}$$

whenever $v(t, x)$ is a solution of

$$\frac{\partial v}{\partial t}(t, x) = \frac{\partial^2 v}{\partial x^2}(t, x) - v^3(t, x) + \int_{-\infty}^{t} k(t, s, x) v(s, x) ds$$

$$+h(t,x) + r(t,x), \quad t > \sigma, \ 0 < x < \pi,$$

$$\frac{\partial v}{\partial x}(t,0) = \frac{\partial v}{\partial x}(t,\pi) = 0, \quad t > \sigma,$$

where $r(t,x) \in C([\sigma,\infty) \times [0,\pi])$ with $\sup_{t \geq \sigma, 0 \leq x \leq \pi} |r(t,x)| < \varepsilon/2$ and $\sup_{\theta \leq \sigma, 0 \leq x \leq \pi} |u(\theta,x) - v(\theta,x)| < \varepsilon$. Set $w(t,x) = u(t,x) - v(t,x)$. Then $w(t,x)$ is a (mild) solution of

$$\frac{\partial w}{\partial t}(t,x) = \frac{\partial^2 w}{\partial x^2}(t,x) - w(u^2(t,x) + u(t,x)v(t,x) + v^2(t,x))$$

$$+ \int_{-\infty}^{t} k(t,s,x)w(s,x)ds - r(t,x), \quad t > \sigma, \ 0 < x < \pi,$$

$$\frac{\partial w}{\partial x}(t,0) = \frac{\partial w}{\partial x}(t,\pi) = 0, \quad t > \sigma.$$

Assume (3.75) is not true. Then there exists an $(t_3, x_3) \in (\sigma, \infty) \times [0, \pi]$ such that $|w(t,x)| < \varepsilon$ on $[\sigma, t_3] \times [0, \pi]$ and $|w(t_3, x_3)| = \varepsilon$, say $w(t_3, x_3) = \varepsilon$. Consider a function $V(t,x)$ defined by $V(t,x) = w(t,x) - \varepsilon$ for $(t,x) \in [\sigma, t_3] \times [0, \pi]$. Clearly, $V(t,x)$ is a (mild) solution of

$$\frac{\partial V}{\partial t}(t,x) = \frac{\partial^2 V}{\partial x^2}(t,x) - (V + \varepsilon)(u^2(t,x) + u(t,x)v(t,x) + v^2(t,x))$$

$$+ \int_{-\infty}^{t} k(t,s,x)w(s,x)ds - r(t,x), \quad \sigma < t \leq t_3, \ 0 < x < \pi,$$

$$\frac{\partial V}{\partial x}(t,0) = \frac{\partial V}{\partial x}(t,\pi) = 0, \quad \sigma < t \leq t_3.$$

If $V(t,x)$ is smooth, then

$$\frac{\partial^2 V}{\partial x^2}(t,x) - \frac{\partial V}{\partial t}(t,x) - (u^2(t,x) + u(t,x)v(t,x) + v^2(t,x))V(t,x)$$

$$= \varepsilon(u^2(t,x) + u(t,x)v(t,x) + v^2(t,x)) - \int_{-\infty}^{t} k(t,s,x)w(s,x)ds + r(t,x)$$

$$\geq \varepsilon\{(v(t,x) + (1/2)u(t,x))^2 + (3/4)u^2(t,x)\} - \varepsilon \int_{-\infty}^{t} k(t,s,x)ds - \varepsilon/2$$

$$> 3\varepsilon/4 - \varepsilon/4 - \varepsilon/2 = 0$$

on $(\sigma, t_3] \times [0, \pi]$. Then, repeating the same argument as in the proof of Lemma 3.8, we can get a contradiction by applying the strong maximum principle. In the case $V(t,x)$ is not smooth, we can get a contradiction again by approximating $V(t,x)$ by some smooth functions as in the proof of Lemma 3.8. Thus we must have (3.75).

3.6.3. Partial Functional Differential Equation

We consider the scalar partial functional differential equations

$$\frac{\partial u}{\partial t} = \Delta u + \mathcal{F}(t, x, u_t(\cdot, x)) \tag{3.76}$$

and

$$\frac{\partial u}{\partial t} = \Delta u + \mathcal{F}(t, x, u_t(\cdot, x)) + \bar{f}(t, x) \tag{3.77}$$

in $\mathbf{R} \times \Omega$ together with the boundary condition

$$\frac{\partial u}{\partial \mathbf{n}} + \kappa(x)u = 0 \quad \text{on } \mathbf{R} \times \partial\Omega, \tag{3.78}$$

where Ω is a bounded domain in $\mathbf{R}^J$ with smooth boundary $\partial\Omega$ (e.g., $\partial\Omega \in C^{2+\alpha}$ for some $\alpha \in (0,1)$), and Δ and $\partial/\partial\mathbf{n}$ respectively denote the Laplacian operator in $\mathbf{R}^J$ and the exterior normal derivative at $\partial\Omega$, and moreover $\kappa \in C^{1+\alpha}(\partial\Omega)$ with $\kappa(x) \geq 0$ on $\partial\Omega$. As the space $\mathbf{X}$, we take

$$\mathbf{X} = C(\bar{\Omega}),$$

where $\bar{\Omega}$ is $\Omega \cup \partial\Omega$ and $C(\bar{\Omega})$ is the Banach space of all real-valued continuous functions on $\bar{\Omega}$ with the supremum norm, and we define the operator A by the closed extension of the operator Δ with domain $\mathcal{D}(\Delta) = \{\xi \in C^2(\bar{\Omega}) : \partial\xi/\partial n + \kappa\xi = 0 \text{ on } \partial\Omega\}$. By [212, Theorem 2], A generates an analytic compact semigroup $\{T(t)\}_{t\geq 0}$ on $\mathbf{X}$. In fact, $\{T(t)\}_{t\geq 0}$ is represented as

$$[T(t)\phi](x) = \int_\Omega U(t; x, y)\phi(y)dy, \quad x \in \bar{\Omega}, \tag{3.79}$$

where $U(t; x, y)$ is the fundamental solution of the heat equation with the boundary condition (3.78) (cf. [68], [116], [108]).

On $\mathcal{F}$ and $\bar{f}$ we impose the following conditions:

(H1') $\mathcal{F} : \mathbf{R} \times \bar{\Omega} \times C_g^0(\mathbf{R}) \mapsto \mathbf{R}$ is continuous in $(t, x, \phi) \in \mathbf{R} \times \bar{\Omega} \times C_g^0(\mathbf{R})$ and linear in ϕ; here g is a (fixed) continuous nonincreasing function such that $g(0) = 1$ and $g(s) \to \infty$ as $s \to -\infty$;

(H2') $\bar{f} : \mathbf{R} \times \bar{\Omega} \mapsto \mathbf{R}$ is continous in $(t, x) \in \mathbf{R} \times \bar{\Omega}$;

(H3') $\mathcal{F}(t, x, \phi)$ is almost periodic in t uniformly for $(x, \phi) \in \bar{\Omega} \times C_g^0(\mathbf{R})$, and $\bar{f}(t, x)$ is almost periodic in t uniformly for $x \in \bar{\Omega}$, where g is the function in (H1').

Set

$$\mathcal{B} = C_g^0(C(\bar{\Omega})),$$

and define $F : \mathbf{R} \times \mathcal{B} \mapsto \mathbf{X}$ and $f : \mathbf{R} \mapsto \mathbf{X}$ by

$$F(t, \phi)(x) = \mathcal{F}(t, x, \phi(\cdot, x)), \quad (t, x, \phi) \in \mathbf{R} \times \bar{\Omega} \times \mathcal{B},$$

and
$$f(t)(x) = \bar{f}(t,x), \quad (t,x) \in \mathbf{R} \times \bar{\Omega},$$
where $\phi(\cdot, x)$ is an element in $C_g^0(\mathbf{R})$ defined by $\phi(\theta, x) = \phi(\theta)(x)$ for $\theta \in \mathbf{R}^-$, and g is the function in (H1'). Observe that if the set W is compact in $\mathcal{B}$, then the set $\{\phi(\cdot, x) : \phi \in W, \ x \in \bar{\Omega}\}$ is compact in $C_g^0(\mathbf{R})$. From this fact, we can easily see that (Hi') implies (Hi) for $i = 1, 2, 3$. Then (3.77) together with (3.78) is represented as the (abstract) equation (3.61). From (3.79) we can see that the (mild) solution $u(\cdot, \sigma, \phi, F + f)$ of (5.7) through (σ, ϕ) satisfies the relation

$$\begin{aligned} u(t,x) &= \int_\Omega U(t-\sigma;x,y)\phi(y)dy \\ &\quad + \int_\sigma^t \int_\Omega U(t-s;x,y)\{\mathcal{F}(s,y,u(s+\cdot,y)) + \bar{f}(s,y)\}dyds \end{aligned} \tag{3.80}$$

for $(t,x) \in [\sigma, \infty) \times \bar{\Omega}$, where $u(t,x) = u(t, \sigma, \phi, F + f)(x)$. Henceforth, the function u satisfying (3.80) will be called a (mild) solution of (3.77) and (3.78).

In what follows, in order to ensure the BC-TS of the null solution of (3.77) and (3.78), we consider the following condition;

(H4') there exist positive constants a and b, $a < b$, such that
$$\sup_{t,\ x} |\mathcal{F}(t,x,\xi) + b\xi(0)| \leq a|\xi|_{BC}, \quad \xi \in BC(\mathbf{R}^-; \mathbf{R}).$$

It is easy to check that if $\mathcal{F}(t,x,\xi) = -b\xi(0) + \int_{-\infty}^0 k(t,s,x)\xi(s)ds$ with $\sup_{t,\ x} \int_{-\infty}^0 |k(t,s,x)|ds \leq a < b$, then (H4') is satisfied.

Lemma 3.9 *Assume* (H1'), (H3') *and* (H4'). *Then the null solution of* (3.77) *and* (3.79) *is BC-TS.*

Proof. Let $\varepsilon > 0$ and $\sigma \in \mathbf{R}$ be given. To prove Lemma 3.9, it suffices to certify that $\phi \in BC(\mathbf{R}^-; \mathbf{X})$ with $|\phi|_{BC} < \varepsilon$ and $h \in BC([\sigma, \infty); \mathbf{X})$ with $\sup_{t \geq \sigma} |h(t)|_\mathbf{X} < (b-a)\varepsilon$ imply $|u(t, \sigma, \phi, h)|_\mathbf{X} < \varepsilon$ for $t \geq \sigma$. If this is not the case, then there exists a $\tau > \sigma$ such that $|u(\tau, \sigma, \phi, h)|_\mathbf{X} = \varepsilon$ and $|u(t, \sigma, \phi, h)|_\mathbf{X} < \varepsilon$ for $t < \tau$. Set $q(t) = F(t, v_t) + bv(t) + h(t)$ for $t \in [\sigma, \tau]$, where $v(t) := u(t, \sigma, \phi, F + f + h)$. The function $q : [\sigma, \tau] \mapsto \mathbf{X}$ is continuous, and it satisfies $|q(t)|_\mathbf{X} < b\varepsilon$ on $[\sigma, \tau]$ by (H4'). Take a sequence $\{\mathcal{Q}^n\} \subset C^1([\sigma, \tau] \times \bar{\Omega}; \mathbf{R})$ with the properties that $Q := \sup\{|\mathcal{Q}^n(t,x)| : t \in [\sigma, \tau],\ x \in \bar{\Omega},\ n = 1, 2, \ldots\} < b\varepsilon$ and that $\sup_{\sigma \leq t \leq \tau} |q^n(t) - q(t)|_\mathbf{X} \to 0$ as $n \to \infty$, where $q^n(t)(x) = \mathcal{Q}^n(t,x)$, $x \in \bar{\Omega}$. Let $V^n(t,x)$ be the (classical) solution of linear partial differential equation

$$\frac{\partial u}{\partial t} = \Delta u - bu + \mathcal{Q}^n(t,x) \quad \text{in } (\sigma, \tau] \times \Omega$$

which satisfies the boundary condition (3.78) on $(\sigma, \tau] \times \partial\Omega$ and the initial condition $V^n(\sigma, x) = \phi(0)(x)$, $x \in \bar{\Omega}$. Set $v^n(t)(x) = V^n(t,x)$ for $x \in \bar{\Omega}$. Then $v^n(t) =$

$T(t-\sigma)\phi(0) + \int_\sigma^t T(t-s)\{-bv^n(s) + q^n(s)\}ds$ for $t \in [\sigma, \tau]$ (cf. [116, Theorems 2.9.1 and 2.9.2]). Applying Gronwall's lemma, one can see that $v^n(t)$ tends to $T(t-\sigma)\phi(0) + \int_\sigma^t T(t-s)\{-bv(s) + q(s)\}ds = v(t)$ uniformly on $[\sigma, \tau]$ as $n \to \infty$. Set $k = \max(|\phi(0)|_\mathbf{X}, Q/b)$, and take a positive integer n_0 so that $k + 1/n_0 < \varepsilon$. We claim that
$$|V^n(t, x)| < k + 1/n, \quad (t, x) \in [\sigma, \tau] \times \bar{\Omega}, \tag{3.81}$$
for all $n \geq n_0$. If this claim is not true, then there exists some $n \geq n_0$ and $(t_0, x_0) \in (\sigma, \tau] \times \bar{\Omega}$ such that $|V^n(t_0, x_0)| = k + 1/n$ (say $V^n(t_0, x_0) = k + 1/n$) and $|V^n(t, x)| < k + 1/n$ on $[\sigma, t_0) \times \bar{\Omega}$. Set $W(t, x) = V^n(t, x) - k - 1/n$ for $(t, x) \in [\sigma, \tau] \times \bar{\Omega}$. Then $W(t_0, x_0) = 0$ and $W(t, x) < 0$ for all $(t, x) \in [\sigma, t_0) \times \bar{\Omega}$. Moreover, we get

$$\begin{aligned}\frac{\partial W}{\partial t}(t, x) &= \frac{\partial V^n}{\partial t}(t, x) \\ &= \Delta W(t, x) - b(W(t, x) + k + 1/n) + Q^n(t, x),\end{aligned}$$

and hence $\Delta W(t, x) - (\partial/\partial t)W(t, x) - bW(t, x) = b(k + 1/n) - Q^n(t, x) \geq Q - Q^n(t, x) \geq 0$ on $(\sigma, t_0] \times \Omega$. If $x_0 \in \Omega$, then $W(t, x) \equiv 0$ on $[\sigma, t_0] \times \bar{\Omega}$ by the strong maximum principle (e.g., [183, Theorems 3.3.5, 3.3.6 and 3.3.7]), which is a contradiction because of $V^n(\sigma, x) = \phi(0)(x) < k + 1/n$. We thus obtain $x_0 \in \partial\Omega$ and $W(t, x) < 0$ on $[\sigma, t_0] \times \Omega$, and hence $\partial W/\partial \mathbf{n} > 0$ at (t_0, x_0) by the strong maximum principle, again. However, this is impossible because of $(\partial W/\partial \mathbf{n})(t_0, x_0) = (\partial V^n/\partial \mathbf{n})(t_0, x_0) = -\kappa(x_0)V^n(t_0, x_0) \leq 0$. Consequently, the claim (3.81) must hold true. Letting $n \to \infty$ in (3.81), we get $|v(t)|_\mathbf{X} \leq k < \varepsilon$ for all $t \in [\sigma, \tau]$, which is a contradiction to $|v(\tau)|_\mathbf{X} = \varepsilon$. This completes the proof.

Combining Theorem 3.19 with Lemma 3.9, we can immediately obtain the following result.

Proposition 3.6 *Assume* (H1')-(H4'). *Then there exists an almost periodic (mild) solution of* (3.77) *and* (3.78).

Under the additional assumption on $\mathcal{F}(t, x, \xi)$ and $\bar{f}(t, x)$, we can deduce the existence of an almost periodic classical solution of (3.77) and (3.78).

Proposition 3.7 *In addition to* (H1')-(H4'), *assume that* $\mathcal{F}(t, x, \xi)$ *and* $\bar{f}(t, x)$ *are locally Hölder continuous in* $(t, x) \in \mathbf{R} \times \bar{\Omega}$ *uniformly for* $\xi \in C_g^0(\mathbf{R})$ *in bounded sets. Then there exists an almost periodic classical solution of* (3.77) *and* (3.78).

Proof. Combining Theorem 3.19 with Lemma 3.9, we see that (3.77) has an almost periodic solution (say, $p(t)$), which is continuously differentiable and satisfies (3.78) on $\mathbf{R}$. Set $P(t, x) = p(t)(x)$ for $(t, x) \in \mathbf{R} \times \bar{\Omega}$. Clearly $P(t, x)$ is almost periodic in t uniformly for $x \in \bar{\Omega}$. It remains to prove that P is a classical solution of (3.77) and (3.78). Set $Q(t, x)(= Q(t)(x)) = \mathcal{F}(t, x, P(t+\cdot, x)) + \bar{f}(t, x)$ for $(t, x) \in \mathbf{R} \times \bar{\Omega}$. The function $Q : \mathbf{R} \times \bar{\Omega} \mapsto \mathbf{R}$ is continuous in (t, x), and moreover it is locally Hölder continuous in t uniformly for $x \in \bar{\Omega}$ as seen in the proof of Theorem 3.20. We assert that $Q(t, x)$ is Hölder continuous in $x \in \bar{\Omega}$ uniformly for t in bounded sets. If the assertion is true, then $Q(t,x)$ is Hölder continuous on $[s,r] \times \bar{\Omega}$ for any $r > s$; consequently $P(t,x) = \int_\Omega U(t-s;x,y)P(s,y)dy + \int_s^t \int_\Omega U(t-\tau;x,y)Q(\tau,y)dyd\tau$ is a classical solution of (3.77) and (3.78), as required (cf. [116, Theorem 2.7.1]). Now, in order to prove the assertion we first show that the function $x \in \bar{\Omega} \mapsto P(t+\cdot,x) \in C_g^0(\mathbf{R})$ is Lipschitz continuous, uniformly for $t \in \mathbf{R}$. By the same reason as in the proof of Theorem 3.20, we can assume that the semigroup $\{T(t)\}_{t \geq 0}$ satisfies (3.62). Take a constant α with $1/2 < \alpha < 1$. Recall that $(-A)^\alpha$, the fractional power of $-A$, is a closed linear operator. It is known (e.g., [90, Theorem 1.6.1]) that the Banach space $\mathcal{D}((-A)^\alpha)$, which is the space of the definition domain of $(-A)^\alpha$ equipped with the graph norm, is continuously imbedded to the space $C^1(\bar{\Omega})$. If $t > \sigma$, then

$$\begin{aligned}|(-A)^\alpha p(t)|_\mathbf{X} &= |(-A)^\alpha (T(t-\sigma)p(\sigma) + \int_\sigma^t T(t-s)Q(s)ds)|_\mathbf{X} \\ &\leq c_\alpha (t-\sigma)^{-\alpha} e^{-\lambda(t-\sigma)} |p(\sigma)|_\mathbf{X} + \\ &\quad + \int_\sigma^t c_\alpha (t-s)^{-\alpha} e^{-\lambda(t-s)} ds \cdot \sup_{s \in \mathbf{R}} |Q(s)|_\mathbf{X}.\end{aligned}$$

Passing to the limit as $\sigma \to -\infty$, we get $\sup_{t \in \mathbf{R}} |(-A)^\alpha p(t)|_\mathbf{X} < \infty$; hence $\sup_{t \in \mathbf{R}} |P(t,\cdot)|_{C^1(\bar{\Omega})} (=: C) < \infty$. Then

$$\sup\{|x-y|^{-1}|P(t+\theta,x) - P(t+\theta,y)| : t \in \mathbf{R},\ \theta \leq 0,\ x \neq y\} \leq C,$$

and consequently

$$\sup\{|x-y|^{-1}|P(t+\cdot,x) - P(t+\cdot,y)|_{C_g^0(\mathbf{R})} : t \in \mathbf{R},\ x \neq y\} \leq CJ$$

by (3.25). Thus the function $x \in \bar{\Omega} \mapsto P(t+\cdot,x) \in C_g^0(\mathbf{R})$ is Lipschitz continuous, uniformly for $t \in \mathbf{R}$. Therefore, in virtue of the Hölder continuity of $\mathcal{F}(t,x,\xi)$ and $\bar{f}(t,x)$, we can see that $Q(t,x)$ is Hölder continuous in $x \in \bar{\Omega}$ uniformly for t in bounded set, as required.

3.7. NOTES

The concept of processes discussed by Dafermos [53], [54], Hale [77], [78] and Kato, Marttynyuk and Shestakov [120] is a useful tool in the study of mathematical analysis for some phenomena whose dynamics is described by equations that contain the derivative with respect to the time variable. Indeed, Hale [77] derived some stability properties for processes and applied those to get stability results for some kinds of equations, including functional differential equations, partial differential equations and evolution equations. These equations have the same property that the space of initial functions is identical with the phase space and solutions take values at time t in the phase space.

Hino and Murakami [101], [103] generalize the concept of processes and get a more extended concept which is called quasi-processes. Furthermore, they discuss some stability properties for quasi-processes and obtain some equivalence relations concerning with stabilities for quasi-processes in connection with those for "limiting" quasi-processes. Theorems 3.1 and 3.2 due to Hino and Murakami [101], [103] are generalizations of [77, Theorem 3.7.4]. Furthermore, Hino and Murakami [102], [104] have given existence theorems of almost periodic integrals for almost periodic processes and almost quasi-processes those include Dafermos's result [53]. That is, Theorems 3.3–3.7 and Theorem 3.8 are due to [102], [104], respectively.

For functional differential equations on a uniform fading memory space with $\mathbf{X} = \mathbf{R}^n$, an axiomatic approach and $\mathcal{B}$-stabilities were given by Hale and Kato [79] and Hino, Murakami and Naito [107]. Henríquez [91] extended fundamental theories on functional differential equations with infinite delay to the case where $\mathbf{X}$ is a general Banach space. Theorem 3.9 is given by Hino and Murakami [101], [103], which is an extension of results for $\mathbf{X} = \mathbf{R}^n$ by Hale and Kato [79], Hino [94] and Murakami [155], [156] to the case of dim $\mathbf{X} = \infty$. For functional differential equations on a fading memory space with $\mathbf{X} = \mathbf{R}^n$, ρ-stabilities properties were discussed by Hamaya [81], [82], Hamaya and Yoshizawa [83], Murakami [157] and Yoshizawa [232]. Theorem 3.10 is given by Hino and Murakami [101]. Proposition 3.2 is an extension of Murakami and Yoshizawa [161] for $\mathbf{X} = \mathbf{R}^n$ to the case of dim$\mathbf{X} = \infty$. As pointed out in [15], some integrodifferential equation can be set up as functional differential equations on a fading memory space (not uniform) and BC-stability is more practical. Thus the BC-stabilities would seem to be more usual than the ρ-stabilities. This fact would produce some difficulties when one tries to discuss the existence of periodic solutions or almost periodic solutions under some BC-stability assumption. Theorems 3.11–3.13 and 3.14 are due to Hino and Murakami [105] and [106], respectively. Theses theorems show that equivalent relations established by Murakami and Yoshizawa [161] for $\mathbf{X} = R^n$ hold even for functional differential equations with infinite delay on a fading memory space $\mathcal{B} = \mathcal{B}((\mathbf{R}^-; \mathbf{X})$ with a general Banach space $\mathbf{X}$.

Theorems 3.15 and 3.16 are due to Hino and Murakami [100]. For a linear ordinary differential equations, Massera [147] proved that the null solution is TS if and only if it is UAS. For functional differential equations on a uniform fading memory space $\mathcal{B}$ with $\mathbf{X} = \mathbf{R}^n$, Hino [93] proved that the null solution of homogeneous systems is $\mathcal{B}$-TS if it is $\mathcal{B}$-UAS, and that the converse also holds under a mild assumption. Moreover, Hino and Murakami [98] have established a similar relation for BC-TS and BC-UAS in linear Volterra integrodifferential equations.

For functional differential equations on a uniform fading memory space $\mathcal{B}$ with $\mathbf{X} = \mathbf{R}^n$, Hino [92] obtained a result on the existence of almost periodic solutions by assuming the existence of a bounded solution which is $\mathcal{B}$-TS or $\mathcal{B}$-UAS. Hino and Murakami [96], Hino, Murakami and Naito [107] and Sawano [194] also obtain results concerning with the existence of almost periodic solutions of nonhomogeneous lineat systems under some stability properties of the null solution of homogeneous systems. Murakami and Yoshizawa [161] have discussed the existence of an almost periodic solution for functional differential equations on a fading memory space with $\mathbf{X} = \mathbf{R}^n$ in the context of BC-stability. For nonhomogeneous systems, see Hino [95] and Hino and Murakami [98].

Hino and Murakami [100] have established a result on the existence of an almost periodic solution of nonhomogeneous linear systems on a fading memory space $\mathcal{B}(\mathbf{R}^-;\mathbf{X})$ with a general Banach space $\mathbf{X}$. Theorems 3.17 and 3.18 due to Hino, Murakami and Yoshizawa [108] would be considered as some extensions of Hino [94, Theorem 4] and Murakami and Yoshizawa [161, Corollary 1] to the case where X is a general Banach space, and of Hino and Murakami [97, Theorem 1] to a nonlinear equation. Proposition 3.4 is due to Hino, Murakami and Yoshizawa [109] and Theorem 3.19 is due to Hino and Murakami [97].

Applications 3.6.1 and 3.6.3 are given by Hino and Murakami [102] and [100], respectively. Application 3.6.2 is given by Hino and Murakami [99].

CHAPTER 4

APPENDICES

4.1. FREDHOLM OPERATORS AND CLOSED RANGE THEOREMS

Let X be a Banach spaces, and T a linear operator from a linear subspace $D(T)$ of X to X, where $D(T)$ denotes the domain of T. Denote by $R(T), N(T)$ the range of T and the null space of T, respectively. For a complex number λ, set $T_\lambda := \lambda I - T$, where I is the identity. The resolvent set, $\rho(T)$, is the set of λ such that $N(T_\lambda) = \{0\}$, $\overline{R(T_\lambda)} = X$, and T_λ^{-1} is continuous, where $\overline{R(T_\lambda)}$ denotes the closure of $R(T_\lambda)$. The set $\sigma(T) := \mathbb{C} \setminus \rho(T)$ is called the spectrum of T. It is classified into three disjoint subsets: the continuous spectrum $C_\sigma(T) = \{\lambda \in \sigma(T) : N(T_\lambda) = \{0\}, \overline{R(T_\lambda)} = X$, and T_λ^{-1} is not continuous; the residual spectrum $R_\sigma(T) = \{\lambda \in \sigma(T) : N(T_\lambda) = \{0\}, \overline{R(T_\lambda)} \neq X\}$; the point spectrum $P_\sigma(T) = \{\lambda \in \sigma(T) : N(T_\lambda) \neq \{0\}\}$. If T is a closed operator, $\lambda \in \rho(T)$ if and only if $N(T_\lambda) = \{0\}, R(T_\lambda) = X$.

If λ in $P_\sigma(T)$, λ is called the eigenvalue of T, $N(T_\lambda)$ is called the (geometric) eigenspace of T with respect to λ, and its dimension is called the geometric multiplicity of λ, [223, p.163]. The space $M_\lambda := \cup_{n=1}^\infty N(T_\lambda^n)$ is a linear subspace of X, and is called the generalized eigenspace of T with respect to λ. The dimension of M_λ is called the algebraic multiplicity of λ,[64, p.9], [223, p.163]. If there is a positive integer k such that $N(T_\lambda^k) = N(T_\lambda^{k+j})$ for all $j = 1, 2, \cdots$, then λ is said to have finite index and the smallest such k for which this is true is called the index, or the ascent of λ.

Let T be a compact, linear operator on X: that is, $T(B)$ is relatively compact in X for every bounded set B of X. Then the Fredholm alternative [195, pp.94] says that $R(I - T)$ is closed and $\dim N(I - T) = \dim N(I - T^*)$ is finite, where T^* denotes the dual operator of T. In particular, either $R(I - T) = X, N(I - T) = \{0\}$, or $R(I - T) \neq X, N(I - T) \neq \{0\}$. This implies that, for $\lambda \neq 0$, $R(T_\lambda)$ is closed, and either $R(T_\lambda) = X, N(T_\lambda) = \{0\}$, that is, $\lambda \in \rho(T)$, or $N(T_\lambda) \neq \{0\}, R(T_\lambda) \neq X$, that is, $\lambda \in P_\sigma(T)$. Furthermore, the spectrum $\sigma(T)$ is a nonempty set, and is either a finite set or a countable set with the only limit point $\lambda = 0$, which belongs to $\sigma(T)$. cf. [240, pp.296-298]. Hence, if $\lambda \in \sigma(T)$ and $\lambda \neq 0$, then $\lambda \in P_\sigma(T), R(T_\lambda)$ is closed, $\dim(X/R(T_\lambda)) = \dim N(T_\lambda) = \dim N(T_\lambda^*)$ is finite and $y \in R(T_\lambda)$ if and only if $(y, x^*) = 0$ for all $x^* \in N(T_\lambda^*)$. Notice that, if $\dim X = \infty$, then $0 \in \sigma(T)$, see [64, p.5].

Let T be a bounded linear operator from X to a Banach space Y. It is said to be a Fredholm operator from X to Y if

i) $\dim N(T)$ is finite,

ii) $R(T)$ is closed,

iii) $\dim N(T^*)$ is finite.

Recall the following two facts.

iv) $N(T^*) = R(T)^\perp$.

v) If M is a closed subspace of Y, then $\mathrm{codim} M = \dim M^\perp$, $\mathrm{codim} M^\perp = M$, cf. [122, p.140], where $M^\perp := \{y* \in Y^* :< y, y^* >= 0 \text{ for } y \in M\}$, and $\mathrm{codim} M = \dim Y/M, \mathrm{codim} M^\perp = \dim Y^*/M^\perp$.

Hence $\dim N(T^*) = \mathrm{codim} R(T)$, that is, the condition iii) is equivalent that

iii)' $\mathrm{codim} R(T)$ is finite.

If T satisfies the condition ii) and at least one of the conditions i) and iii), it is called a semi-Fredholm operator. The set of Fredholm operators from X to Y is denoted by $\Phi(X, Y)$, whereas $\Phi_+(X, Y)(\Phi_-(X, Y))$ denotes the set of semi-Fredholm operators which satisfies ii) and i)(iii)). If $X = Y$, $\Phi(X) := \Phi(X, X), \Phi_\pm(X) := \Phi_\pm(X, X)$. If T is a compact operator on X, $\lambda I - T$ is a Fredholm opertor for $\lambda \neq 0$ whose index is zero. These definitions are adopted for closed linear operators provided its domain is dense.

We will use the closed range property of an operator having a form $I - T$ in applications of the Chow-Hale fixed point therem for the existence of periodic solutions of periodic linear functional differential equaitons. For this purpose we recall perturbation results of semi-Fredholm operators. If the perturbation is compact, we have the following result [195, p.128, p.183].

Theorem 4.1 *If $T \in \Phi_+(X, Y)$ and if $K : X \to Y$ is a compact operator, then $T + S \in \Phi_+(X, Y)$.*

Theorem 4.2 *Let T be an injective, closed linear operator from X to Y. Then $R(T)$ is closed if and only if there exists a constant $c > 0$ such that $|x| \leq c|Tx|$ for $x \in D(T)$.*

Proof. T^{-1} is a closed operator whose domain is $R(T)$. If $R(T)$ is closed, then T^{-1} becomes a bounded linear operator from the closed graph theorem. Hence, there exists a constant $c \geq 0$ such that $|T^{-1}y| \leq c|y|$ for $y \in R(T)$; or $|x| \leq c|Tx|$ for $x \in D(T)$.

The converse implication is proved easily.

If T is not injective in the above theorem, we take the quotient space $X/N(T)$. Set $[x] := x + N(T)$ for $x \in X$. Since $N(T)$ is a closed subspace, this becomes a Banach space with the norm $|[x]| = \inf\{|x + y| : y \in N(T)\}$. Define $[T] : D(T)/N(t) \to Y$ by $[T]([x]) = T(x), x \in D(T)$. Then $R(T) = R([T])$ and T is a closed linear operator and $R(T) = R([T])$, see [64, pp.28-29]. Hence we have the following corollary.

Corollary 4.1 *Let T be a closed linear operator from X to Y. Then $R(T)$ is closed if and only if there exists a constant $c > 0$ such that $\|[x]\| \leq c|Tx|$ for $x \in D(T)$. In particular, if there exists a constant $c > 0$ such that $|x| \leq c|Tx|$ for $x \in D(T)$, $R(T)$ is closed.*

If T is a closed linear operator such that $D(T) = X$, it is a bounded linear operator. Suppose that $\dim N(T)$ is finite. Then there exists a closed subspace Z of X such that $X = N \oplus Z$. Now we define a map $\pi : Z \to X/N$ by $\pi(z) = [z]$ for $z \in Z$. Since π is a continuous isomorphism, π^{-1} is also continuous from the closed graph theorem. Therefore, there is a constant k_Z such that

$$\|[z]\| \leq |z| \leq k_Z |z| \quad \text{for } z \in Z.$$

Let P_Z be the projection onto N along Z.

Lemma 4.1 *Let T be a bounded linear operator from X to Y, and N, Z, P_Z, k_Z be defined as in the above. Then the following statements are equivalent.*

i) *$R(T)$ is closed.*

ii) *There exists a constant c such that $\|[x]\| \leq c|Tx|$ for $x \in X$.*

iii) *There exists a constant c_Z such that $|(I - P_Z)x| \leq c_Z |Tx|$ for $x \in X$.*

In fact, ii) implies iii) as $c_Z = ck_Z$, while iii) implies ii) as $c = c_Z$.

Proof. It suffices to prove that the conditions ii) and iii) are equivalent. If the condition ii) holds, then

$$|(I - P_Z)x| \leq k_Z \|[(I - P_Z)x]\| \leq ck_Z |T(I - P_Z)x| = ck_Z |Tx|.$$

Conversely, if the condition iii) holds, then

$$\|[x]\| = \|[(I - P_Z)x]\| \leq |(I - P_Z)x| \leq c_Z |Tx|.$$

We refer the following result [195, Therem 6.4, p.128] as a perturbation of semi-Fredholm operators.

Theorem 4.3 *Let $T \in \Phi_+(X, Y)$, and c_Z be the constant appearing in the condition (3) in the foregoing lemma. If $S : X \to Y$ is a bounded linear operator satisfying $\|S\| \leq 1/2c_Z$, then $T + S \in \Phi_+(X, Y)$ and*

$$\dim N(T + S) \leq \dim N(T).$$

The best possible estimate of k_Z is known as follows, cf. [64, p.6] and [181, B.4.9, pp.29-30, 28.2.6, p.386].

Theorem 4.4 *Let N be an n-dimensional subspace of X. Then there exists a projection P such that $R(P) = N$ and $\|P\| \leq \sqrt{n}$.*

Summarizing these results, we have the folowing perturbation theorem for the application to the existence of periodic solutions later.

Theorem 4.5 *Suppose that $T \in \Phi_+(X,Y)$. Let $\dim N(T) = n$, and c be a positive constant such that $|[x]| \leq c|Tx|$ for $x \in X$. If $S : X \to Y$ is a bounded linear operator such that $\|S\| \leq 1/2c(1+\sqrt{n})$, then $T + S \in \Phi_+(X,Y)$, and $\dim N(T+S) \leq n$.*

Proof. Take a closed subspace Z such that $X = N \oplus Z$ and that $\|P_Z\| \leq \sqrt{n}$. Let $x \in X$, and set $(I - P_Z)x = z$. Suppose $y \in [z]$. Then $y = y - z + z$ implies that $y - z = P_Z(y)$. Hence

$$|z| \leq |z - y| + |y| \leq \|P_Z\||y| + |y| \leq (1 + \sqrt{n})|y|.$$

Since this inequality holds for all $y \in [z]$, we have that $|z| \leq (1 + \sqrt{n})|[z]|$. Since $[z] = [x]$, it follows that $|(I-P_Z)x| \leq (1+\sqrt{n})|[x]|$. Hence $|(I-P_Z)x| \leq c(1+\sqrt{n})|Tx|$ for $x \in X$: we can take $c_Z = c(1 + \sqrt{n})$.

Theorem 4.6 *Suppose that $T \in \Phi_+(X,Y)$. Set $\dim N(T) = n$. Then there exists a closed subspace Z of X such that $X = N(T) \oplus Z$, and $Q := T|Z$, the restriction of T to Z, has a bounded inverse, and $\|Q^{-1}\| \leq c(1+\sqrt{n})$, where c is the constant in the previous theorem.*

Proof. Take the subspace Z as in the proof of the previous theorem. Take a $z \in Z$. It is represented by $z = (I - P_Z)x$ for some $x \in X$. Since $TP_Zx = 0$, we have

$$|z| = |(I - P_z)x| \leq c(1+\sqrt{n})|Tx| = c(1+\sqrt{n})|T(I-P_Z)x| = c(1+\sqrt{n})|Tz|.$$

This implies the assertion.

4.2. ESSENTIAL SPECTRUM AND MEASURES OF NONCOMPACTNESS

Let A be a closed linear operator from X to X. We will use the definition of the essential spectrum $E_\sigma(A)$ by Browder [39]. The complex number ζ lies in $E_\sigma(A)$ whenever at least one of the following conditions holds:

i) $R(\zeta I - A)$ is not closed;

ii) $\cup_{n \geq 0} N((\zeta I - A)^n)$ is of infinite dimension;

iii) The point ζ is a limit point of $\sigma(A)$.

The point $\lambda \notin E_\sigma(A)$ is called a normal point of A. If it lies in $\sigma(A)$, $\lambda I - A \in \Phi_+(X)$ and we call it the normal eigenvalue of A.

Let T be a bounded linear operator from X to X. The spectral radius $r_\sigma(T)$ of T is defined as $r_\sigma(T) = \sup\{|\lambda| : \lambda \in \sigma(T)\}$, and the following Gelfand formula is well known

$$r_\sigma(T) = \lim_{n \to \infty} \|T^n\|^{1/n}.$$

Similary the radius of the essential spectral radius $r_e(T)$ is defined for $E_\sigma(T)$, and the corresponding formula is given by Nussbaum [176]. To present it, we introduce the Kuratowskii measure α of noncompactness of bounded sets of X, cf. [80], [215]. Let B be a bounded set of X. We define

$$\alpha(B) = \inf\{d > 0 : B \text{ has a finite cover of diameter } < d\}.$$

For the bounded linear operator T, we define $\alpha(T)$ as the infimum of k such that $\alpha(TB) \leq k\alpha(B)$ for all bounded sets $B \subset X$. Properties of $\alpha(B)$ and $\alpha(T)$ are found in these books. In particular, $\alpha(B) = 0$ if and only if $\overline{B}$, the clolsure of B, is compact; $\alpha(T) = 0$ if and only if T is compact; the Nussbaum formula is given as

$$r_e(T) = \lim_{n \to \infty} \alpha(T^n)^{1/n}. \tag{4.1}$$

If $\alpha(T) < 1$, then T is called α-contraction. Let $C(I, E)$ be the Banach space of continuous functions from a compact subset I of R to E with the supremum norm. If H is a bounded subset of $C(I, E)$, its α-measure is computed as follows, Set $H(t) = \{\phi(t) : \phi \in H\}, H(I) = \cup_{t \in I} H(t)$. From the definition of α-measure, it is clear that

$$\sup\{\alpha(H(t)) : t \in I\} \leq \alpha(H). \tag{4.2}$$

If H is equicontinuous, the following result holds, [133, Lemma 1.4.2, p.20].

Lemma 4.2 *If H is a bounded, equicontinuous set of the space $C(I, E)$, I being a comapct interval, then*

$$\alpha(H) = \alpha(H(I)) = \sup\{\alpha(H(t)) : t \in I\}.$$

For a bounded set $H \subset C([a, b], E)$ and for $t \in [a, b]$ we use the following notations:

$$\omega(\delta; t, H) = \sup\{|\phi(r) - \phi(s)| : r, s \in [t - \delta, t + \delta], \phi \in H\}$$

$$\omega(t, H) = \inf\{\omega(\delta; t, H) : \delta > 0\} = \lim_{\delta \to 0+} \omega(\delta; t, H)$$

and

$$\omega(H) = \sup\{\omega(t, H) : a \leq t \leq b\}.$$

The following lemma can be found in [177] which generalizes the above lemma.

Lemma 4.3 *Let H be a bounded subset of $C([a, b], E)$. Then*

$$\max\{(1/2)\omega(H), \sup_{a \leq t \leq b} \alpha(H(t))\} \leq \alpha(H) \leq 2\omega(H) + \sup_{a \leq t \leq b} \alpha(H(t)).$$

In what follows we will use the measure of noncompactness to verify the compactness of convolution operators. Let $T(t)$ be a C_0-semigroup on E. For $u \in C([a,b], E)$, we put

$$T * u(t) = \int_a^t T(t-s)u(s)ds \quad \text{for } t \in [a,b],$$

For a subset $\mathcal{U} \subset C([a,b], E)$ we put $T * \mathcal{U} = \{T * u : u \in \mathcal{U}\}$, which is a subset of $C([a,b], E)$, and $(T * \mathcal{U})(t) = \{(T * u)(t) : u \in \mathcal{U}\}$, which is a subset of E. If $T(t)$ is compact for $t > 0$, we can prove directly that $H = T * \mathcal{U}$ is equicontinuos. However, we have a more general results in this case.

Let us define

$$\gamma_T := \lim_{t \to 0+} \sup\{\|T(s)\| : 0 < s \leq t\}.$$

Then, we have the following theorem.

Theorem 4.7 *Let $\mathcal{U}$ be a bounded set in $C([a,b], E)$, and $T(t)$ a C_0-semigroup on E. Then, for $a < t \leq b$,*

$$\omega(t, T * \mathcal{U}) \leq 2\gamma_T \sup\{\alpha(T * \mathcal{U}(s)) : a \leq s \leq t\}.$$

Proof. Suppose that $a < c < t$. Set $M = \sup\{\alpha(T * \mathcal{U}(s)) : a \leq s \leq t\}$. Let $\varepsilon_1 > 0$. Then the set $T * \mathcal{U}(c)$ is covered by a finite set $B_i, i = 1, 2, \cdots, n$ of E with diameter less than $M + \varepsilon_1$. For $i = 1, 2, \cdots, n$, take $v_i \in \mathcal{U}$ such that $T * v_i(c) \in B_i$. Let $T * u \in T * \mathcal{U}$. Then there exists an index i such that $T * u(c) \in B_i$. Set $f = T * u, g_i = T * v_i$. If $c < t_1 \leq t_2 \leq b$, we have that

$$|f(t_1) - f(t_2)| \leq |f(t_1) - g_i(t_1)| + |f(t_2) - g_i(t_2)| + |g_i(t_1) - g_i(t_2)|.$$

We estimate the right side. Set

$$N = \sup\{|T * u(t)| : t \in [a,b], u \in \mathcal{C}\}, S = \sup\{\|T(t)\| : t \in [a,b]\}.$$

Since $T * u(c), T * v_i(c) \in B_i$, we have that

$$\begin{aligned}
|f(t_1) - g_i(t_1)| &\leq \|T(t_1 - c)\| \left| \int_a^c T(c-s)(u(s) - v_i(s))ds \right| \\
&\quad + \int_c^{t_1} \|T(t_1 - s)\| |u(s) - v_i(s)| ds \\
&\leq \|T(t_1 - c)\| |T * u(c) - T * v_i(c)| + 2NS(t_1 - c) \\
&\leq \|T(t_1 - c)\|(M + \varepsilon_1) + 2NS(t_1 - c).
\end{aligned}$$

The similar estimate is valid for $|f(t_2) - g(t_2)|$. As a result we have that

$$\begin{aligned}
|f(t_1) - f(t_2)| &\leq (\|T(t_1 - c)\| + \|T(t_2 - c)\|)(M + \varepsilon_1) \\
&\quad + 2NS(t_1 - c + t_2 - c) + |g_i(t_1) - g_i(t_2)|.
\end{aligned}$$

Let $\varepsilon_2 > 0$. Since $g_1, g_2, \cdots, g_n$ are continuous, there exists a $\delta(\varepsilon_2)$ such that, if $\delta < \delta(\varepsilon_2)$, then $|g_i(t_1) - g_i(t_2)| < \varepsilon_2$ for $i = 1, 2, \cdots$ and for t_1, t_2 such that $|t_1 - t|, |t_2 - t| \leq \delta$. Thus we have that, if $|t_1 - t|, |t_2 - t| \leq \delta < \delta(\varepsilon_2)$, then

$$|f(t_1) - f(t_2)| \leq (\|T(t_1 - c)\| + \|T(t_2 - c)\|)(M + \varepsilon_1)$$
$$+ 2NS(t_1 - c + t_2 - c) + \varepsilon_2.$$

Since $f \in T * \mathcal{U}$ is arbitrary, it follows that, if $\delta < \delta(\varepsilon_2)$ and $t - \delta > c$,

$$\omega(\delta; t, T * \mathcal{U}) \leq 2\sup\{\|T(s)\| : 0 < s \leq t + \delta - c\}(M + \varepsilon_1)$$
$$+ 4(t + \delta - c)NS + \varepsilon_2.$$

Taking the limit as $\delta \to 0+$, we have that

$$\omega(t, T * \mathcal{U}) \leq 2(M + \varepsilon_1)\overline{\lim_{\delta \to 0+}}\sup\{\|T(s)\| : 0 < s \leq t + \delta - c\}$$
$$+ 4(t - c)NS + \varepsilon_2.$$

Since ε_2 is arbitrary,

$$\omega(t, T * \mathcal{U}) \leq 2(M + \varepsilon_1)\overline{\lim_{\delta \to 0+}}\sup\{\|T(s)\| : 0 < s \leq t + \delta - c\} + 4(t - c)NS.$$

Let $c \to t - 0$. Then

$$\omega(t, T * \mathcal{U}) \leq 2(M + \varepsilon_1)\overline{\lim_{r \to 0+}}\sup\{\|T(s)\| : 0 < s \leq r\}.$$

Since ε_1 is arbitrary, we obtain that $\omega(t, T * \mathcal{U}) \leq 2M\gamma_T$, as desired.

Theorem 4.8 *Let $\mathcal{U}$ be a bounded set in $C([a, b], E)$, and $T(t)$ be a C_0-semigroup on E. Then*

$$\alpha(T * \mathcal{U}) \leq \gamma_T \sup\{\alpha((T * \mathcal{U})(t)) : a \leq t \leq b\}.$$

In particular, if $T(t)$ is a C_0-contraction semigroup, then

$$\alpha(T * \mathcal{U}) = \sup\{\alpha((T * \mathcal{U})(t)) : a \leq t \leq b\}.$$

Proof. Set $M := \sup\{\alpha((T * \mathcal{U})(t)) : a \leq t \leq b\}$. Take $\varepsilon > 0$ and a sequence $a = t_1 < t_2 < \cdots < t_n < t_{n+1} = b$. Then every set $B_j := (T * \mathcal{U})(t_j), 1 \leq j \leq n$ has a finite cover $\cup_i B(i, j), 1 \leq i \leq k_j$, such that $\text{diam} B(i, j) < M + \varepsilon$. For a multi-index $\iota := (i_1, \cdots, i_n)$, where $1 \leq i_j \leq k_j, j = 1, \cdots n$, we set $V_\iota := \{T * u : (T * u)(t_j) \in B(i_j, j)\}$. Then $T * \mathcal{U} \subset \cup_\iota V_\iota$. Suppose that $T * u, T * v \in V_\iota$, and that $t_j < t \leq t_{j+1}, 1 \leq j \leq n$. Since

$$T * u(t) = T(t - t_j) \int_a^{t_j} T(t_j - s)u(s)ds + \int_{t_j}^t T(t - s)u(s)ds,$$

we have that

$$|T * u(t) - T * v(t)| \leq \|T(t - t_j)\| |T * u(t_j) - T * v(t_j)| +$$
$$+ \int_{t_j}^t \|T(t - s)\| |u(s) - v(s)|ds.$$

Set $\delta := \max\{t_{j+1} - t_j : 1 \leq j \leq n\}, \gamma_\delta = \sup\{\|T(s)\| : 0 < s \leq \delta\}, N = \text{diam} T * \mathcal{U}$. If $a \leq t \leq b, t \neq t_j, 1 \leq j \leq n$, then

$$|T * u(t) - T * v(t)| \leq \gamma_\delta (M + \varepsilon) + \delta \gamma_\delta N.$$

Since $|T * u(t) - T * v(t)|$ is continuous, this inequality holds for all $t \in [a, b]$; hence $\text{diam} V_\iota \leq \gamma_\delta(B + \varepsilon) + \delta \gamma_\delta N$, and $\alpha(T * \mathcal{U}) \leq \gamma_\delta(M + \varepsilon) + \delta \gamma_\delta N$. Taking a limit as $\delta \to 0+$, we have that $\alpha(T * \mathcal{U}) \leq \lim_{\delta \to 0+} \gamma_\delta(M + \varepsilon)$. Since $\varepsilon > 0$ is arbitrary small, we obtain the first estimate in the lemma.

Since $\gamma_T \leq 1$ for a contraction semigroup $T(t)$, in view of the inequality 4.2 we have at once the second assertion in the lemma.

Let $H = \{f_1, f_2, \cdots\}$ be a sequence of strongly measurable functions on $I := [a, b]$ to E, and set $H(t) := \{f_n(t) : n \geq 1\}, t \in I$, and

$$S := \left\{ \int_a^b f_n(t) dt : n \geq 1 \right\}.$$

Is the behaviour of measure of noncompactness like the one of the norm in E with respect to integration? Heinz [89] gives answers to this problem for the Kuratowskii measure α and the Hausdorff measure β, where

$$\beta(B) := \inf\{r > 0 : B \text{ has a finite cover of balls of radius } < r\}.$$

for bounded sets B. Since the discussion is rather simple for the case of β, we present here his result for β in the very restricted style, and this result is sufficient for our use. Before proceeding we remark that $\beta(B) \leq \alpha(B) \leq 2\beta(B)$. Suppose that B is contained in a closed subspace F of E. If we compute $\alpha(B), \beta(B)$ by taking the covering of B in the set F only, we denote $\alpha(B, F), \beta(B, F)$, respectively. However, it is clear that $\alpha(B) = \alpha(B, F), \beta(B) = \beta(B, F)$.

We recall defintions of measurability. A function $\phi : I \to E$ is said to be finitely-valued if it is constant on each finite number of disjoint measurable subset of I. A function f is said to be strongly measurable if there exists a sequence of finitely-valued functions $\{\phi_n(t)\}$ strongly convergent to $f(t)$ for $t \in I \setminus J_0$ for some $J_0 \subset I$ such that $\mu(J_0) = 0$, where μ is the Lubesgue measure. Then the set $\{f(t) : t \in I \setminus J_0\}$ is separable.

Let $J_n, n \geq 1$, be the null set such that $\{f_n(t) : t \in I \setminus J_n\}$ is separable. Set $J = \cup_{n \geq 1} J_n$, Then J is a null set, and $H(t)$ is separable for $t \notin J$: that is, there exists a countable set $D \subset E$ such that $H(t) \subset \overline{D}$. Let E_1 be the smallest closed subset of E which contains D. It is the closure of the set of finite linear combinations $c_1 x_1 + c_2 x_2 + \cdots + c_k x_k, x_i \in D, 1 \leq i \leq k$. Hence E_1 is a separble Banach space, and $H(t) \subset E_1$ for $t \notin J$.

Put
$$h(t) := \sup\{|f_n(t)| : n \geq 1\}.$$

If $h(t)$ is integrable on I, the set S defined in the above is a bounded set of E.

Theorem 4.9 *Suppose that E is a separable Banach space. Let*

$$H = \{f_1, f_2, \cdots\}$$

be a sequence of measurable functions from $I = [a, b]$ to E for which $h(t)$ is integrable on I. Then $\beta(H(t))$ is measurable, and

$$\beta\left(\left\{\int_a^b f_n(t)dt : n \geq 1\right\}\right) \leq \int_a^b \beta(\{f_n(t) : n \geq 1\})\, dt.$$

Proof. First, let us approximate $\beta(H(t))$ by a sequence of suitable funcions $r_k(t), k = 1, 2, \cdots$, as follows. Let D be a countable set such that $\overline{D} = E$, and let $\mathcal{C}$ be the the family of all closed balls with centers in D and with rational radii. Then $\mathcal{C}$ is a countable set. The set of all finite sequences $\{A_1, A_2, \cdots, A_q\}$ of sets in $\mathcal{C}$ is again countable and can thus be written as a sequence $\{C_k\}, k = 1, 2 \cdots$, where

$$C_k = \{A_1^k, A_2^k, \cdots, A_{q(k)}^k\}.$$

Put

$$V_k := \cup_{j=1}^{q(k)} A_j^k, \quad k \geq 1,$$

and define nonnegative simple functions R_k on I by

$$R_k(t) := \begin{cases} \beta(V_k) & \text{if } H(t) \subset V_k \\ \infty & \text{otherwise} \end{cases}$$

for $k \geq 1$, and $R_0 = h$. Since

$$\{t \in I : H(t) \subset V_k\} = \cap_{n=1}^\infty f_n^{-1}(V_k),$$

and since $f_n^{-1}(V_k)$ are measurable set, R^k is measurble. Finally, set

$$r_k(t) := \min\{R_0(t), R_1(t), \cdots, R_k(t)\}, \quad t \in I,$$

for $k \geq 1$. Then cleary $\beta(H(t)) \leq r_k(t) \leq R_0(t) = h(t)$ for $t \in I$. Moreover, we have

$$\beta(H(t)) = \lim_{k \to \infty} r_k(t)$$

a.e. on I. In fact, since $h(t)$ is integrable on I, the set $I_\infty := \{t \in I : h(t) = \infty\}$ is a set of measure zero. Suppose that $t \notin I_\infty$. Then $\beta(H(t)) \leq h(t) < \infty$. Set $\rho = \beta(H(t))$, and let $\varepsilon > 0$. Then there exists a finite number of balls B_j with radii less than $\rho + \varepsilon/3$ such that the union of the balls covers the set $H(t)$. Let B be one of the balls, and let c be its center. Take a point $c' \in D$ such that $|c - c'| \leq \varepsilon/3$. For $x \in B$, we have $|x - c'| \leq |x - c| + |c - c'| \leq \rho + 2\varepsilon/3$. Take a rational number ρ' such that $\rho + 2\varepsilon/3 < \rho' < \rho + \varepsilon$, and let B' be the ball with the center c' and with the radius ρ'. Then $B \subset B'$. If we replace each B_j by such a ball, we have a member C_m of $\mathcal{C}$ such that $H(t) \subset V_m$ and $R_m(t) = \beta(V_m) \leq \beta(H(t)) + \varepsilon$. Since $\beta(H(t)) \leq r_k(t) \leq R_m(t)$ for $k \geq m$, it follows that

$$\beta(H(t)) \leq r_k(t) < \beta(H(t)) + \varepsilon$$

for $k \geq m$. Hence we have proved $\lim_{k \to \infty} r_k(t) = \beta(H(t))$ a.e. on I.

Thus the proof of the theorem is complete if we prove that, for $k \geq 1$,

$$\beta(S) \leq \int_a^b r_k(t)dt.$$

Let k be fixed. Set

$$J_i = \{t \in I \setminus I_\infty : r_k(t) = R_i(t)\},$$

for $i = 0, 1, 2, \cdots, k$, and $I_0 = J_0, I_i = J_i \setminus \cup_{0 \leq j \leq i-1} J_j, 1 \leq i \leq k$. Then $I = I_0 \cup I_1 \cup \cdots \cup I_k \cup I_\infty$ is a decomposition by a disjoint measurable sets. Set

$$S_i = \left\{\int_{I_i} f_n(t)dt : n \geq 1\right\}.$$

Since $S \subset S_0 + S_1 + \cdots + S_k + S_\infty$, $\beta(S) \leq \beta(S_0) + \beta(S_1) + \cdots + \beta(S_k) + \beta(S_\infty)$. Since $\mu(I_\infty) = 0$, we have $S_\infty = \{0\}$, and $\beta(S_\infty) = 0$. If we prove

$$\beta(S_i) \leq \int_{I_i} r_k(t)dt$$

for $0 \leq i \leq k$, we complete the proof. This is obvious for $i = 0$ since

$$\left|\int_{I_0} f_n(t)de\right| \leq \int_{I_0} |f_n(t)|dt \leq \int_{I_0} h(t)dt = \int_{I_0} r_k(t)dt.$$

Let $1 \leq i \leq k$, and $t \in I_i$. Then $H(t) \subset V_i$ and $R_i(t) = \beta(V_i)$; othewise, $R_i(t) = \infty$, which is impossible since $R_i(t) \leq R_0(t) = h(t) < \infty$ for $t \in I_i$. As is well known from the integration theory, this implies that $S_i \subset \mu(I_i)\overline{co}V_i$, where $\overline{co}V_i$ is the convex closure of V_i. Hence we have

$$\beta(S_i) \leq \beta(\mu(I_i)\overline{co}V_i) = \mu(I_i)\beta(V_i) = \int_{I_i} R_i(t)dt = \int_{I_i} r_k(t)dt.$$

Corollary 4.2 *Let K be a countable or uncountable family of measurable functions on $I = [a, b]$ to E such that $h(t) := \sup\{|f(t)| : f \in K\}$ is integrable on I. Assume that $K(t) := \{f(t) : f \in K\}$ is relatively compact in E for t a.e. on I. Then the set*

$$T = \left\{\int_a^b f(t)dt : f \in K\right\}$$

is relatively compact in E.

Proof. It suffices to prove that any sequence of T has a convergent subsequence. Let S be a sequence of T made of a sequence $H = \{f_1, f_2, \cdots\}$ in K. As ramarked before the previous theorem, there exists a separable closed subset E_1 of E and null set $J \subset I$ such that $H(t) \subset E_1$. For $n = 1, 2, \cdots$, Redefine $f_n(t) = 0$ for $t \in J$. Then the integral of f_n does not change, $f_n(t) \in E_1$ for $t \in E_1$ and $S \subset E_1$. Hence we have

$$\beta(S, E_1) \leq \int_a^b \beta(H(t), E_1) dt = \int_{I \setminus J} \beta(H(t), E_1) dt.$$

Since $\beta(H(t), E_1) = \beta(H(t), E)$, and since $\beta(H(t), E) = 0$ a.e. on I, the last integral is zero. Hence $\beta(S, E) = \beta(S, E_1) = 0$, which implies S is relatively compact.

4.3. SUMS OF COMMUTING OPERATORS

In this section we collect some known notions and results concerning spectral properties of sums of commuting operators. Throughout the paper we always assume that A is a given operator on **X** with $\rho(A) \neq \emptyset$, (and thus it is closed).

We recall now the notion of two commuting operators.

Definition 4.1 Let A and B be operators on a Banach space G with non-empty resolvent set. We say that A *and* B *commute* if one of the following equivalent conditions hold:

i) $R(\lambda, A) R(\mu, B) = R(\mu, B) R(\lambda, A)$ for some (all) $\lambda \in \rho(A), \mu \in \rho(B)$,

ii) $x \in D(A)$ implies $R(\mu, B) x \in D(A)$ and $A R(\mu, B) x = R(\mu, B) A x$ for some (all) $\mu \in \rho(B)$.

For $\theta \in (0, \pi), R > 0$ we denote $\Sigma(\theta, R) = \{z \in \mathbf{C} : |z| \geq R, |arg z| \leq \theta\}$.

Definition 4.2 Let A and B be commuting operators. Then

i) A is said to be of class $\Sigma(\theta + \pi/2, R)$ if there are positive constants θ, R such that $\theta < \pi/2$, and

$$\Sigma(\theta + \pi/2, R) \subset \rho(A) \quad \text{and} \quad \sup_{\lambda \in \Sigma(\theta + \pi/2, R)} \|\lambda R(\lambda, A)\| < \infty,$$

ii) A and B are said to satisfy condition P if there are positive constants $\theta, \theta', R, \theta' < \theta$ such that A and B are of class $\Sigma(\theta + \pi/2, R), \Sigma(\pi/2 - \theta', R)$, respectively.

If A and B are commuting operators, $A + B$ is defined by $(A + B)x = Ax + Bx$ with domain $D(A + B) = D(A) \cap D(B)$.

In the sequel we will use the following topology, defined by A on the space **X**, $\|x\|_{\mathcal{T}_A} := \|R(\lambda, A)x\|$, where $\lambda \in \rho(A)$. It is seen that different $\lambda \in \rho(A)$ yields equivalent norms. We say that an operator C on **X** is A-closed if its graph is closed with respect to the topology induced by $\mathcal{T}_A$ on the product $\mathbf{X} \times \mathbf{X}$. It is easily seen that C is A-closable if $x_n \to 0, x_n \in D(C), Cx_n \to y$ with respect to $\mathcal{T}_A$ in **X** implies $y = 0$. In this case, A-closure of C is denoted by $\overline{C}^A$.

Theorem 4.10 *Assume that A and B commute. Then the following assertions hold:*

i) If one of the operators is bounded, then
$$\sigma(A+B) \subset \sigma(A) + \sigma(B).$$

ii) If A and B satisfy condition P, then $A + B$ is A-closable, and
$$\sigma(\overline{(A+B)}^A) \subset \sigma(A) + \sigma(B).$$

In particular, if $D(A)$ is dense in $\mathbf{X}$ then $\overline{(A+B)}^A = \overline{A+B}$, where $\overline{A+B}$ denotes the usual closure of $A+B$.

Proof. For the proof we refer the reader to [11, Theorems 7.2, 7.3]. ∎

Note that the assertion ii) of Theorem 4.10 can be improved a little.

Proposition 4.1 *Let A and B be commuting operators such that there are positive constants $R, \gamma, \delta, 0 < \theta' < \theta < \pi, 1 < \gamma + \delta$ and*

i)
$$\Sigma(\theta + \pi/2, R) \subset \rho(A) \quad \text{and} \quad \sup_{\lambda \in \Sigma(\theta+\pi/2,R)} |\lambda|^\gamma \|R(\lambda, A)\| < \infty,$$

ii)
$$\Sigma(\pi/2 - \theta', R) \subset \rho(B) \quad \text{and} \quad \sup_{\lambda \in \Sigma(\pi/2-\theta',R)} |\lambda|^\delta \|R(\lambda, B)\| < \infty.$$

Then the assertion ii) of Theorem 4.10 holds also true.

Proof. The proof of the proposition can be taken from that of [11, Theorem 7.3] by taking into account the convergence of all the integrals used in the proof of [11, Theorem 7.3]. ∎

4.4. LIPSCHITZ OPERATORS

Let $\mathbf{X}$ and $\mathbf{Y}$ be given Banach spaces over the same field $\mathbf{R}, \mathbf{C}$. An operator $A : \mathbf{X} \to \mathbf{Y}$ is called *Lipschitz continuous* if there is a positive constant L such that
$$\|Ax - Ay\|_\mathbf{Y} \leq L\|x - y\|_\mathbf{X}, \ \forall x, y \in \mathbf{X}.$$

For a Lipschitz continuous operator A the following
$$\|A\| := \sup_{x,y \in \mathbf{X}, \ x \neq y} \|Ax - Ay\|/\|x-y\|$$

is finite and is called the *Lipschitz constant* of A. The set of all Lipschitz continuous operators from $\mathbf{X}$ to $\mathbf{Y}$ is denoted by $\mathcal{L}ip(\mathbf{X}; \mathbf{Y})$ and $\mathcal{L}ip(\mathbf{X}; \mathbf{X}) = \mathcal{L}ip(\mathbf{X})$ for short. A member $A \in \mathcal{L}ip(\mathbf{X})$ is said to be *invertible* if there is a $B \in \mathcal{L}ip(\mathbf{X})$ such that $A \cdot B = B \cdot A = I$. B is called the *inverse* of A and is denoted by A^{-1}

Theorem 4.11 *(Lipschitz Inverse Mapping) Let* $\mathbf{X}$ *be a Banach space,* A *is an invertible member of* $\mathcal{L}ip(\mathbf{X})$ *and* B *is a member of* $\mathcal{L}ip(\mathbf{X})$ *such that* $\|B\| \cdot \|A^{-1}\| < 1$. *Then* $A + B$ *is invertible in* $\mathcal{L}ip(\mathbf{X})$ *and*

$$\|(A+B)^{-1}\| \leq \|A^{-1}\|(1 - \|B\| \cdot \|A^{-1}\|)^{-1}.$$

Proof. We first prove the following assertion: If $A \in \mathcal{L}ip(\mathbf{X})$ such that $\|A\| < 1$. Then $(I - A)$ is invertible in $\mathcal{L}ip(\mathbf{X})$ and

$$\|(I-A)^{-1}\| \leq (1 - \|A\|)^{-1}. \tag{4.3}$$

In fact, for $x, y \in \mathbf{X}$

$$\|(I-A)x - (I-A)y\| \geq \|x-y\| - \|Ax - Ay\| \geq (1 - \|A\|)\|x-y\|.$$

Thus $I - A$ is injective. If $z, w \in R(I - A)$, then

$$\|(I-A)^{-1}z - (I-A)^{-1}w\| \leq (1 - \|A\|)^{-1}\|z - w\|.$$

For $x \in \mathbf{X}$ by induction we can prove that

$$\|B_{n+1}x - B_n x\| \leq \|A\|^n \|Ax\|, \quad \forall n = 0, 1, 2, \ldots$$

where by induction we define $B_0 := I$, $B_n := I + AB_{n-1}$, $\forall n = 1, 2, \ldots$. Indeed, this holds true for $n = 0$, so if we assume it to be true for $n - k$, then

$$\begin{aligned}
\|B_{k+1}x - B_k x\| &= \|AB_k x - AB_{k-1} x\| \\
&\leq \|A\| \cdot \|B_k x - B_{k-1} x\| \\
&\leq \|A\| \cdot \|A\|^{k-1} \|Ax\|,
\end{aligned}$$

so the assertion follows by induction. For any positive integer p,

$$\begin{aligned}
\|B_{n+p}x - B_n x\| &= \left\| \sum_{k=0}^{p-1} (B_{n+k+1}x - B_{n+k}x) \right\| \\
&\leq \sum_{k=0}^{p-1} \|(B_{n+k+1}x - B_{n+k}x)\| \\
&\leq \sum_{k=0}^{p-1} \|A\|^{n+k}\|Ax\| \leq \|A^n\|\|Ax\|(1 - \|A\|)^{-1}.
\end{aligned}$$

Since $\|A\| < 1$ and $\mathbf{X}$ is a Banach space, $Cx = \lim_{m \to \infty} B_m x$ exists for all $x \in \mathbf{X}$ and

$$\|Cx - B_n x\| = \lim_{p \to \infty} \|B_{n+p}x - B_n x\| \leq \|A\|^n \|Ax\|(1 - \|A\|)^{-1}.$$

Since A is continuous,

$$Cx = \lim_{n \to \infty} B_n x = \lim_{n \to \infty} (I - AB_{n-1})x = x + ACx.$$

This shows that $C = I + AC$, so C is a right inverse of $I - A$, i.e. $(I - A)C = I$. Finally, this shows the surjectiveness of $I - A$, proving the assertion that $I - A$ is invertible in $\mathcal{L}ip(\mathbf{X})$.

We are now in a position to prove the theorem. In fact, we have $(A + B) = (I + BA^{-1})A$ and $\|BA^{-1}\| \leq \|B\| \cdot \|A^{-1}\| < 1$. By the above assertion, $(I + BA^{-1})^{-1}$ exists as an element of $\mathcal{L}ip(\mathbf{X})$. Hence $(A + B)^{-1} = A^{-1}(I + BA^{-1})^{-1}$. and

$$\|(I + BA^{-1})^{-1}\| \leq (1 - \|B\| \cdot \|A^{-1}\|)^{-1}.$$

A modification of the above theory for Lipschitz continuous operators from a Banach space $\mathbf{X}$ to another Banach space $\mathbf{Y}$ can be easily made. For instance, the following is true:

Theorem 4.12 *Let A be an invertible member of $\mathcal{L}ip(\mathbf{X}, \mathbf{Y})$. Then for sufficiently small positive k, the operator $A+B$ is an invertible member of $\mathcal{L}ip(\mathbf{X}, \mathbf{Y})$ if $\|B\| < k$.*

Proof. Set $C = A^{-1}(A + B) - I$. Then C is a member of $\mathcal{L}ip(\mathbf{X})$, and for all $x, y \in \mathbf{X}$,

$$\begin{aligned}\|Cx - Cy\|_{\mathbf{X}} &= \|(A^{-1}(A+B) - A^{-1}A)x - (A^{-1}(A+B) - A^{-1}A)y\| \\ &\leq \|A^{-1}\| \cdot \|B\| \cdot \|x - y\|.\end{aligned}$$

Thus for sufficietly small positive k, $I + C$ is invertible, so is $A + B$.

REFERENCES

1. E. Ait Dads, K. Ezzinbi, O. Arino, Periodic and almost periodic results for some differential equations in Banach spaces, *Nonlinear Anal.* **31** (1998), no. 1-2, 163–170.
2. E. Ait Dads, K. Ezzinbi, O. Arino, Pseudo almost periodic solutions for some differential equations in a Banach space, *Nonlinear Anal.* **28** (1997), no. 7, 1141–1155.
3. S. Aizicovici, N. Pavel, Anti-periodic solutions to a class of nonlinear differential equations in Hilbert space. *J. Funct. Anal.* **99** (1991), no. 2, 387–408.
4. H. Amann, "Ordinary Differential Equations", de Gruyter Studies in Math. **13**, Walter de Gruyter, Berlin 1990.
5. H. Amann, Periodic solutions of semi-linear parabolic equations, *in* "Nonlinear Analysis", A Collection of Papers in Honor of Erich Roth, Academic Press, New York, 1978, 1-29.
6. A. Ambrosetti, Un theorema di esistenza per le equazioni differenziali negli spazi di Banach, *Rend Sem. Math. Univ. Padova*, **39**(1967), 349-360.
7. L. Amerio, G. Prouse, "Almost Periodic Functions and Functional Equations", Van Nostrand Reinhold, New York, 1971.
8. W. Arendt, C.J.K. Batty, Almost periodic solutions of first and second oder Cauchy problems, *J. Diff. Eq.* **137**(1997), N.2, 363-383.
9. W. Arendt, C.J.K. Batty, Tauberian theorems and stability of one-parameter semigroups, *Trans. Amer. Math. Soc.* **306**(1988), 837-852.
10. W. Arendt, C.J.K. Batty, Asymptotically almost periodic solutions of inhomogeneous Cauchy problem on the half-line, *Bull. London Math. Soc.* **31**(1999), 291-304.
11. W. Arendt, F. Räbiger, A. Sourour, Spectral properties of the operators equations $AX + XB = Y$, *Quart. J. Math. Oxford (2)*, **45**(1994), 133-149.
12. W. Arendt, S. Schweiker, Discrete spectrum and almost periodicity, *Ulmer Seminare über Funktionalanalysis und Differentialgleichungen*, no. 2, 1997, pp. 59-72.
13. O. Arino, G. Ladas, Y.G. Sficas, On oscillations of some retarded differential equations, *SIAM J. Math. Anal.* **18** (1987), no. 1, 64–73.
14. O. Arino, I. Gyori, Necessary and sufficient condition for oscillation of a neutral differential system with several delays, *J. Diff. Eq.* **81** (1989), no. 1, 98–105.
15. F. V. Atkinson, J. R. Haddock, On determining phase spaces for functional differential equations, *Funkcial. Ekvac.* **31** (1988), 331-347.
16. B. Aulbach, N.V. Minh, Nonlinear semigroups and the existence, stability of semilinear nonautonomous evolution equations, *Abstract and Applied Analysis* **1**(1996), 351-380.

17. B. Aulbach, N.V. Minh, The concept of spectral dichotomy for difference equations. II., *Journal of Difference Equations and Applications* **2**(1996), N.3, 251-262.
18. B. Aulbach, N.V. Minh, Almost periodic mild solutions of a class of partial functional differential equations, *Abstract and Applied Analysis*. To appear.
19. B. Aulbach, N.V. Minh, A sufficient condition for almost periodicity of solutions of nonautonomous nonlinear evolution equations. Submitted.
20. B. Aulbach, N.V. Minh, P.P. Zabreiko, A generalization of monodromy operator and applications in ordinary differential equations, *Differential Equations and Dynamical Systems* **1**(1993), N.3, 211-222.
21. M.E. Ballotti, J.A. Goldstein, M.E. Parrott, Almost periodic solutions of evolution equations, *J. Math. Anal. Appl.* **138** (1989), no. 2, 522–536.
22. V. Barbu, "Nonlinear semigroups and differential equations in Banach spaces", Noordhoff, Groningen, 1976.
23. V. Barbu, Optimal control of linear periodic resonant systems in Hilbert spaces, *SIAM J. Control Optim.* **35** (1997), no. 6, 2137–2156.
24. V. Barbu, N.H. Pavel, Periodic solutions to one-dimensional wave equation with piecewise constant coefficients, *J. Diff. Eq.* **132** (1996), no. 2, 319–337.
25. V. Barbu, N.H. Pavel, Periodic optimal control in Hilbert space, *Appl. Math. Optim.* **33** (1996), no. 2, 169–188.
26. B. Basit, Harmonic analysis and asymptotic behavior of solutions to the abstract Cauchy problem, *Semigroup Forum* **54**(1997), 58-74.
27. B. Basit, Some problems concerning different types of vector almost periodic periodic functions, *Dissertaiones Math.* **338**(1995).
28. B. Basit, H. Günzler, Asymptotic behavior of solutions of systems of neutral and convolution equations, *J. Diff. Eq.* **149** (1998), no. 1, 115–142.
29. A.G. Baskakov, Semigroups of difference operators in the spectral analysis of linear differential operators, *Funct. Anal. Appl.* **30**(1996), no. 3, 149–157 (1997).
30. C.J.K. Batty, Asymptotic behaviour of semigroups of operators, *in* "Functional Analysis and Operator Theory", Banach Center Publications volume 30, Polish Acad. Sci. 1994, 35 - 52.
31. C.J.K. Batty, W. Hutter, F. Räbiger, Almost periodicity of mild solutions of inhomogeneous periodic Cauchy problems, *J. Diff. Eq.* **156**(1999), 309-327.
32. M. Biroli, A. Haraux, Asymptotic behavior for an almost periodic, strongly dissipative wave equation, *J. Diff. Eq.* **38** (1980), no. 3, 422–440.
33. J. Blot, Oscillations presque-periodiques forcees d'equations d'Euler-Lagrange, *Bull. Soc. Math. France* **122** (1994), no. 2, 285–304.
34. J. Blot, Almost periodically forced pendulum, *Funkcial. Ekvac.* **36** (1993), no. 2, 235–250.
35. J. Blot, Almost-periodic solutions of forced second order Hamiltonian systems, *Ann. Fac. Sci. Toulouse Math.* (5) **12** (1991), no. 3, 351–363.
36. J. Blot, Une approche variationnelle des orbites quasi-periodiques des systemes hamiltoniens, *Ann. Sci. Math. Quebec* **13** (1990), no. 2, 7–32.

REFERENCES

37. J. Blot, Trajectoires presque-periodiques des systemes lagrangiens convexes, *C. R. Acad. Sci. Paris Ser. I Math.* **310** (1990), no. 11, 761–763.

38. J. Blot, P. Cartigny, Bounded solutions and oscillations of concave Lagrangian systems in presence of a discount rate, *Z. Anal. Anwendungen* **14** (1995), no. 4, 731–750.

39. F.E. Browder, On the spectral theory of elliptic differential operators. I, *Math. Annalen*, **142**(1961), 22-130.

40. T. Burton, "Stability and Periodic Solutions of Ordinary and Functional Differential Equations", Academic Press, Orlando, Florida. 1985.

41. T. Burton, T. Furumochi, Periodic solutions of Volterra equations and attractivity, *Dynamical Sysytems and Applications* **3**(1994), 583-598.

42. T. Burton, T. Furumochi, Almost periodic solutions of Volterra equations and attractivity, *J. Math. Anal. Appl.* 198 (1996), no. 2, 581–599.

43. B. Calvert, A. G. Kartsatos, On the compactness of the nonlinear evolution operator in a Banach space, *Bull. London Math. Soc.* **19** (1987), no. 6, 551–558.

44. R. Chill, "Fourier Transforms and Asymptotics of Evolution Equations". PhD dissertation, University of Ulm, 1998.

45. S.N. Chow, J.K. Hale, Strongly limit-compact maps, *Funkc. Ekvac.* **17**(1974), 31-38.

46. S.N. Chow, K. Lu, J. Mallet-Paret, Floquet theory for parabolic differential equations I: the time-periodic case, *Center for Dynamical Systems and Nonlinear Studies*, Report No. 58, Georgia Institute of Technology, 1991.

47. C.V. Coffman, J.J. Schäffer, Linear differential equations with delays: Admissibility and conditional exponential stability, *J. Diff. Eq.* **9**(1971), 521-535.

48. J. Conway, "Functions of One Complex Variable", Springer, Berlin, 1973.

49. W.A. Coppel, "Dichotomies in Stability Theory", Lecture Notes in Math. vol. 629, Springer- Verlag, Berlin - New York, 1978.

50. C. Corduneanu, "Almost periodic functions" Interscience Tracts in Pure and Applied Mathematics, No. 22. Interscience Publishers [John Wiley & Sons], New York-London-Sydney, 1968.

51. M.G. Crandall, T.M. Liggett, Generation of nonlinear transformations of on general Banach spaces, *Amer. J. Math.* **93**(1971), 265-298.

52. M.G. Crandall, A. Pazy Nonlinear evolution equations in Banach spaces, *Israel J. Math.* **11**(1972), 57-94.

53. C. M. Dafermos, An invariance principle for compact processes, *J. Diff. Eq.* **9** (1971), 239–252.

54. C.M. Dafermos, Almost periodic processes and almost periodic solutions of evolution equations, *in* "Dynamical Systems, Proceedings of a University of Florida International Symposium, 1977"Academic Press, pp. 43-57.

55. Ju. L. Daleckii, M.G. Krein, "Stability of Solutions of Differential Equations in Banach Space", Amer. Math. Soc., Providence, RI, 1974.

56. D. Daners, P.K. Medina, "Abstract Evolution Equations, Periodic Problems and Applications", Pitman Research Notes in Math. Ser. volume 279, Longman. New York 1992.

57. G. Da Prato, Synthesis of optimal control for an infinite-dimensional periodic problem, *SIAM J. Control Optim.* **25** (1987), no. 3, 706–714.
58. G. Da Prato, P. Grisvard, Sommes d'operateurs lineares et equations differentielles operationelles, *J. Math. Pures Appl.* **54**(1975), 305-387.
59. G. Da Prato, A. Ichikawa, Quadratic control for linear periodic systems, *Appl. Math. Optim.* **18** (1988), no. 1, 39–66.
60. G. Da Prato, A. Ichikawa, Optimal control of linear systems with almost periodic inputs, *SIAM J. Control Optim.* **25**(1987), no. 4, 1007–1019.
61. E.B. Davies, "One-parameter Semigroups", Academic Press, London, 1980.
62. G. Dore, A. Venni, On the closedness of the sum of two closed operators, *Math. Z.* **196**(1987), 189-201.
63. N. Dunford, J.T. Schwartz, "Linear Operators, Part 1", Wiley-Interscience, New York, 1988.
64. D.E. Edmunds, W.D. Evans, "Spectral Theory and Differential Operators", Oxford Univ. Press, New York, 1987.
65. K.J. Engel, R. Nagel, "One-parameter Semigroups for linear Evolution Equations". Springer, Berlin, 1999.
66. M. Farkas, "Periodic motions", Applied Mathematical Sciences, 104. Springer-Verlag, New York, 1994.
67. A.M. Fink, "Almost Periodic Differential Equations", Lecture Notes in Math., **377**, Springer Verlag, Berlin, 1974.
68. A. Friedman, "Partial Differential Equations of Parabolic Type", Englewood Cliffs, N. J. Prentice-Hall, 1964.
69. T. Furumochi, Almost periodic solutions of integral equations. Proceedings of the Second World Congress of Nonlinear Analysts, Part 2 (Athens, 1996). *Nonlinear Anal.* **30** (1997), no. 2, 845–852.
70. T. Furumochi, T. Naito, Nguyen Van Minh, Boundedness and almost periodicity of solutions of partial functional differential equations. Preprint.
71. J.A. Goldstein, "Semigroups of Linear Operators and Applications", Oxford Mathematical Monographs, Oxford University Press, Oxford 1985.
72. K. Gopalsamy, Oscillations in linear systems of differential-difference equations, *Bull. Austral. Math. Soc.* **29** (1984), no. 3, 377–387.
73. K. Gopalsamy, "Stability and oscillations in delay differential equations of population dynamics". Mathematics and its Applications, **74**. Kluwer Academic Publishers Group, Dordrecht, 1992.
74. K. Gopalsamy, G. Ladas, Oscillations of delay differential equations, *J. Austral. Math. Soc. Ser.* B **32** (1991), no. 4, 377–381.
75. I. Gyori, T. Krisztin, Oscillation results for linear autonomous partial delay differential equations, J. Math. Anal. Appl. **174** (1993), no. 1, 204–217.
76. J. Haddock, Liapunov functions and boundedness and global existence of solutions, *Applicable Anal.* **1**(1972), 321-330.
77. J.K. Hale, "Asymptotic Behavior of Dissipative Systems", Amer. Math. Soc., Providence, RI, 1988.

REFERENCES

78. J.K. Hale, "Theory of Functional Differential Equations", Springer-Verlag, New York - Berlin 1977.

79. J. K. Hale, J. Kato, Phase space space for retarded equation with infinite delay, *Funkcial. Ekvac.* **21** (1978), 11-41.

80. J.K. Hale, S.M. Verduyn-Lunel, "Introduction to Functional Differential Equations", Springer-Verlag, Berlin-New york, 1993.

81. Y. Hamaya, Periodic solutions of nonlinear integrodifferential equations, *Tohoku Math. J.* **41** (1989), 105–116.

82. Y. Hamaya, Total stability property in limiting equations of integrodifferential equations, *Funkcial. Ekvac.* **33** (1990), 345–362.

83. Y. Hamaya, T. Yoshizawa, Almost periodic solutions in integrodifferential equations, *Proc. Royal Soc. Edinburgh*, **114 A** (1990), 151–159.

84. A. Haraux, "Nonlinear evolution equations - global behavior of solutions". Lecture Notes in Mathematics, **841**.Springer-Verlag, Berlin - Heidelberg - New York, 1981.

85. A. Haraux, A simple almost periodicity criterion and applications, *J. Diff. Eq.* **66**(1987), 51-61.

86. A. Haraux, Asymptotic behavious of trajectories for some nonautonomous, almost periodic processes, *J. Diff. Eq.* **49**(1983), 473-483.

87. A. Haraux, M. Otani, Quasi-periodicity of bounded solutions to some periodic evolution equations, *J. Math. Soc. Japan* **42** (1990), no. 2, 277–294.

88. L. Hatvani, T. Kristin, On the existence of periodic solutions for linear inhomogeneous and quasilinear functional differential equations, *J. Diff. Eq.* **97**(1992), 1-15.

89. H.P. Heinz, On the behavior of measures of noncompactness with respect to differentiation and integration of vector-valued function, *Nonlinear Analysis*, **7**(1983), 1351-1371.

90. D. Henry, "Geometric Theory of Semilinear Parabolic Equations", Lecture Notes in Math., Springer-Verlag, Berlin-New York, 1981.

91. H.R. Henriquez, Periodic solutions of quasi-linear partial functional differential equations with unbounded delay, *Funkcial. Ekvac.* **37**(1994), 329-343.

92. Y. Hino, Stability and existence of almost periodic solutions of some functional differential equations *Tohoku Math. J.* **28** (1976), 389–409.

93. Y. Hino, Total stability and uniformly asymptotic stability for linear functional differential equations with infinite delay, *Funkcial. Ekvac.* **24** (1981), 345–349.

94. Y.Hino, Stability properties for functional differential equations with infinite delay, *Tohoku Math. J.* **35** (1983), 597–605.

95. Y. Hino, Almost periodic solutions of a linear Volterra system, *Differential and Integral Equations*, **3** (1990), 495-501.

96. Y. Hino, S. Murakami, Favard's property for linear retarded equations with infinite delay, *Funkcial. Ekvac.* **29** (1986), 11–17.

97. Y. Hino, S. Murakami, Stability properties of linear Voltera equations, *J. Diff. Eq.* **59** (1991), 121-137.

98. Y. Hino, S. Murakami, Total stability and uniform asymptotic stability for linear Volterra equations, *J. London Math. Soc.* (2) **43** (1991), 305-312.

REFERENCES

99. Y. Hino, S. Murakami, Stability properties in abstract linear functional differential equations with infinite delay, *Dynamical Systems and Applications*, (1995), 329–341.

100. Y. Hino, S. Murakami, Existence of an almost periodic solutions of abstract linear functional differential equations with infinite delay, *Dynamic Systems and Applications*, (1996), 385–398.

101. Y. Hino, S. Murakami, A genelarization of processes snd stabilities in abstract functional differential equations, *Funkcial. Ekvac.* **41** (1998), 235–255.

102. Y. Hino, S. Murakami, Almost periodic processes and the existence of almost periodic solutions *Electric J. of Qualitative Theory of Differential Equations*, **1** (1998) 1–19.

103. Y. Hino, S. Murakami, Skew product flows of quasi-processes and stabilities, *Fields Institute Communications*, **21** (1999), 329–341.

104. Y. Hino, S. Murakami, Almost periodic quasi-processes and the existence of almost periodic solutions. *Proceedings of international conference of biomathematics, bioinformatics and applications of functional differential equations, Akdeniz University*, 94–103.

105. Y. Hino, S. Murakami, Total Stability in abstract functional differential equations with infinite delay. To appear in *Electric J. of Qualitative Theory of Differential Equations*.

106. Y. Hino, S. Murakami, Quasi-processes and stabilities in functional differential equations. To appear.

107. Y. Hino, S. Murakami, T. Naito, "Functional Differential Equations with Infinite Delay", Lect. Notes Math. 1473, Springer-Verlag, 1991.

108. Y. Hino, S. Murakami, T. Yoshizawa, Existence of almost periodic solutions of some functional-differential equations with infinite delay in a Banach space, *Tohoku Math. J.* (2) **49** (1997), no. 1, 133–147

109. Y. Hino, S. Murakami, T. Yoshizawa, Almost periodic solutions of abstract functional differential equations with infinite delay, *Nonlinear Analysis*, **30** (1997), 853–864.

110. W.A. Horn, Some fixed point theorems for compact maps and flows in Banach spaces, *Trans. Amer. Math. Soc.* **149**(1970), 391–404.

111. J.S. Howland, Stationary scattering theory for time-dependent Hamiltonians, *Math. Ann.* **207** (1974), 315–335.

112. Z. S. Hu, A. B. Mingarelli, On a question in the theory of almost periodic differential equations, *Proc. Amer. Math. Soc.* **127** (1999), pp. 2665-2670.

113. W. Hutter, Hyperbolicity of almost periodic evolution families, *Tübinger Berichte*, 1997, pp. 92-109.

114. W. Hutter, "Spectral Theory and Almost Periodicity of Mild Solutions of Nonautonomous Cauchy problems", PhD Thesis, University of Tübingen (1998).

115. H. Ishii, On the existence of almost periodic complete trajectories for contractive almost periodic processes, *J. Diff. Eq.* **43** (1982), no. 1, 66–72.

116. S. Itô, "Diffusion Equations", Kinokuniya, Tokyo, 1979 (in Japanese).

117. R. Johnson, A linear almost periodic equation with an almost automorphic solution, *Proc. Amer. Math. Soc.* **82** (1981), no. 2, 199–205.

118. M.A. Kaashoek, S.M. Verduyn-Lunel, An integrability condition on the resolvent for hyperbolicity of the semigroup, *J. Diff. Eq.* **112**(1994), 374-406.

119. A. G. Kartsatos, On the compactness of the evolution operator generated by certain nonlinear Ω-accretive operators in general Banach spaces, *Proc. Amer. Math. Soc.* **123** (1995), no. 7, 2081–2091.

120. J. Kato, A.A. Martynyuk, A.A. Shestakov, "Stability of motion of nonautonomous systems (method of limiting equations)". Stability and Control: Theory, Methods and Applications, 3. Gordon and Breach Publishers, Amsterdam, 1996.

121. J. Kato, Uniformly asymptotic stability and total stability, *Tohoku Math. J.* **22** (1970), 254-269.

122. T. Kato, "Perturbation Theory for Linear Operators", Springer-Verlag, 1966.

123. S. Kato, Almost periodic solutions of functional differential equations with infinite delays in a Banach space, *Hokkaido Math. J.* **23**(1994), 465-474.

124. S. Kato, M. Imai, Remarks on the existence of almost periodic solutions of systems of nonlinear differential equations, *Nonlinear Anal.* **25**(1995), N. 4, 409-415.

125. S. Kato, Y. Sekiya, Existence of almost periodic solutions of nonlinear differential equations in a Banach space, *Math. Japon.* **43** (1996), no. 3, 563–568.

126. Y. Katznelson, "An Introduction to Harmonic Analysis", Dover Publications, New York, 1968.

127. N. Kenmochi, M. Otani, Nonlinear evolution equations governed by subdifferential operators with almost periodic time-dependence, *Rend. Accad. Naz. Sci. XL Mem. Mat.* (5) **10** (1986), no. 1, 65–91.

128. H. Kielhöfer, Global solutions of semilinear evolution equations satisfying an energy inequality, *J. Diff. Eq.* **36**(1980), 188-222.

129. H. Komatsu (Ed.), "Functional Analysis and Related Topics, 1991", Lecture Notes in Math. vol. 1540 Springer, Berlin - New York 1993.

130. Kreulich, J., Eberlein-weakly almost periodic solutions of evolution equations in Banach spaces, *Differential and Integral Equations* **9**(1996), N.5, 1005-1027.

131. P. Kuchment, "Floquet Theory For Partial Differential Equations", Birkhauser Verlag, Basel, 1993.

132. P. Kuchment, Floquet theory for partial differential equations, Russian Math. Surveys, v.37, no.4,1982, p. 1-60.

133. V. Lakshmikantham, S. Leela, "Nonlinear Differential Equations in Abstract Spaces", Pergamon Press, Oxford-New York, 1981.

134. C. Langenhop, Periodic and almost periodic solutions of Volterra integral differential equations with infinite memory, *J. Diff. Eq.* **58**(1985),391-403.

135. Yu. Latushkin, S. Monthomery-Smith, Evolutionary semigroups and Lyapunov theorems in Banach spaces, *J. Func. Anal.* **127**(1995), 173-197.

136. Yu. Latushkin, S. Montgomery-Smith,T. Randohlph, Evolutionary semigroups and dichotomy of linear skew-product flows on locally compact spaces, *J. Diff. Eq.* **125** (1996), 73-116.

137. B.M. Levitan, V.V. Zhikov, "Almost Periodic Functions and Differential Equations", Moscow Univ. Publ. House 1978. English translation by Cambridge University Press 1982.

138. J. Liu, Bounded and periodic solutions of differential equations in Banach space,*Differential equations and computational simulations, I* (Mississippi State, MS, 1993),*Appl. Math. Comput.* **65** (1994), no. 1-3, 141–150.

139. Y. Li, Z. Lin, Z. Li, A Massera type criterion for linear functional differential equations with advanced and delay, *J. Math. Anal. Appl.* **200**(1996), 715-725.

140. A. Lunardi, "Analytic Semigroups and Optimal Regularity in Parabolic Problems", Birhauser, Basel, 1995.

141. Yu. I. Lyubich, "Introduction to the Theory of Banach Representations of Groups". Birkhäuser, Basel, Boston, Berlin, 1988.

142. R.Martin, "Nonlinear operators and differential equations in Banach spaces", Wiley-Interscience, New York 1976.

143. A.A. Martynyuk, A theorem on instability under small perturbation, *Dokl. Akad. Nauk Ukrainy* 1992, no. 6, 14–16.

144. A.A. Martynyuk, Stability theorem for nonautonomous equations with small perturbations, *Dokl. Akad. Nauk Ukrainy 1992*, no. 4, 8–10.

145. A.A. Martynyuk, Boundedness of solutions of nonlinear systems with small perturbations. (Russian) *Dokl. Akad. Nauk SSSR* **317** (1991), no. 5, 1055–1058; translation in *Soviet Math. Dokl.* **43** (1991), no. 2, 591–594.

146. A.A. Martynyuk, V.V. Shegai, On the theory of stability of autonomous systems. (Russian) *Prikl. Mekh.* **22** (1986), no. 4, 97–102, 135.

147. J.L. Massera, The existence of periodic solutions of systems of differential equations, *Duke Math. J.***17**, (1950). 457–475.

148. J. L. Massera, Contributions to stability theory, *Ann. of Math.* **64** (1956), 182-206.

149. N.V. Medvedev, Certain tests for the Existence of bounded solutions of systems of differential equations, *Differential'nye Uravneniya* 4(1968), 1258-1264.

150. M. Memory, Stable and unstable manifolds for partial functional differential equations, *Nonlinear Anal.* **16**(1991), 131-142.

151. N.V. Minh, Semigroups and stability of differential equations in Banach spaces, *Trans. Amer. Math. Soc.* **345**(1994), 322-341.

152. N.V. Minh, On the proof of characterisations of the exponential dichotomy, *Proc. Amer. Math. Soc.* **127**(1999), 779-782.

153. N.V. Minh, F. Räbiger, R. Schnaubelt, On the exponential stability, exponential expansiveness and exponential dichotomy of evolution equations on the half line, *Int. Eq. and Oper. Theory* **32**(1998), 332-353.

154. Yu. A. Mitropolsky, A.M. Samoilenko, D.I. Martynyuk, "Systems of evolution equations with periodic and quasiperiodic coefficients" Mathematics and its Applications (Soviet Series), 87. Kluwer Academic Publishers Group, Dordrecht, 1993.

155. S. Murakami, Perturbation theorems for functional differential equations with infinite delay via limiting equations, *J. Diff. Eq.* **59** (1985), 314-335.

156. S. Murakami, Stability in functional differential equations with infinite delay, *Tohoku Math. J.* **37** (1985), 561–570.

157. S. Murakami, Almost periodic solutions of a system of integrodifferential equations, *Tohoku Math. J.* **37** (1985), 561–570.

158. S. Murakami, Linear periodic functional differential equations with infinite delay, *Funkcial. Ekvac.* **29**(1986), N.3, 335-361.

159. S. Murakami, T.Naito, Fading memory spaces and stabilities for functional differential equations with infinite delay, *Funkcial. Ekvac.*, **32** (1989), 91–105.

160. S. Murakami, T. Naito, N.V. Minh, Evolution semigroups and sums of commuting operators: a new approach to the admissibility theory of function spaces, *J. Diff. Eq.*. To appear.

161. S. Murakami, T. Yoshizawa, Relationships between BC-stabilities and ρ-stabilities in functional differential equations with infinite delay, *Tohoku Math. J.* **44** (1992), 45-57.

162. S. Murakami, T. Yoshizawa, Asymptotic behavior in a system of Volterra integrodifferential equations with diffusion, *Dynamic Systems Appl.* **3** (1994), 175–188.

163. R. Nagel (Ed.), "One-parameter Semigroups of Positive Operators", Lec. Notes in Math. 1184, Springer-Verlag, Berlin, 1986.

164. T. Naito, J.S. Shin, Evolution equations with infinite delay, *RIMS Kokyuroku*, **984**(1997), 147-160.

165. T. Naito, J.S. Shin, S. Murakami, On solution semigroups of general functional differential equations, *Nonlinear Analysis*, **30**(1997), 4565-4576.

166. T. Naito, J.S. Shin, S. Murakami, The generator of the solution semigroup for the general linear functional differential equation, *Bulletin of the University of Electro-Communications*, **11** (1998), 29-38.

167. T. Naito, N.V. Minh, Evolution semigroups and spectral criteria for almost periodic solutions of periodic evolution equations, *J. Diff. Eq.* **152**(1999), 358-376.

168. T. Naito, N.V. Minh, R. Miyazaki, J.S. Shin, A decomposition theorem for bounded solutions and the existence of periodic solutions of periodic differential equations, *J. Diff. Eq.* **160**(2000), pp. 263-282.

169. T. Naito, N.V. Minh, J. S. Shin, New spectral criteria for almost periodic solutions of evolution equations. Submitted.

170. T. Naito, N.V. Minh, R. Miyazaki, Y. Hamaya, Boundedness and almost periodicity in dynamical systems, *J. Diff. Eq. Appl.*. To appear.

171. M. Nakao, Existence of an anti-periodic solution for the quasilinear wave equation with viscosity, *J. Math. Anal. Appl.* **204** (1996), no. 3, 754–764.

172. M. Nakao, H. Okochi, Anti-periodic solution for $u_{tt} - (\sigma(u_x))_x - u_{xxt} = f(x,t)$, *J. Math. Anal. Appl.* **197** (1996), no. 3, 796–809.

173. J. M. A. M. van Neerven, "The asymptotic Behaviour of Semigroups of Linear Operator", Birkhaüser Verlag. Basel. Boston. Berlin, Operator Theory, Advances and Applications Vol.88 1996.

174. J. M. A. M. van Neerven, Characterization of exponential stability of a semigroup of operators in terms of its action by convolution on vector-valued function spaces over $\mathbf{R}_+$ *J. Diff. Eq.* **124** (1996), no. 2, 324–342.

175. G. Nickel, "On evolution semigroups and well posedness of non-autonomous Cauchy problems", PhD Thesis, Tübingen University, Tübingen 1996.

176. R.D. Nussbaum, The radius of the essential spectrum, *Duke Math. J.* **37**(1970), 473-478.

177. R.D. Nussbaum, A generalization of the Ascoli theorem and an application to functional differential equations, *J. Math. Anal. Appl.* **35**(1971), 600-610.

178. M. Otani, Almost periodic solutions of periodic systems governed by subdifferential operators, *Proc. Amer. Math. Soc.* **123** (1995), no. 6, 1827–1832.

179. A. Pazy, "Semigroups of Linear Operators and Applications to Partial Differential Equations", Applied Math. Sci. 44, Spriger-Verlag, Berlin-New York 1983.

180. G. Pecelli, Dichotomies for linear functional differential equations, *J. Diff. Eq.* **9**(1971), 555-579.

181. A. Pietsch, " Operator Ideals", North-Holland Math. Lib. Vol.20, North-Holland Pub. Company, Amsterdam, New York, Oxford, 1980.

182. G. Prouse, Periodic or almost-periodic solutions of a nonlinear functional equation. I. *Atti Accad. Naz. Lincei Rend. Cl. Sci. Fis. Mat. Natur.* (8) **43** (1967), 161–167.

183. M. H. Protter, H. F. Weinberger, "Maximum Principles in Differential Equations", Springer-Verlag, New York, 1984.

184. J. Prüss, On the spectrum of C_0-semigroups, *Trans. Amer. Math. Soc.* **284**(1984), 847-857.

185. J. Prüss, "Evolutionary Integral Equations and Applications", Birkhäuser, Basel, 1993.

186. J. Prüss, Bounded solutions of Volterra equations, *SIAM Math. Anal.* **19**(1987), 133-149.

187. R. Rau, Hyperbolic evolution semigroups on vector valued function spaces *Semigroup Forum* **48** (1994), no. 1, 107–118.

188. F. Räbiger, R. Schnaubelt, The spectral mapping theorem for evolution semigroups on spaces of vector valued functions, *Semigroup Forum*, **48**(1996), 225-239.

189. W. M. Ruess, Almost periodicity properties of solutions to the nonlinear Cauchy problem in Banach spaces *in* "Semigroup theory and evolution equations" (Delft, 1989), 421–440, Lecture Notes in Pure and Appl. Math., 135, Dekker, New York, 1991.

190. W. M. Ruess, W.H. Summers, Weak almost periodicity and strong ergodic limit theorem for periodic evolution systems, *J. Func. Anal.* **94** (1990), 177-195.

191. W. M. Ruess, W.H. Summers, Almost periodicity and stability for solutions to functional-differential equations with infinite delay, *Differential Integral Equations*, **9** (1996), no. 6, 1225–1252.

192. W.M. Ruess, Q.P. Vu, Asymptotically almost periodic solutions of evolution equations in Banach spaces, *J. Diff. Eq.* **122**(1995), 282-301.

193. W. Rudin, "Functional Analysis", Second edition, McGraw-Hill Inc., New York 1991.

194. K. Sawano, Exponential asymptotic stability for functional differential equations with infinite retardations, *Tohoku Math. J.* **31** (1979), 363-382.

195. M. Schechter, "Principles of functional Analysis", Academic Press, New York and London, 1971.

196. R. Schnaubelt, "Exponential bounds and hyperbolicity of evolution semigroups", PhD Thesis, Tübingen University, Tübingen 1996.

197. G. Seifert, Nonlinear evolution equations with almost periodic time dependence, *SIAM J. Math. Anal.* **18**(1987), 387-392.

REFERENCES

198. G. Seifert, Almost periodic solutions for a certain class of almost periodic systems, *Proc. Amer. Math. Soc.* **84**(1982), 47-51.

199. G. Seifert, Almost periodicity in semiflows, *Proc. Amer. Math. Soc.* **123**(1995), 2895-2899.

200. G. Seifert, Almost periodic solutions for delay-differential equations with infinite delays, *J. Diff. Eq.* **41**(1981), 416-425.

201. J.S. Shin, Uniqueness and global convergence of successive approximations for solutions of functional integral equations with infinite delay, *J. Math. Anal. Appl.* **120**(1986), 71-88.

202. J.S. Shin, An existence theorem of functional differential equations with infinite delay in a Banach sapce, *Funkcial. Ekvac.* **30**(1987), 19-29.

203. J.S. Shin, On the uniqueness of solutions for functional differential equations, *Funkcial. Ekvac.* **30**(1987), 227-236.

204. J.S. Shin, Existence of solutions and Kamke's theorem for functional differential equations in Banach space, *J. Diff. Eq.* **81**(1989), 294-312.

205. J.S. Shin, Comparison theorems and uniqueness of mild solution to semilinear functional differential equations in Banach spaces, *Nonlinear Analysis*, **23**(1994), 825-847.

206. J.S. Shin, T. Naito, Semi-Fredholm operators and periodic solutions for linear functional differential equations, *J. Diff. Eq.* **153**(1999), 407-441.

207. J.S. Shin, T. Naito, Existence and continuous dependence of mild solutions to semilinear functional differential equations in Banach spaces, *Tohoku Math. J.* **51**(1999), 555-583.

208. J.S. Shin, T. Naito, N.V. Minh, On stability of solutions in linear autonomous functional differential equations, *Funkcial. Ekvac.*. To appear.

209. J.S. Shin, T. Naito, N.V. Minh, Existence and uniqueness of periodic solutions to periodic linear functional differential equations with finite delay, *Funkcial. Ekvac.*. To appear.

210. E. Sinestrari, On the abstract Cauchy problem of parabolic type in spaces of continuous functions, *J. Math. Anal. Appl.* **107**(1985), 16-66.

211. V.E. Sljusarcuk, Estimates of spectra and the invertibility of functional operators. (Russian) *Mat. Sb. (N.S.)* **105**(147) (1978), no. 2, 269–285.

212. H. B. Stewart, Generation of analytic semigroups by strongly elliptic operators under general boundary conditions, *Trans. Amer. Math. Soc. 259* (1980), 299–310.

213. H. Tanabe, "Equations of evolution". Pitman, Boston, Mass.-London, 1979.

214. H. Tanabe, "Functional analytic methods for partial differential equations". Monographs and Textbooks in Pure and Applied Mathematics, 204. Marcel Dekker, Inc., New York, 1997.

215. J.M.A. Toledano, T.D. Benavides, G.L. Acedo, "Measures of Noncompactness in Metric Fixed Point Theory", Birkhäuser, 1991.

216. C.C. Travis, G.F. Webb, Existence and stability for partial functional differential equations, *Trans. Amer. Math. Soc.*, **200**(1974), 394-418.

217. Q.P. Vu, Stability and almost periodic of trajectories of periodic processes, *J. Diff. Eq.* **115**(1995), 402-415.

218. Q.P. Vu, The operator equation $AX - XB = C$ with unbounded operator A and B and related abstract Cauchy problems, *Math. Z.* **208**(1991), 567-588.

219. Q.P. Vu, On the spectrum, complete trajectories and asymptotic stability of linear semi-dynamical systems, *J. Diff. Eq.* **105**(1993), 30-45.

220. Q.P. Vu, Almost periodic and strongly stable semigroups of operators, *in* "Linear Operators", Banach Center Publications, Volume 38, Polish Acad. Sci., Warsaw 1997, p. 401 - 426.

221. Q.P. Vu, Almost periodic solutions of Volterra equations, *Diff. Int. Eq.* **7**(1994), 1083-1093.

222. Q.P. Vu, E. Schüler, The operator equation $AX - XB = C$, stability and asymptotic behaviour of differential equations, *J. Diff. Eq.* **145**(1998), 394-419.

223. G.F. Webb, "Theory of Nonlinear Age-Dependent Population Dynamics", Marcell Dekker, Inc., 1986.

224. G.F. Webb, An operator theoretic formulation of asynchronous exponential growth, *Trans. Am. Math. Soc.* **303**(1987),751-763.

225. G.F. Webb, Continuous nonlinear perturbations of linear accretive operators in Banach spaces, *J. Func. Anal.* **10** (1972), 191-203.

226. J. Wu, "Theory and Applications of Partial Functional Differential Equations ", Applied Math. Sci. **119**, Springer, Berlin- New york, 1996.

227. X. Xiang, N. Ahmed, Existence of periodic solutions of semilinear evolution equations with time lags, *Nonlinear Analysis*, **11**(1992), 1063-1070.

228. M. Yamaguchi, Almost periodic solutions of one dimensional wave equations with periodic coefficients, *J. Math. Kyoto Univ* **29-3**(1989), 463-487.

229. M. Yamaguchi, Nonexistence of bounded solutions of one-dimensional wave equations with quasiperiodic forcing terms, *J. Diff. Eq.* **127** (1996), no. 2, 484–497.

230. M. Yamaguchi, Quasiperiodic motions of vibrating string with periodically moving boundaries, *J. Diff. Eq.* **135** (1997), no. 1, 1–15.

231. T. Yoshizawa, "Stability Theory and the Existence of Periodic Solutions and Almost Periodic Solutions", Applied Math. Sciences 14, Springer-Verlag, New York, 1975.

232. T. Yoshizawa, Some results on Volterra equations with infinite delay, in Oualitative Theory of Differential Equations, (B. Sz-Nazy and L. Hatvani, eds.), *Colloq. Math. Soc. Janos Bolyai*, **53**, (1990) 653–660.

233. K. Yosida, "Functional Analysis", 4th edition, Springer. Berlin New York, 1974.

234. P.P. Zabreiko, N.V. Minh, The group of characterisatic operators and applications in the theory of linear ordinary differential equations, *Russian Acad. Sci. Dokl. Math.* **45**(1992), N.3, 517-521.

235. S. Zaidman, "Topics in abstract differential equations", Pitman Research Notes in Mathematics Series, **304**, Longman Scientific & Technical, New York, 1994.

236. S. Zaidman, Almost-periodicity of solutions of the abstract wave equation, *Libertas Math.* **14** (1994), 85–89.

237. S. Zaidman, Existence of asymptotically almost-periodic and of almost-automorphic solutions for some classes of abstract differential equations, *Ann. Sci. Math. Quebec* **13** (1989), no. 1, 79–88.

REFERENCES

238. S. Zaidman, A nonlinear abstract differential equation with almost-periodic solution, *Riv. Mat. Univ. Parma* (4) **10** (1984), 331–336.

239. S. Zaidman, "Almost-periodic functions in abstract spaces", Research Notes in Mathematics, **126**. Pitman (Advanced Publishing Program), Boston, Mass.-London, 1985.

240. E. Zeidler, "Applied Functional Analysis", Applied Math Sciences, 109, Springer-Verlag, Berlin, 1991.

241. V.V. Zhikov, On the theory of admissibility of pairs of function spaces, *Soviet Math. Dokl.* **13**(1972), N.4, 1108-1111.

242. V.V. Zhikov, Some questions of admissibility and dichotomy. The averaging principle. *Izv. Akad. Nauk SSSR* Ser. Mat. 40 (1976), no. 6, 1380–1408, 1440.(Russian)

Index

BC_ρ^U-stability, 188
$(\rho, \mathbf{X})$-stability, 189
$(\rho, \mathbf{X})$-total stability, 196
$(V\psi)(\cdot)$, 117
$\alpha(B)$, 225
α-contraction, 225
δ_T, 120
ε-period, 26
ε-translation, 26
γ_T, 120
$\Lambda(\mathbf{X})$, 26
$\Lambda(\mathbf{Z}, \mathbf{X})$, 134
$\Lambda_{AP}(\mathbf{X})$, 105
$\overline{C}^A$, 231
$\Phi_+(X,Y)(\Phi_-(X,Y))$, 222
$\Phi_\pm(X)$, 222
ρ-total stability, 195
ρ-totally stable, 195
ρ-Uniform Asymptotic Stability, 192
ρ-stabilities, 188
ρ-stability, 188
$\sigma(g)$, 133
$\sigma_\Gamma(P)$, 78
$\sigma_b(f)$, 27
τ-anti-periodic, 141
$\mathcal{B}$-stability, 185
$\mathcal{S}_L(\omega)$, 115
$\mathcal{T}_A$-topology, 231
A-closure of C, 231
$AP(\mathbf{X})$, 27
$APS(\mathbf{X})$, 139
$BUC(\mathbf{R}, \mathbf{X})$, 15
$BUC^1(\mathbf{R}, \mathbf{X})$, 48
C_0-semigroup, 7
$k(t)$, 112
$K(t,\sigma)\phi$, 110

$r_\sigma(T)$, 224
$r_e(T)$, 225
S_M, 117
$sp(u)$, 24
$U(t,\sigma)$, 110

A and B commute, 231
abstract functional differential equations with infinite delay, 176
accretive, 17
admissibility of a function space, 48
admissibility of an operator, 153
almost periodic, 168
almost periodic function, 26
almost periodic integrals, 168
almost periodic solution, 202
anti-periodic, 45
approximate point spectrum, 14
Approximation Theorem, 27
asymptotic smoothness, 164
asymptotically almost periodic, 168
autonomous functional operator, 62

B-class, 59
BC-stability, 190
BC-totally stable, 190
BC-Uniform Asymptotic Stability, 192
Beurling spectrum, 24
Bochner's criterion, 27
Bohr spectrum, 27

class $\Sigma(\theta + \pi/2, R)$, 54
commuting operators, 231
compact semigroup, 10
condition C, 89
condition H, 35
condition H1, 47

condition H2, 47
condition H3, 47
condition H4, 144
condition H5, 144
condition H6, 147
condition P, 54

decomposition of solution operators, 110

equation in the hull, 179
essential spectral radius, 225
essential spectrum, 224
evolution semigroup, 31
evolutionary process, 32

fading memory space, 177
Fourier- Carleman transform, 25

generalized solution, 156

integer and finite basis, 86
integral of the quasi-process, 164

Kuratowski's measure of noncompactness, 181

limiting equation, 179
limiting quasi-processes, 164
Lipschitz Inverse Mapping, 233

mild solution of functional evolution equation, 62
mild solution of higher order equations, 56
mild solution on $\mathbf{R}$, 48
mildly admissible, 49
monodromy operator, 35

normal point, 224
Nussbaum formula, 225

operator $L_{\mathcal{M}}$, 49

point spectrum, 14
processes, 163

quasi-process, 163

regular, 180
relatively dense, 26
residual spectrum, 14

semi-Fredholm operator, 222
semigroup of type ω, 17, 144
separation condition, 172
skew product flow, 164
solution on $\mathbf{R}$, 48
solution operator $U(t,\sigma)$, 110
Spectral Decomposition Theorem, 82
spectral inclusion, 13
spectral mapping theorem, 11
spectral radius, 224
spectral separation condition, 85
spectrum of a function, 25
spectrum of a sequence, 133
strongly asymptotically smooth, 164
strongly continuous group, 14
strongly continuous semigroup, 7

totally ergodic, 29
trigonometric polynomial, 26
type number, 122

ultimate boundedness, 151
uniform fading memory space, 177

weak spectral mapping theorem, 15
weakly admissible, 48

Other titles in the Stability and Control: Theory, Methods and Applications series.

Volume 13

Advances in Stability Theory at the end of the 20th Century
edited by A.A. Martynyuk

Volume 14

Dichotomies and Stability in Nonautonomous Linear Systems
Yu.A. Mitropolskii, A.M. Samoilenko and V.L. Kulik

Volume 15

Almost Periodic Solutions of Differential Equations in Banach Spaces
Yoshiyuki Hino, Toshiki Naito, Nguyen Van Minh and Jong Son Shin

This book is part of a series. The publisher will accept continuation orders which may be cancelled at any time and which provide for automatic billing and shipping of each title in the series upon publication. Please write for details.